Statistical Methods and Applications Using Quantiles

This book presents a unified and modern treatment of statistical methods based on quantiles, bridging classical regression, distributional modelling, and contemporary data analysis. Moving beyond mean-based approaches, it develops a coherent framework for modelling conditional and marginal distributions through quantile functions, with particular attention to interpretation, inference, and practical implementation. The book combines theoretical developments with applications across the health, social, environmental, and ecological sciences. Its aim is to provide both a conceptual foundation and a practical toolkit for researchers seeking robust, flexible, and interpretable methods for analysing complex data.

Key Features:

- A coherent treatment of statistical modelling through the lens of quantile functions
- Integration of conditional and unconditional quantiles within a single framework, bridging two strands of the literature that are typically treated separately
- Emphasis on distributional thinking, focusing on the entire outcome distribution rather than summary measures
- Coverage of advanced topics including mixed-effects and nonlinear quantile models
- Quantile-based tools for distributional comparison, including differences, ratios, and tail summaries
- Fully reproducible examples with R code using real and simulated datasets

The book is intended for graduate students, researchers, and practitioners in statistics, biostatistics, econometrics, and related fields. It is suitable for advanced courses on regression modelling, distributional methods, or applied data analysis, and can also serve as a reference for methodological research. Applied scientists working with heterogeneous or non-Gaussian data will find practical guidance for implementation and interpretation. A working knowledge of regression methods is assumed, while more advanced topics are developed progressively, allowing readers to engage with both foundational concepts and current research directions.

Marco Geraci is Professor of Statistics and Medical Statistics at Sapienza University of Rome and an internationally recognised expert in quantile-based statistical methods. His research focuses on the development of flexible modelling approaches for complex data, with applications in the health and social sciences. He has held academic positions at University College London, the University of Manchester, and the University of South Carolina. His work has been widely cited, with several publications recognised as highly cited contributions in statistics. Professor Geraci was Statistical Editor of the *Journal of Child Health Care* and has served on the editorial boards of leading statistical journals, including the *Journal of the Royal Statistical Society Series A*. He is a Fellow of the Royal Statistical Society and a member of the editorial board of *Significance*. He is also the author of widely used R packages for quantile modelling, which have contributed to the uptake of modern statistical methods in applied research.

Statistical Methods and Applications Using Quantiles

Marco Geraci

CRC Press
Taylor & Francis Group
Boca Raton London New York

CRC Press is an imprint of the
Taylor & Francis Group, an **informa** business
A CHAPMAN & HALL BOOK

First edition published 2027
by CRC Press
2385 NW Executive Center Drive, Suite 320, Boca Raton, FL 33431

and by CRC Press
4 Park Square, Milton Park, Abingdon, Oxon, OX14 4RN

CRC Press is an imprint of Taylor & Francis Group, LLC

ISBN: 978-1-032-35330-2 (hbk)
ISBN: 978-1-032-35334-0 (pbk)
ISBN: 978-1-003-32641-0 (ebk)

DOI: 10.1201/9781003326410

Typeset in Latin Modern font
by KnowledgeWorks Global Ltd.

Publisher's note: This book has been prepared from camera-ready copy provided by the authors.

In memory of my mother, my father, and Dario

Contents

Preface

In the past few centuries, remarkable scholars attempted to address important scientific questions in their own fields of study. The flourishing body of statistical knowledge from which today's statisticians draw has its roots in those endeavours. For one, modern statistics is greatly indebted to astronomy in the 18th century. Among other sciences, astronomy was, as it has become today, a field of experimental observations driven by mathematical theories. In his *Nouvelles méthodes pour la détermination des orbites des comètes*, Legendre introduced one of the most widely known statistical methods to 'deduce the most accurate possible results from observational measurements' (Stigler, 1986, p. 13): the method of least squares.

All statistical methods face two challenges when used in applied research: the first challenge is to give the right answer to the question asked by the researcher; the second challenge is to do so efficiently. If, rather simplistically, one defines the 'big data' problem as the inadequacy of computational resources relative to the size of the problem to be solved, then it would be fair to say that scientists have been experiencing this issue long before the present time. Some sort of computational 'Darwinism' might explain how estimation challenges in data analysis have triggered a natural selection that has favoured some methods over others. However, the practical advantages of the *means* do not necessarily lead to the *intended end*, unless the end is adapted to the means. Adolphe Quetelet, mathematician and social statistician, introduced the notion of the average man, an egalitarian idea but also an easy approach to data synthesis in an era when modern computers were not available. Quetelet's average man, defined to be 'all that is great, good, or beautiful' ('tout ce qu'il y a de grand, de beau et de bien') (Quetelet, 1835, p. 276; see also Stigler, 1986, p. 171), was at that time an imaginary but attractive idea of the *normal individual* as 'that which is such that it ought to be' (Canguilhem, 1989). However, this idea of mediocrity was soon strongly criticised from a moral point of view (Stigler, 1986, p. 171). Today, we can equally reject that notion from social, biological, and medical standpoints. Diversity is arguably a driving force in shaping modern societies that acknowledge demographic, ethnic, and cultural differences among their constituent members. Increasingly, public policies, tax reforms, and poverty reduction interventions are designed to account for such differences. Precision medicine, the tailoring of medical treatment to the characteristics of individual patients and their diseases, is another striking example of the importance of biological variation. In a setting where deviations from normality are meaningful, the notion of mediocrity is no longer defensible.

This book is intended to give a general perspective on the many facets of random variables that statisticians should consider when analysing their data. I believe that the study of *quantiles* is well suited to providing a useful and comprehensive view of these many facets. This book is not intended to be a critique of 'competing' approaches; on the contrary, it is my belief that quantiles are just the other side of the same coin, the result of looking at the same problem from a different perspective.

May 2, 2026

Acknowledgements

My sincere thanks go to the R Foundation for Statistical Computing and the R Development Core Team, without whom statistics would have far less joy. I am very much indebted to Mario Cortina Borja (UCL Great Ormond Street Institute of Child Health) for his invaluable comments on several drafts of this book. I am grateful to Rob Calver at Chapman & Hall/CRC for his guidance and support during the publication process.

I thank several colleagues for the never-dull collaborations and discussions on topics related to this book, including Matteo Bottai (Karolinska Institutet), Chris Jones (Open University), Alessio Farcomeni (University of Rome 'Tor Vergata'), Lea Petrella (Sapienza University of Rome), and Nansi Boghossian (University of South Carolina).

I am grateful to the UCL Partners Biostatistics Network, the UCL Department of Statistical Science, the UCL Great Ormond Street Institute of Child Health, and the Royal Statistical Society for hosting my lectures and workshops on quantile regression, and to the students of the University of South Carolina for their engagement and feedback during my course *Introduction to Quantile Regression*. Invitations to present my work at a wide range of institutions across Europe and North America have led to many stimulating exchanges, and I thank the colleagues and students who took part in them.

Most of all, my heartfelt gratitude to my family for their encouragement and support, and for always asking the question 'Is the book finished yet?'.

Overview

Organisation of the book

This book is organised to develop a unified view of statistical modelling through the lens of quantiles, progressing from foundational concepts to advanced methodological and applied topics. It is structured in four parts: Part I introduces the principles of statistics, Part II focuses on unconditional quantiles, Part III develops methods for conditional quantiles, and Part IV covers selected further topics.

The first part of the book (Part I) introduces basic concepts that will recur throughout. Chapter 1 presents the main motivation for the book. It discusses why the study of distributions, rather than summary measures alone, is central to statistical analysis, why normality should not be taken as the default paradigm, and how quantiles emerged historically as useful tools for describing variation and heterogeneity. Chapter 2 presents the basic probabilistic framework, including random variables, distribution functions, quantiles, moments, and core ideas in statistical modelling and inference. These concepts provide the formal foundation for the rest of the book. Chapter 3 discusses key features of a distribution, namely location, scale, and shape, with particular emphasis on quantile-based measures. It also introduces the distinction between unconditional and conditional modelling, which underpins much of the subsequent development.

Chapters 4 to 7 focus on unconditional quantiles (Part II). Chapter 4 introduces sample quantiles, order statistics, empirical distributions, and mid-functions. Chapter 5 deals with the computation of quantiles, including sorting, linear programming, nonsmooth optimisation, and smooth approximations. Chapter 6 discusses topics in inference, such as likelihood-based approaches, auxiliary distributions, and the method of quantiles. Chapter 7 covers inferential methods for constructing standard errors, confidence intervals, and tests of hypotheses for one or more samples and one or more quantiles.

Part III, comprising Chapters 8 to 10, turns to conditional quantiles. Chapter 8 introduces linear quantile regression, including modelling, interpretation, estimation, inference, choice of quantiles, and practical pitfalls. Chapter 9 extends the discussion to nonlinear quantile regression, covering nonlinear parametric models, transformation models, nonparametric models, and distribution-based approaches. Chapter 10 addresses discrete responses, including jittering, conditional mid-quantiles, and latent-response formulations.

The final part of the book (Part IV) is devoted to further topics. Chapter 11 examines quantile-based distributional comparisons, including quantile differences, quantile ratios, location, scale, and shape comparisons, conditional comparison functions, and L-statistics. Chapter 12 considers regression models for distribution functions, including binomial regression, distributional regression, and unconditional quantile regression. Chapter 13 deals with conditional quantile estimation in the presence of dependence, with emphasis on clustered and longitudinal data, as well as time series.

Throughout the book, selected theoretical results are presented in dedicated boxes so as to avoid interrupting the flow of the main exposition, while still providing rigorous statements and derivations for the interested reader. In addition, R code chunks and snippets are used extensively to illustrate practical implementation, using both simulated and real data examples.

The overall aim is to provide a coherent framework in which quantiles serve as a unifying tool for modelling, inference, and interpretation across a wide range of statistical problems.

Data and software

This section provides an overview of the datasets and software used throughout the book. The examples span a range of application domains, including epidemiology, public health, socio-economic studies, and environmental science, reflecting the broad applicability of the methods presented.

- **Childhood Respiratory Disease Study**. Data on lung function in children from East Boston, recorded in the 1970s.
- **City Populations in West Virginia**. Demographic data (2000 Census) on the population size in different West Virginia cities. Data source: https://www.census.gov.
- **Crohn's Disease Patients and First-Degree Relatives**. Data on clinical and familial characteristics of Crohn's disease patients. Data source: Sorrentino et al. (2014).
- **EU Statistics on Income and Living Conditions**. A harmonised survey providing information on income, poverty, and social exclusion across European countries. Data source: European Union Statistics on Income and Living Conditions (EU-SILC), https://ec.europa.eu/eurostat/web/microdata/european-union-statistics-on-income-and-living-conditions.
- **Staghorn Corals**. Ecological data used to study depth distributions and species-specific heterogeneity in marine environments. Data source: Muir et al. (2015), https://datadryad.org/dataset/doi:10.5061/dryad.3b568.
- **UK Mathematical Development in Children from the Millennium Cohort Study**. Longitudinal data used to study cognitive development and educational outcomes. Data source: UK Data Service, http://ukdataservice.ac.uk.
- **UK Quarterly Gas Consumption**. Quarterly UK gas consumption from 1960 to 1986. Available in `R` via `datasets::UKGas`.
- **US Environmental Protection Agency**. Environmental measurements of tetrachlorobenzene concentration levels. Available in `R` via `EnvStats::EPA.94b.tccb.df`.
- **US Fatality Analysis Reporting System**. A nationwide database maintained by the National Highway Traffic Safety Administration (NHTSA), containing detailed information on fatal motor vehicle crashes in the United States. Data source: https://www.nhtsa.gov/research-data/fatality-analysis-reporting-system-fars.
- **US Integrated Public Use Microdata Series**. Harmonised individual-level data on pre-tax wage and salary income, along with demographic and socio-economic characteristics, from US Census and American Community Survey (ACS) samples. Data source: Ruggles et al. (2025).

- **US National Center for Health Statistics on Births and Infant Mortality.** Vital national statistics data providing detailed information on births and infant deaths in the United States, widely used in perinatal and population health research. Data source: National Center for Health Statistics (NCHS), http://www.cdc.gov/nchs.
- **US National Health and Nutrition Examination Survey.** A nationally representative survey combining interviews and physical examinations. Data source: National Health and Nutrition Examination Survey (NHANES), available in `R` via the `NHANES` package.
- **World Bank Internet Adoption.** Country-level panel data on internet usage rates over time. Data source: https://data.worldbank.org/indicator/IT.NET.USER.ZS.

Some material reproduced or adapted from United States government sources, including the Centers for Disease Control and Prevention (CDC), is included for scholarly and illustrative purposes only. Such use does not imply endorsement by the CDC, the Department of Health and Human Services, or the United States Government.

All analyses are carried out in `R` (R Core Team, 2026), using a combination of base functionality and contributed packages. In particular, the `quantreg` (Koenker, 2025), `Qtools` (Geraci, 2026, 2016b), and `lqmm` (Geraci, 2014) packages are used extensively for estimation, along with `ggplot2` (Wickham, 2016) for visualisation. Code snippets are provided throughout to facilitate reproducibility and to illustrate practical implementation of the methods discussed. Additional material, including `R` packages developed by the author, datasets used in the examples, and the `R` code used for the analyses and figures in this book, is available on the author's GitHub page at `https://github.com/marco-geraci`. Some of the packages are also available on CRAN, although GitHub provides the latest pre-submission versions.

Notation

This section gives an overview of notational conventions used in the present book.

The symbol '$\equiv$' means 'equal by definition'. Lower- and uppercase letters (e.g., x, X, and ω) denote scalar quantities, either fixed or random. Lowercase bold letters (e.g., $\boldsymbol{x}$ and $\boldsymbol{\omega}$) are used for vectors. Uppercase bold fonts (e.g., $\boldsymbol{X}$ and $\boldsymbol{\Omega}$) denote vectors of random variables or matrices. If necessary, the dimension of a matrix with n rows and m columns is indicated using subscripts (e.g., $\boldsymbol{A}_{n \times m}$). The entries of a vector or matrix use the same letter as the vector or matrix itself but are not bold. Thus, $\boldsymbol{x} = \begin{bmatrix} x_1 \\ \vdots \\ x_n \end{bmatrix}$ and $\boldsymbol{A} = \begin{bmatrix} a_{11} & a_{12} \\ a_{21} & a_{22} \end{bmatrix}$.

The transpose of $\boldsymbol{A}$ is denoted by $\boldsymbol{A}^\top$. If $\boldsymbol{A}$ is an invertible square matrix, then $\boldsymbol{A}^{-1}$ denotes its inverse. Any vector is assumed to be a column, so its transpose is a row. The symbols $\boldsymbol{I}$, $\boldsymbol{0}$, $\boldsymbol{1}$ denote the identity matrix, a vector of zeros and a vector of ones, respectively, with dimension specified by a subscript (e.g., $\boldsymbol{1}_n$ for an n-dimensional vector). The *norm* of a vector $\boldsymbol{x}$ is denoted by $\|\boldsymbol{x}\|$, that is $\|\boldsymbol{x}\| = \sqrt{\boldsymbol{x}^\top \boldsymbol{x}}$.

Pr	Probability
$\Pr(\cdot \mid \cdot)$	Conditional probability

E	Expected value
var	Variance (scalar) or variance-covariance (matrix)
cov	Covariance
cor	Correlation
$\stackrel{d}{=}$ or $\sim$	Identically distributed as
$\xrightarrow{p}$	Converges in probability
$\xrightarrow{d}$	Converges in distribution
$\dot{\sim}$	Approximately distributed as
$\perp\!\!\!\perp$	Independent
IID	Independent and identically distributed
NID	Independent but not identically distributed
$\stackrel{\text{IID}}{\sim}$	Independent and identically distributed as
$\stackrel{\text{NID}}{\sim}$	Independent but not identically distributed as
$\mathcal{N}\left(\mu, \sigma^2\right)$	Normal distribution with mean μ and variance σ^2
Φ	Distribution function of the standard normal distribution $\mathcal{N}(0, 1)$
ϕ	Density function of the standard normal distribution $\mathcal{N}(0, 1)$
$\mathcal{L}(\mu, \sigma)$	Laplace distribution with location parameter μ and scale parameter σ
$\mathcal{AL}(\mu, \sigma, \tau)$	Asymmetric Laplace distribution with location parameter μ, scale parameter σ, and shape parameter τ
$\mathcal{U}(a, b)$	Uniform distribution on (a, b)
$\mathcal{B}(a, b)$	Beta distribution with parameters a, b
χ^2_ν	Chi-squared distribution with ν degrees of freedom
t_ν	Student's t distribution with ν degrees of freedom

$L(\theta)$	Likelihood of θ
$\ell(\theta)$	Log-likelihood of θ
$\hat{\theta}$	Estimate or estimator of θ
$\hat{\theta}_n$	Estimator of θ as a function of the sample size n
log	Natural logarithm, unless stated otherwise
I	Indicator function: $I(A) = \begin{cases} 1 & \text{if } A \text{ is true} \\ 0 & \text{if } A \text{ is false} \end{cases}$
$\operatorname{sgn}(x)$	Sign function: $\operatorname{sgn}(x) = \begin{cases} -1 & \text{if} \quad x < 0 \\ 0 & \text{if} \quad x = 0 \\ 1 & \text{if} \quad x > 0 \end{cases}$
$(x)_+$	Plus function: $(x)_+ = \max(x, 0)$
$\lfloor x \rfloor$	Floor function: the largest integer not greater than x
$\lceil x \rceil$	Ceiling function: the smallest integer not less than x
$B(a, b, x)$	Incomplete beta function: $B(a, b, x) = \int_0^x t^{a-1} (1-t)^{b-1} \, \mathrm{d}t$
$\Gamma(a, x)$	Incomplete gamma function: $\Gamma(a, x) = \int_x^\infty t^{a-1} e^{-t} \, \mathrm{d}t$
$\mathbb{R}$	Real numbers
$\mathbb{R}_c$	Real numbers restricted to condition c (e.g., $\mathbb{R}_{>0} = \{x \in \mathbb{R} : x > 0\}$)
$\mathbb{R}^d$	d-dimensional Euclidean space
$\mathbb{N}$	Natural numbers (nonnegative)
$\emptyset$	Empty set
inf	Infimum of a set (the greatest lower bound)
sup	Supremum of a set (the least upper bound)
$\alpha(F) = \inf\{x : F(x) > 0\}$	Left endpoint of the distribution function F
$\omega(F) = \sup\{x : F(x) < 1\}$	Right endpoint of the distribution function F

Part I

Principles of statistics

1

Introduction

A new scientific truth does not triumph by convincing its opponents and making them see the light, but rather because its opponents eventually die and a new generation grows up that is familiar with it.

Max Planck

1.1 What's in a distribution?

Rarely is a child happy with a tiny slice of chocolate cake if the whole cake is right in front of them. Likewise, it is unthinkable that a scientist will be content with only one piece of information when mouth-watering, rich (and costly) data are available to them. Yet, it is not uncommon to see research studies, expensively funded by governmental agencies or charities, draw their conclusions based on results where hundreds or thousands of observations have been crunched down to a few summary statistics. Clearly, there is no risk for the scientist to end up with a stomach ache if they gorge on all the available data. At worst, they will not extract any more information than already assimilated through those few summary statistics. So, when is a 'minimalist' approach justifiable from a scientific (and economic) point of view? This question has become particularly important in an era when more and more data are at our fingertips.

When presented with sample data, two seemingly contrasting needs arise. One is to find a model that captures as much variation as possible, which often increases complexity. The other is to achieve the same goal as parsimoniously as possible. Only when the balance between the two is struck can a scientist be truly satisfied. The task requires statistical knowledge and expertise combined with sound judgement about the problem at hand. The application of statistics becomes both art and science, guided by experience and good practice.

The following example outlines some of the principles we have briefly introduced. Consider Figure 1.1, which depicts the distribution of mathematical development (MD) scores in children aged five participating in the Millennium Cohort Study, a United Kingdom birth cohort longitudinal survey begun at the start of the 21st century. Higher scores correspond to higher levels of MD. The mean and standard deviation are equal to 23.0 and 4.7, respectively. Would these two numbers alone be sufficient to describe satisfactorily the distribution of MD scores? Of course not. There is an infinite number of distributions with the same mean and standard deviation but completely different shapes. Knowing that the skewness index (third standardised moment) is equal to -1.7 would lead to restrict the focus to left-skewed distributions, though the lack of bounds for this index makes it difficult to judge the strength of such asymmetry. Actually, knowledge of the third standardised moment gives little clue about the shape of the distribution (Mood et al., 1974, p. 75). Finally, the kurtosis index (fourth standardised moment) for the MD scores is equal to 6.4. However, this

DOI: 10.1201/9781003326410-1

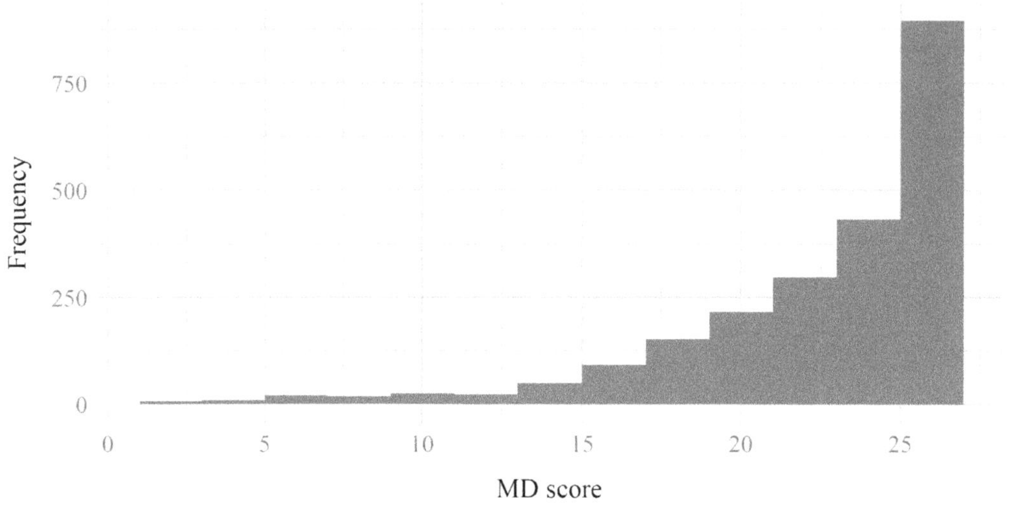

Figure 1.1
Frequency diagram of mathematical development (MD) scores in 5-year-old children of the Millennium Cohort Study, 2006.

information is not particularly helpful here because the distribution is asymmetric (Jones et al., 2011). We could carry on with a long list of moments of higher order, and our intuitive understanding of the MD score distribution would probably not get any better.

Consider the five-number summary (FNS), which consists of the minimum, the first, second and third quartiles, and the maximum of the sample. For the MD scores, we have FNS $= \{0, 21, 25, 27, 27\}$. This simple sequence of numbers provides a rather intuitive and parsimonious snapshot of the empirical distribution. The distance between the minimum and the first quartile, which represents the first striking feature of the distribution, indicates that a quarter of the sample is spread over a relatively long stretch. In contrast, the relatively small difference between the first quartile and the median and that between the median and the third quartile suggest a rapid accumulation of observations. Remarkably, the third quartile is also the sample maximum: since educational attainment scores like the MD scores are bounded, a ceiling effect is to be expected. The left-skewness is apparent from these observations. In addition, one can easily calculate Bowley's coefficient of skewness (Bowley, 1901) as $[(27 - 25) - (25 - 21)]/(27 - 21) = -0.3$. This index is also known as Galton's skewness and is bounded on the interval $(-1, 1)$ (Chapter 3).

Knowledge of the mean and a few higher moments of the distribution does not provide direct information about probabilities except under specific parametric models. Moreover, theoretical studies suggest that moments are not informative about the central part of the distribution (Lindsay and Basak, 2000). In contrast, we can easily obtain an approximate estimate of the moments of the distribution by using weighted averages of sample quantiles. The relevant equation for the calculation of the expected value is introduced in Equation (1.1) further below and generalised to higher moments in Section 2.3.2. For the data plotted in Figure 1.1, we take the midpoints of the intervals between successive quartiles, weighted by the probabilities (i.e., proportions) of observations contained by the intervals themselves. We then estimate the sample mean (Section 2.3.2)

$$\hat{m} = 10.5 \cdot 0.25 + 23.0 \cdot 0.25 + 26.0 \cdot 0.25 + 27.0 \cdot 0.25 = 21.6$$

and variance

$$\hat{s}^2 = (10.5 - \hat{m})^2 \cdot 0.25 + (23.0 - \hat{m})^2 \cdot 0.25 + (26.0 - \hat{m})^2 \cdot 0.25 + (27.0 - \hat{m})^2 \cdot 0.25 = 6.6^2.$$

Better approximations are provided as the number of quantiles is increased, for example, 22.6 and 5.4^2 using deciles, and 22.9 and 4.9^2 using vigintiles. Granted, if we are able to calculate the sample mean and variance directly from the raw data, then we do not need to approximate these estimates by means of quantiles. However, the main message here is that we *can* recover moments from sample quantiles, but not vice versa unless we make some assumptions about the underlying distribution that generated the data (and provided that the estimates of the moments are not severely affected by outliers).

There are three fundamental aspects that govern a distribution: (a) location, (b) scale, and (c) shape. Studies in applied research traditionally over-stress the former and typically provide little information about the other two (perhaps as a consequence of the widespread use of moments and the difficulty of interpretation of those of higher order). This topic will be discussed in more detail in Chapters 3 and 11. For now, we proffer the following principles: (i) our interest lies in phenomena subject to variability; (ii) variability is a source of both uncertainty and information; and (iii) probability distributions provide the basic tool to measure uncertainty. If these premises are agreed upon, we must also agree on the pivotal role of distributions (rather than central tendency alone) in applied statistics.

In particular, the focus of many statistical analyses is to contrast distributions, either as the result of a change in experimental conditions or in the characteristics of the population units. For example, consider the simple case in which the analyst wants to study the random variable Y and compare two groups (e.g., blood pressure in treated and controls, salary for academics and CEOs of blue-chip companies, years of schooling in men and women, etc.). The next step is to decide *how* to compare the groups. Should we compare the distributions as a whole, or would summary statistics alone suffice? Suppose Y is a continuous random variable with support S_Y and distribution function F_Y. Suppose, for simplicity, that we decide to contrast the averages in the two populations. We can first define the dummy variable $X \in \{0, 1\}$ to identify the two populations of interest and then calculate the conditional expectation $\mathrm{E}(Y \mid X = x)$. We now apply a useful formula for the expectation of a random variable, which will be introduced more formally in the next chapter (Section 2.3). The expected value of Y can be defined in terms of F_Y or of its inverse F_Y^{-1} (Parzen, 1979), that is

$$\mathrm{E}(Y) = \int_{S_Y} y \, \mathrm{d}F_Y(y) = \int_0^1 F_Y^{-1}(p) \, \mathrm{d}p. \tag{1.1}$$

Analogously, we have

$$\mathrm{E}(Y \mid X = x) = \int_0^1 F_Y^{-1}(p \mid X = x) \, \mathrm{d}p$$

for the conditional (on X) expectation of Y. Here, $F_Y^{-1}(p \mid X = x)$ denotes the pth quantile of Y given $X = x$, that is, the conditional quantile function. It is easy to verify that

$$\begin{aligned} \mathrm{E}(Y \mid X = x) &= \mathrm{E}(Y \mid X = 0) + \{\mathrm{E}(Y \mid X = 1) - \mathrm{E}(Y \mid X = 0)\} \cdot x \\ &= \int_0^1 F_Y^{-1}(p \mid X = 0) \, \mathrm{d}p + \left\{ \int_0^1 F_Y^{-1}(p \mid X = 1) - F_Y^{-1}(p \mid X = 0) \, \mathrm{d}p \right\} \cdot x. \end{aligned} \tag{1.2}$$

The addend $\mathrm{E}(Y \mid X = 0)$ often plays the role of a baseline, whereas substantive interest frequently centres on the second term, $F_Y^{-1}(p \mid X = 1) - F_Y^{-1}(p \mid X = 0)$. This assumes an important role in the analyst's quest as it contrasts the distributions of the two populations.

The question that the analyst faces is whether it suffices to take a (weighted) average of these differences as in Equation (1.2) or to dig a bit deeper. If, say, $F_Y^{-1}(p \mid X = 1) - F_Y^{-1}(p \mid X = 0) = \beta_1$, that is, the difference is constant over the probability p, then the average gives all the information. Indeed, in the normal case $\mathcal{N}\left(\beta_0 + \beta_1 x, \sigma^2\right)$, we have $F_Y^{-1}(p \mid X = x) = \beta_0 + \beta_1 x + \sigma\Phi^{-1}(p)$. This implies that $\mathrm{E}(Y \mid X = 0) = \beta_0$ and $\mathrm{E}(Y \mid X = 1) - \mathrm{E}(Y \mid X = 0) = \beta_1$. By combining terms that do not depend on x, we have $F_Y^{-1}(p \mid X = x) = \alpha_0(p) + \beta_1 x$, where the intercept $\alpha_0(p) = \beta_0 + \sigma\Phi^{-1}(p)$ is the only parameter that depends on p.

However, as we will see in the next section, distributions may differ not only by location, but also by scale and shape. It is not unusual to come across statements like 'male infants are heavier than females at birth', 'girls do better than boys at school', 'lack of physical activity is associated with obesity', which we often find in newspaper headlines. As soon as these findings are confirmed in study after study, these 'truths' become part of our knowledge. However, common sense will tell everyone that not *all* boys are heavier than girls at birth, that not *all* girls do better than boys at school, and that not *all* sedentary people are obese. We know that those 'truths' concern an *expectation*, a simplified and perhaps convenient representation of a complex reality.

1.2 Is normality the norm or the exception?

Two hundred years ago, a big revolution took place in the scientific world as we know it. Scholars began to look at measurements more and more closely by pursuing what we would call today 'statistical reasoning'. They wanted to make sense out of them, either to verify a theoretical model through predictions or to find new insights into the qualities of natural phenomena. The birth of the method of least squares (LS) in 1805 marked a step change in addressing what was known as the problem of 'the combination of observations' (Stigler, 1986). This new potent and numerically appealing approach sparked the development of statistical theory and soon spread to a vast number of applied fields.

In modern notation, we can state the problem as follows. Consider n pairs of observations $(Y_1, \boldsymbol{x}_1)$, $(Y_2, \boldsymbol{x}_2)$, ..., $(Y_n, \boldsymbol{x}_n)$, where the Y_is are independent copies of the random variable Y with continuous distribution function F and the $\boldsymbol{x}_i$s are p-dimensional vectors of *known* regressors. We want to solve

$$\min_{\boldsymbol{\beta} \in \mathbb{R}^p} \frac{1}{n} \sum_{i=1}^{n} \left(Y_i - \boldsymbol{x}_i^\top \boldsymbol{\beta}\right)^2. \tag{1.3}$$

The $p \times 1$ parameter vector $\boldsymbol{\beta}$, which is fixed and *unknown*, represents the quantity of interest. A special case is given when $\boldsymbol{x}_i^\top \boldsymbol{\beta} = \beta_0$, so that the solution to Equation (1.3) is the arithmetic mean of the Y_is.

Let $Y_i = \boldsymbol{x}_i^\top \boldsymbol{\beta} + \varepsilon_i$, $i = 1, \ldots, n$, where

i. $\mathrm{E}(\varepsilon_i) = 0$, therefore $\mathrm{E}(Y_i) = \boldsymbol{x}_i^\top \boldsymbol{\beta}$ (linearity and additivity assumption),

ii. $\mathrm{var}(\varepsilon_i) = \sigma^2 < \infty$ (homoskedasticity assumption – i.e., constant variance) and

iii. $\mathrm{cor}(\varepsilon_i, \varepsilon_j) = 0$, $i \neq j$ (uncorrelated errors assumption).

If these assumptions hold, then, according to the Gauss-Markov theorem, the LS estimator of $\boldsymbol{\beta}$ has the smallest variance of all linear and unbiased estimators (best linear unbiased estimator – BLUE).

The LS method, therefore, enjoys some optimal properties without the need to introduce additional assumptions on the error distribution. However, if in addition the errors are normally distributed, then the LS estimator coincides with the maximum likelihood estimator and is the most efficient among *all* unbiased estimators, in the sense of attaining the Cramér–Rao lower bound (Cox and Hinkley, 1997). Was this result known in the early 1800s? Almost certainly not. So, where does the normal assumption come from?

In 1809, Carl Friedrich Gauss understood that it was necessary to introduce a measure of accuracy of the estimates to support the validity of the LS method and he addressed the problem in probabilistic terms. He looked at the error law that would satisfy some general conditions (e.g., symmetry and mode at the origin) and that would lead, in a linear model, to the 'most probable' (maximum likelihood in modern jargon) solution equal to that given by solving Equation (1.3). This was clearly a circular argument (Stigler, 1986), an assumption to justify *ex post* (Koenker, 2005) a choice determined by the very argument's conclusion.

If supported by Gauss's reasoning alone, perhaps the normal distribution would not have received the same attention it did in the following years. In fact, a breakthrough came from Pierre-Simon Laplace, almost at the same time of Gauss's result, with what is now called the central limit theorem (CLT).[1] It is useful to state the CLT in its simplest form.[2]

Consider n copies $Y_1, \ldots, Y_n$ of the random variable Y, with $\mathrm{E}(Y_i) = \mu \in \mathbb{R}$ and $\mathrm{var}(Y_i) = \sigma^2 < \infty$. Moreover, let F denote the unknown cumulative distribution function of Y and we write $Y \sim F$. The sample mean $\hat{\theta} \equiv 1/n \sum_i^n Y_i$ provides an estimator of μ that does not require specifying a particular parametric form for the distribution F. Knowledge of the distribution of an estimator, called *sampling distribution*, is of utmost interest since inferential procedures depend on it. The CLT states that the distribution of the standardised sample mean[3] $\sqrt{n}(\hat{\theta}_n - \mu)/\sigma$ will approach a standard normal distribution as the sample size n becomes bigger and bigger. This result, in turn, leads to the well-known approximate $(1-\alpha)$-level confidence interval for θ

$$\hat{\theta} \pm \frac{\hat{\sigma}}{\sqrt{n}} z_{1-\alpha/2},$$

where $\hat{\sigma}/\sqrt{n}$ is the (estimated) standard deviation of the sample mean (standard error) and $z_{1-\alpha/2}$ is the $(1-\alpha/2)$th quantile of a standard normal ('critical value' in test of hypotheses). In the CLT, there is no reference to the parent distribution, that is, the theorem is valid regardless of the particular F. However, if the distribution F is normal, then obviously the sampling distribution of $\hat{\theta}_n$ will be normal for all values of n.

There might be a temptation to consider the CLT as a theoretical justification for introducing normal assumptions on F, which would be of course unreasonable and would distract from the main goal of the analysis. The asymptotic normality of the sampling distribution is not a prerogative of the sample mean. Sample quantiles too, for example, are asymptotically normal (Chapter 4). While we look at sampling distributions to understand the properties of estimators and thus study the comparative performance among them, we should not take our eyes off the prize, that is, the parent distribution. The CLT does not say whether the mean:

- provides all the information about F that we need or that we can extract from the data;
- is an appropriate and meaningful parameter for any distribution F;

[1] Laplace's result was read to the Academy of Sciences on April 9, 1810 and can be described as a major generalisation of De Moivre's limit theorem (Stigler, 1986, p. 136).

[2] There are different variants of the CLT with relaxed assumptions. However, they are not relevant to this particular discussion.

[3] We use the subscript n whenever it is necessary to stress the dependence of an estimator on the sample size n.

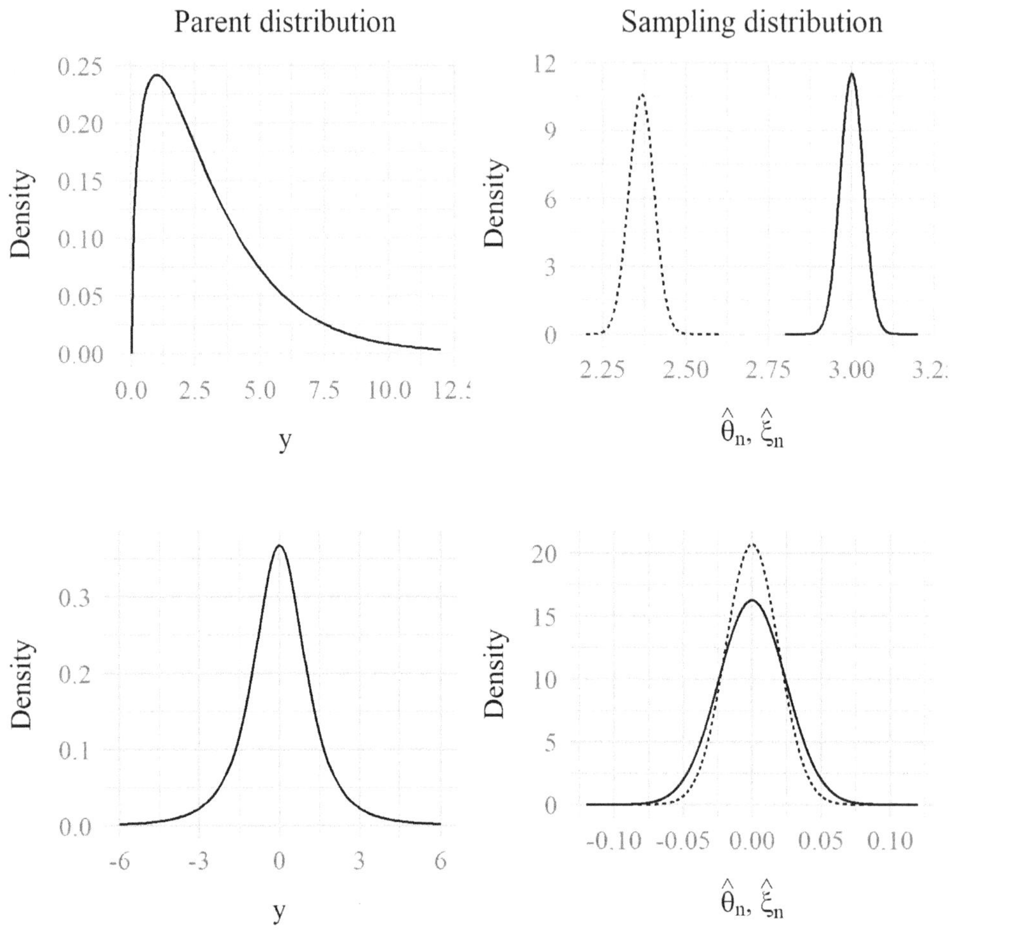

Figure 1.2
The plots on the left-hand side depict the population distribution of a chi-squared (top) and a t-distribution (bottom) with three degrees of freedom. The corresponding sampling distributions of the sample mean (solid line) and sample median (dashed line) with sample size $n = 5000$ are reported on the right-hand side plots.

- is estimated efficiently, that is, with dispersion equal or smaller than that associated with alternative statistics.

Consider this theoretical example. Suppose that Y follows a chi-squared distribution with $\nu = 3$ degrees of freedom (χ^2_3) (Figure 1.2, top left) and let f and F denote its density and distribution functions, respectively. For n sufficiently large, we can approximate the sampling distribution of the sample mean $\hat{\theta}_n$ with $N(\nu, \omega_1)$, where $\nu = 3.0$ and $\omega_1 = 2\nu/n = 6.0/n$. Analogously, we can approximate the sampling distribution of the sample median $\hat{\xi}_{0.5,n}$ by using $N(\xi_{0.5}, \omega_2)$, where $\xi_{0.5} = 2.4$ and $\omega_2 = (0.5 \cdot 0.5)/nf^2(\xi_{0.5}) = 7.1/n$ (see Section 4.3). The two estimators are clearly not directly comparable as they target different values in the distribution (Figure 1.2, top right). It will depend on the particular application whether the mean or the median is the most appropriate location parameter, although the median is often preferred when the distribution is asymmetric, as its interpretation as the value that splits the distribution into two equal halves applies regardless of the shape. Clearly, when assuming a chi-squared model for the population, it would be straightforward (and, generally, more efficient) to calculate the median from the fitted distribution with $\hat{\nu} \equiv \hat{\theta}$ degrees of freedom.

Suppose now that Y follows a Student's t-distribution with $\nu = 3$ degrees of freedom (Figure 1.2, bottom left). We have that $\hat{\theta}_n$ is approximately $N(0, \omega_1)$, where $\omega_1 = \nu/n(\nu - 2) = 3.0/n$. Analogously, we can approximate the sampling distribution of the sample median $\hat{\xi}_{0.5,n}$ with $N(0, \omega_2)$, where $\omega_2 = (0.5 \cdot 0.5)/nf^2(\xi_{0.5}) = 1.8/n$. In this case, both mean and median target the same value (Figure 1.2, bottom right), yet the sample median is asymptotically more efficient (smaller asymptotic variance) than the sample mean. It is interesting to note that the asymptotic variance of the sample median is smaller for the Student's t than for the chi-squared population since the density f at the median is larger for the former distribution. This topic is discussed in more detail in Section 6.2.

When Laplace worked on the CLT, he had very much in mind the problem of combining observations that was encountered in fields like, astronomy and geodesy, whence the LS method had sprung. He considered the sum of a large number of (independent) random variables representing measurement errors about the same true parameter. Needless to say, this parameter was the arithmetic mean. The idea that the sum of many factors produces a symmetric, bell-shaped curve was subsequently extended to describe some natural phenomena, especially in biology. For example, the distribution of human birth weights in a population has been traditionally described by the normal distribution, as the result of many combined influences. Somewhat simplistically, we could write $Y = \sum_{j=1}^{m} U_j$, where Y denotes the population birth weight and the U_js are m unobserved contributing factors. The latter play a similar role as the errors in a linear model. However, in practical applications these factors are not considered noise to be smoothed out, as one would with measurement errors. On the contrary, it is of medical interest to identify those factors that intervene in or are associated with the development of the foetus.

Figure 1.3 shows the distribution of birth weights (g) recorded for about three million singletons born at term (i.e., after 37 weeks of gestation) among the United States (US)

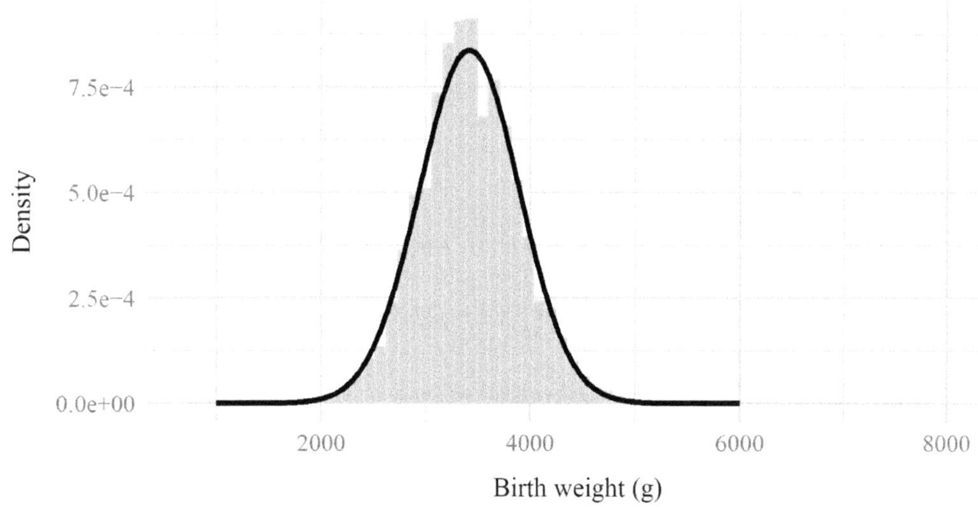

Figure 1.3
Histogram of birth weights (g) for singletons born at term, United States, 2001.

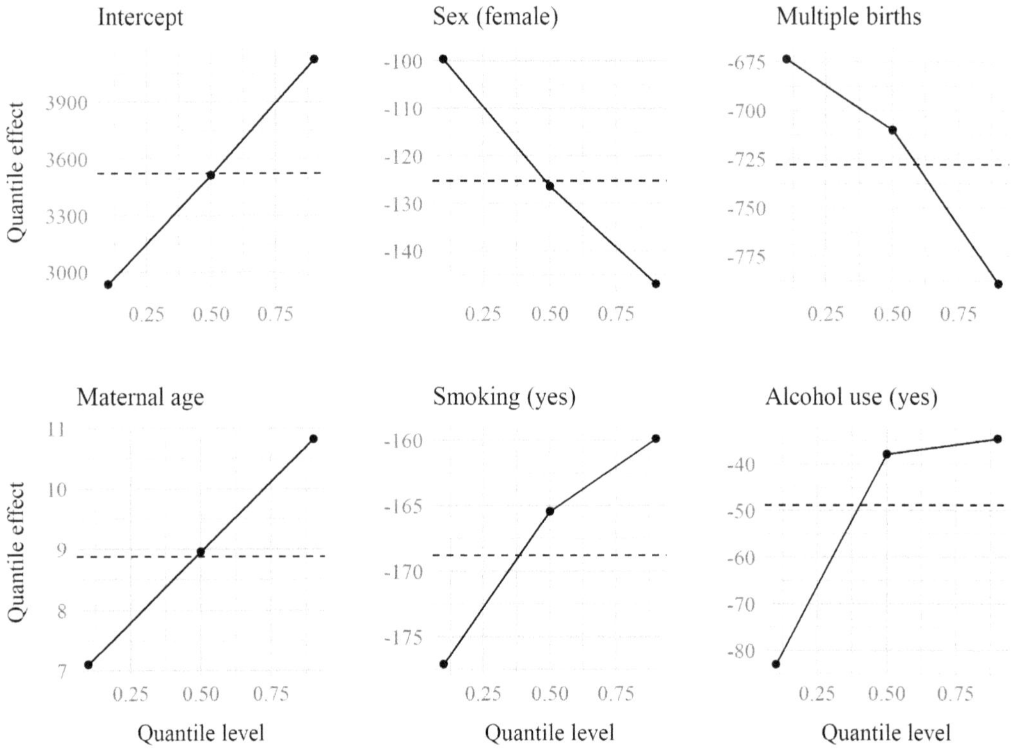

Figure 1.4
Birth weight quantile effects (g) for infants born at term, United States, 2001.

population in 2001 using the NCHS Births dataset. The solid black line represents the density of the theoretical normal curve with mean (3416 g) and standard deviation (477 g) estimated from the same data, which offers a reasonable first approximation to the observed distribution. At first glance, this might tempt one to use mean regression to investigate factors affecting birth weight. However, an attentive analysis would reveal that there is more than the mean (Abrevaya, 2001; Koenker and Hallock, 2001; Abrevaya and Dahl, 2008; Chernozhukov and Fernández-Val, 2011; Geraci, 2016a), despite the approximate normality of the birth weight distribution.

In the following example, we look at US birth weights by sex of the child, multiplicity of birth, maternal age at birth (years), and tobacco and alcohol use of the mothers during pregnancy, which are well-known birth weight determinants. Conditional modelling is carried out using the linear regression specification

$$\texttt{birth weight} \sim \texttt{female} + \texttt{multiple births} + (\texttt{maternal age} - 30)$$
$$+ \texttt{smoking} + \texttt{alcohol}.$$

All the regressors are binary, with the exception of maternal age, which is continuous. Furthermore, maternal age is centred at 30 years to aid interpretation.

Figure 1.4 graphically summarises the estimated effects for the mean and three selected quantiles of the conditional distribution of birth weight. The numerical values and corresponding standard errors are reported in Table 1.1. Here we focus on the mean, first

Table 1.1
Birth weight mean and quantile effects (g), together with standard errors (SE), for infants born at term, United States, 2001.

	Mean		1st decile		Median		9th decile	
	Est.	SE	Est.	SE	Est.	SE	Est.	SE
Intercept	3522.9	0.4	2934.4	0.8	3513.9	0.5	4126.8	0.8
Sex (female)	−125.3	0.5	−99.6	1.0	−126.3	0.6	−147.0	1.0
Multiple births	−727.8	2.2	−673.5	4.1	−710.1	2.5	−789.2	3.0
Maternal age	8.9	0.0	7.1	0.1	9.0	0.1	10.8	0.1
Smoking (yes)	−168.8	0.8	−177.1	1.6	−165.4	1.0	−159.9	1.6
Alcohol use (yes)	−48.8	3.5	−83.0	7.8	−38.0	4.1	−34.6	6.9

decile, median and ninth decile of the conditional birth weight distribution. These summary statistics are given as intercepts for the reference group, that is male singletons born to 30-year-old mothers who did not smoke or drink alcohol during pregnancy, while each of the other coefficients measures the changes in birth weights associated with a particular regressor as compared to the reference, *ceteris paribus*. For example, girls are on average 125 grams lighter than boys. However, small girls are 100 grams lighter than small boys ('small' here refers to the first decile of the distribution). In contrast, large girls are 147 grams lighter than large boys ('large' here refers to the ninth decile of the distribution). In other words, the difference between the birth weight distributions in girls and boys, conditional on the other variables, is not constant and equal to the mean difference, but depends on the particular *quantile* being compared (see Section 8.3.4 for a more in-depth analysis of these data).

1.3 The firmament, ellipticity of the Earth, and Galtonian ogives: Quantiles in the making

The median is often cited as a special case of *quantiles* and is arguably the first quantile that has ever been used in applied statistics. Historically, its appearance in data analysis dates back to the 17th and 18th centuries when, however, it was not yet known as the *median* – the value in a set of ranked observations that divides the data into two parts of equal size (Everitt and Skrondal, 2010) – but, still, as a value at the centre of symmetrically distributed observations. The analysis of astronomical data by Galileo Galilei in 1632 and geodetic measurements by Roger Joseph Boscovich (Ruđer Josip Bošković) in 1757 are presumably the earliest examples of application of the least absolute deviations (LAD) estimator in its, respectively, unconditional and conditional forms. Galileo considered the problem of determining the distance from Earth of a newly discovered star based on the observations of its altitude and the altitude of the polar star (Galilei, 1632) – for a detailed account, see Hald (2003, p. 149). Boscovich, a polymath born in Croatia of Italian heritage who was educated at the *Collegium Romanum* in Rome (Heyde and Seneta, 2001), adopted a constrained minimisation of absolute residuals to determine the ellipticity of the Earth (Boscovich, 1757) – see Koenker and Bassett (1985) and Koenker (2005, p. 2) for details.

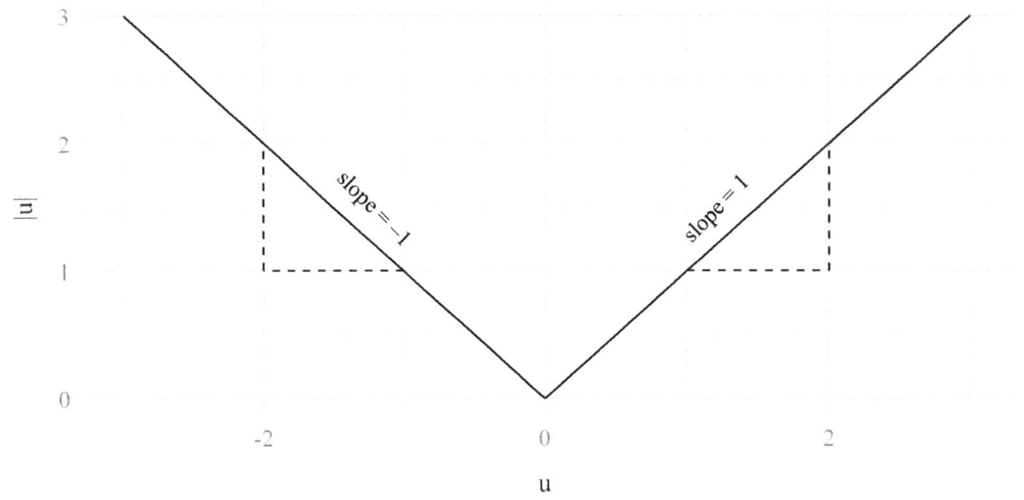

Figure 1.5
Absolute value function. This function is convex. The derivative (i.e., the slope) is 1 if $u \geq 0$ and -1 if $u < 0$.

As noted above, the solution to the LS problem provides the sample mean. Similarly, the median of a sample $Y_1, \ldots, Y_n$ with common distribution F can be calculated as *a* solution[4] to the following LAD problem:

$$\min_{\xi \in \mathbb{R}} \frac{1}{n} \sum_{i=1}^{n} |Y_i - \xi|, \tag{1.4}$$

where the absolute value of a real number u is defined as

$$|u| = \begin{cases} u & \text{if} \quad u \geq 0 \\ -u & \text{if} \quad u < 0. \end{cases}$$

Figure 1.5 plots the absolute value function for $-3 \leq u \leq 3$. It is immediate to see that at 0 this function has no derivative (differentiability in the strict sense). Going from just below 0 to just above 0, there is a sharp change of the slope from -1 to 1. (We use the convention that the derivative at zero is positive.) However, the convexity of the absolute value provides scope for a heuristic demonstration of how the LAD estimate and the sample median are related.

For $\xi \neq Y_i$ for all i, the derivative of the objective function in (1.4) is

$$\frac{\partial}{\partial \xi} \frac{1}{n} \sum_{i=1}^{n} |Y_i - \xi| = \frac{1}{n} \sum_{i:Y_i - \xi < 0} 1 - \frac{1}{n} \sum_{i:Y_i - \xi > 0} 1. \tag{1.5}$$

By equating the expression above to zero, we cannot obtain a closed-form solution as in the case of the LS problem. However, summing ones is equivalent to counting. Therefore, after we collect n observations y_i, we count the number of observations greater than ξ and those less than ξ. Equivalently, we count the number of negative and positive residuals, given ξ. Suppose that the former count gives n_1 and the latter gives n_2, with $n = n_1 + n_2$

[4]In later chapters, we shall see that the solution to (1.4) need not be unique.

(for simplicity, assume that $\sum_i I(Y_i = \xi) = 0$). From Equation (1.5) we obtain

$$\frac{1}{n} \sum_{i:y_i - \xi < 0} 1 = \frac{n_1}{n},$$
$$\frac{1}{n} \sum_{i:y_i - \xi > 0} -1 = -\frac{n_2}{n}.$$

Without loss of generality, suppose that n is even. If ξ is such that $n_1 = n_2 = n/2$, we have $\frac{n_1}{n} = \frac{n_2}{n} = \frac{1}{2}$ and $\frac{n_1}{n} - \frac{n_2}{n} = 0$. In other words, a value of ξ that minimises the absolute deviations is the median. This property was originally proved by Fechner (1897) who, according to Keynes (1911), first introduced the median into use (see also Kotz et al., 2001). For other proofs of this property, see Joag-Dev (1989).

Before moving onto the formal developments of the theory of quantiles, we need to linger a little while longer in the 18th century to discuss the origins of an important probability distribution closely related to the median and, as we shall see, to quantiles in general. As mentioned previously, the problem of minimising measurement errors as a means to find the true value of experimental observations posed analytical challenges. Particular attention has been traditionally paid to the normal distribution in virtue of its mathematical properties, which were handy for solving analytic problems before the advent of modern computers. This fact was already known to Laplace who, in 1778, derived the normal distribution as the law by which the frequency of the error is an exponential function of the square of the error (Wilson, 1923). This law proved to be far more analytically tractable than another law derived by the same Laplace in earlier work (Laplace, 1774), a law by which the frequency of an error is expressed as an exponential function of the numerical magnitude of the error, disregarding sign (Wilson, 1923). With the aid of modern notation and parameterisation, we can express this law as

$$f(y) = \frac{1}{2\sigma} e^{-\frac{|y-\xi|}{\sigma}},$$

where $|y - \xi|$ is the error 'disregarding sign', ξ is the 'true' value and σ is a positive scale factor. Today we call it the *Laplace distribution* or double exponential distribution. Keynes (1911, p. 328) discussed the relationship between ξ and the median to answer the question 'If the most probable value of the quantity is equal to the median of the measurements, what is the law of error?'. In other words, the Laplace density is maximised when ξ is the median. An excellent monograph about this distribution is given by Kotz et al. (2001).

The notion of quantiles began to develop about a hundred years after Laplace in the writings of a polymath of the Victorian era. Sir Francis Galton, with his remarkable studies on anthropometry, provided a systematic definition of quantiles and popularised terms such as 'median', 'quartile' and 'percentile' (Galton, 1882, 1885b). Galton's interest in the analysis of measurements and proportions of the human body is epitomised in Figure 1.6 reproducing the main table from his *Nature* article (Galton, 1885a), which presents a concise account of ideas developed more fully in his longer paper in the *Journal of the Anthropological Institute of Great Britain and Ireland* (Galton, 1885b). We quote from the latter (page 276) for his original description:

> The meaning of [the Table], and that of the new word "percentile" which is defined in the heading to that Table, will be understood by the help of a single example, for which I will take the line referring to Strength of Squeeze among males. [...] This line of the Table consequently presents an exact and very complete account of the distribution of strength in one respect among the middle 90 per cent.

Galton understood the importance of ordered data as a means to characterise the distribution of a random variable. Given his inclination for graphical representation of data, he also

ANTHROPOMETRIC PER-CENTILES

Values surpassed, and Values unreached, by various percentages of the persons measured at the Anthropometric Laboratory in the late International Health Exhibition

(*The value that is unreached by n per cent. of any large group of measurements, and surpassed by 100−n of them, is called its nth per-centile*)

Subject of measurement	Age	Unit of measurement	Sex	No. of persons in the group	Values surpassed by per-cents as below										
					95	90	80	70	60	50	40	30	20	10	5
					Values unreached by per-cents. as below										
					5	10	20	30	40	50	60	70	80	90	95
Height, standing, without shoes ..	23-51	Inches	M.	811	63·2	64·5	65·8	66·5	67·3	67·9	68·5	69·2	70·0	71·3	72·4
			F.	770	58·8	59·9	61·3	62·1	62·7	63·3	63·9	64·6	65·3	66·4	67·3
Height, sitting, from seat of chair ...	23-51	Inches	M.	1013	33·6	34·2	34·9	35·3	35·4	36·0	36·3	36·7	37·1	37·7	38·2
			F.	775	31·8	32·3	32·9	33·3	33·6	33·9	34·2	34·6	34·9	35·6	36·0
Span of arms ...	23-51	Inches	M.	811	65·0	66·1	67·2	68·2	69·0	69·9	70·6	71·4	72·3	73·6	74·8
			F.	770	58·6	59·5	60·7	61·7	62·4	63·0	63·7	64·5	65·4	66·7	68·0
Weight in ordinary indoor clothes ..	23-26	Pounds	M.	520	121	125	131	135	139	143	147	150	156	165	172
			F.	276	102	105	110	114	118	122	129	132	136	142	149
Breathing capacity	23-26	Cubic inches	M.	212	161	177	187	199	211	219	226	236	248	277	290
			F.	277	92	102	115	124	131	138	144	151	164	177	186
Strength of pull as archer with bow	23 26	Pounds	M.	519	56	60	64	68	71	74	77	88	82	89	96
			F.	276	30	32	34	36	38	40	42	44	47	51	54
Strength of squeeze with strongest hand	23-26	Pounds	M.	519	67	71	76	79	82	85	88	91	95	100	104
			F.	276	36	39	43	47	49	52	55	58	62	67	72
Swiftness of blow.	23-26	Feet per second	M.	516	13·2	14·1	15·2	16·2	17·3	18·1	19·1	20·0	20·9	22·3	23·6
			F.	271	9·2	10·1	11·3	12·1	12·8	13·4	14·0	14·5	15·1	16·3	16·9
Sight, keenness of —by distance of reading diamond test-type	23-26	Inches	M.	398	13	17	20	22	23	25	26	28	30	32	34
			F.	433	10	12	16	19	22	24	26	27	29	31	32

Figure 1.6
Galton's percentiles. Reproduced from Galton (1885a, Main table p. 224).

introduced the 'ogive'[5] (Galton, 1875), which anticipated what we now call the quantile function. Figure 1.7 reproduces a diagram from his 1875 paper *Statistics by intercomparison, with remarks on the law of frequency of error* (Galton, 1875), which he describes as follows:

> Nothing more is required for the due performance of this process than to be able to say which of the two objects, placed side by side, or known by description, has the larger share of the quality we are dealing with. Whenever we possess this power of discrimination, it is clear that we can marshal a group of objects in the order in which they severally possess that quality. [...] Therefore, when the objects are marshalled in the order of their magnitude along a level base at equal distances apart, a line drawn freely through the tops of the ordinates which represent their several magnitudes will form a curve of double curvature. It will be nearly horizontal over a long space in the middle, if the objects are very numerous; it will bend down at one end until it is nearly vertical, and it will rise up at the other end until there also it is nearly vertical. Such a curve is called, in the phraseology of architects, an "ogive", and is represented by OG in the diagram [...], in which the process of statistics by intercomparison is clearly shown.

Over time, the interest in the applications of quantiles has grown in parallel with advances in the corresponding inference and computing algorithms. The theoretical studies on

[5]The term *ogive* was borrowed from architecture by Galton. The term was later adopted in statistics for cumulative frequency curves.

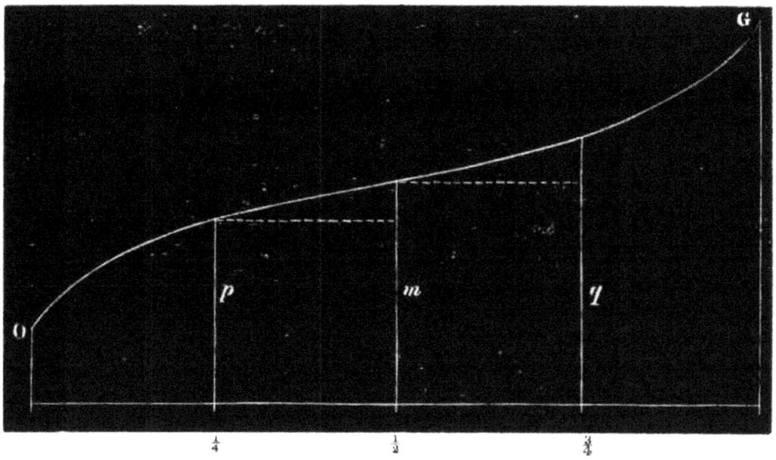

Figure 1.7
Galton's ogive. Reproduced from Galton (1875, Figure 1 p. 36).

unconditional and conditional quantiles of continuous random variables started to appear in the statistical literature of the 20th century. According to David (1995), it seems that the term 'quantile' itself was first used in print by Kendall (1940). In his paper, Kendall showed that sample quantiles of independent and identically distributed observations have an asymptotic normal distribution (see also Hald, 1998, for historical notes on asymptotic studies on unconditional quantiles). Interestingly, the early 1940s (Wilks, 1942) also saw the emergence of the study of *order statistics*, which, as we will see, are quantiles' next of kin. The asymptotic theory of conditional quantile functions is more recent, beginning with the work of Bassett and Koenker (1978) on the median estimator and its generalisation to other quantiles by Koenker and Bassett (1978).

In the case of discrete data, studies have somewhat lagged behind, most probably because of the analytical drawbacks surrounding the discontinuities that characterise discrete quantile functions. Some forms of approximation to continuity have been recently proposed to study the large sample behaviour of quantile estimators. For example, Ma et al. (2011) have demonstrated the asymptotic normality of unconditional sample quantiles based on the definition of mid-distribution function (Parzen, 2004), while Geraci and Farcomeni (2022) extended these results to conditional mid-quantiles. Machado and Santos Silva (2005) proposed inferential approaches to the estimation of conditional quantiles for counts based on data jittering (see, in particular, Chapters 4, 7, and 10 for a formal discussion of quantiles with discrete data). Finally, it is worth mentioning that asymptotic results are also available for unconditional (see, e.g., Oberhofer and Haupt, 2005, and references therein) and conditional (Parente and Santos Silva, 2016) sample quantiles when data are not independent, as well as for nonlinear median functions (Wang, 1995) (these topics are discussed in detail in Chapters 9 and 13).

2

Basic definitions

2.1 Random variables and probability functions

This introductory chapter describes some basic concepts that will recur throughout the book. An excellent treatment of several of the following topics is given in the classic textbook by Mood et al. (1974), from which this chapter draws. We follow a classical approach whereby random variables and distribution functions are introduced first, then followed by quantiles and their conditional counterparts, which form the basis of the modelling framework developed in subsequent chapters.

2.1.1 Univariate

Random variables are the building blocks of statistics. In medical and social sciences, for example, these may represent clinical variables and sociodemographic characteristics of individuals in a given population. A random variable Y is defined as follows.

Definition 1 (Random variable). *Given the probability space* $(\Omega, \mathcal{A}, \Pr(\cdot))$, *a random variable is a function that assigns a real number to outcomes in the sample space* Ω *of a given experiment, namely* $Y : \Omega \to \mathbb{R}$. *Moreover,* $A_y = \{\omega : Y(\omega) \leq y\} \in \mathcal{A}$ *for every* $y \in \mathbb{R}$.

So, for example, if we consider testing a patient for diabetes, the sample space or collection of possible events is given by the set $\Omega = \{\text{no disease, disease}\}$. Then we could define a random variable Y as the function such that $Y : \{\text{no disease, disease}\} \to \{0, 1\}$, which assigns the value 0 to the event 'no disease' and the value 1 to the event 'disease'. Note that these values are completely arbitrary, however they are practical. Analogously, if we consider rolling a die we could define Y : {⚀, ⚁, …, ⚅} $\to \{1, 2, \ldots, 6\}$. And so on. Definition 1, albeit a bit formal, emphasises that our interest lies in the events themselves rather than in the numerical values representing them.

The other key elements of a probability model are an appropriately defined collection of events, typically a σ-algebra $\mathcal{A}$, and a probability function, $\Pr(\cdot)$, that assigns a number between 0 and 1 to the elements of $\mathcal{A}$. The following definition introduces the probabilistic element of a random variable.

Definition 2 (Cumulative distribution function). *The cumulative distribution function (CDF) of a random variable* Y, *denoted by* $F_Y(\cdot)$, *is a function whose domain and codomain are the real line* $\mathbb{R}$ *and the unit interval* $[0, 1]$, *respectively. The CDF of a random variable satisfies* $F_Y(y) = \Pr(Y \leq y) = \Pr(\{\omega : Y(\omega) \leq y\})$ *for every* $y \in \mathbb{R}$.

Based on Definition 2, it is possible to introduce the following properties of a CDF:

i. $F_Y(-\infty) \equiv \lim_{y \downarrow -\infty} F_Y(y) = 0$ and $F_Y(\infty) \equiv \lim_{y \uparrow \infty} F_Y(y) = 1$;

ii. $F_Y(\cdot)$ is nondecreasing;

DOI: 10.1201/9781003326410-2

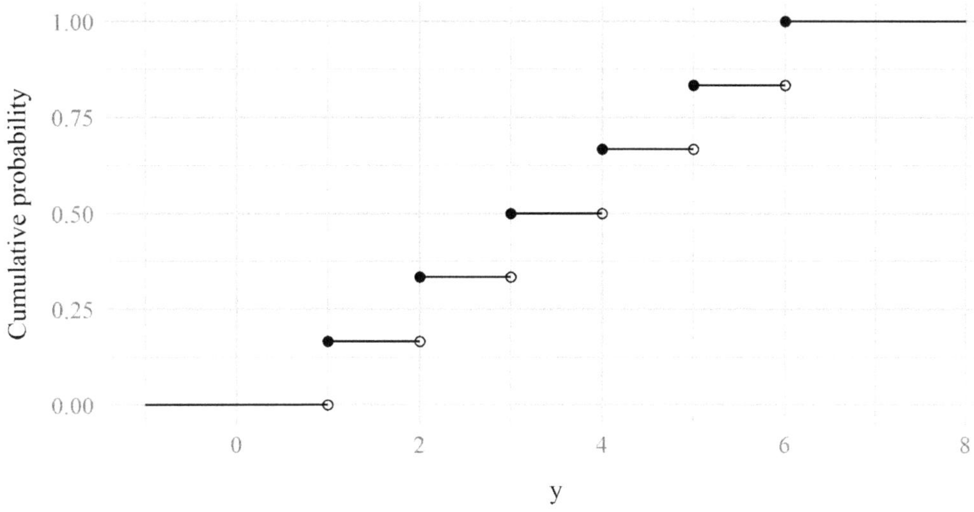

Figure 2.1
Cumulative distribution function for the die-rolling experiment.

iii. $F_Y(\cdot)$ is right-continuous.

The CDF and its properties are illustrated in the following two examples. Let's first consider the experiment with a fair die. Since we defined the variable Y as an integer between 1 and 6, the probability $F_Y(y) = \Pr(Y \leq y)$ for $y < 1$ is identically zero (Figure 2.1). At $y = 1$, the function jumps to $\Pr(Y \leq 1) = \Pr(Y < 1) + \Pr(Y = 1) = 1/6$; at the next integer, the function jumps again to $\Pr(Y \leq 2) = \Pr(Y < 2) + \Pr(Y = 2) = 2/6$; and so forth. Finally, the cumulative probability is equal to 1 and remains constant for any value $y \geq 6$. In Figure 2.1, we used the convention that, at the point of discontinuity or 'jump', the function takes its value corresponding to the ordinate of the filled circle as opposed to that of the hollow circle. This also visually clarifies the meaning of right-continuous. If a point of discontinuity is approached from its right, the value of the function is given by the value of the function *at* that point. The CDF is clearly nondecreasing (i.e., either increasing or constant) as the probability is calculated cumulatively up to the value specified as argument of the function.

Let's now consider another example. Define Y as the birth weight of infants in the 2001 US population born after 37 weeks of gestation. If we cumulate the area under the solid line in Figure 1.3, proceeding from left to right, we obtain the graph in Figure 2.2. The probability that an infant taken at random from the population will have a birth weight less than or equal to, say, 3500 g is easily calculated by looking at the corresponding ordinate, $F_Y(3500) = \Pr(Y \leq 3500) = 0.57$. Also, we have that $F_Y(0) = \Pr(Y \leq 0) = 0$. Although in practice we are not interested in negative or infinite weights (nor have they ever been observed!), the CDF is defined so as to assign probabilities to all relevant events in the algebra $\mathcal{A}$, including the empty set $\{\omega : Y(\omega) \leq 0\} = \emptyset$.

It can be seen that the graphs in Figures 2.1 and 2.2 are conceptually similar, but qualitatively different. This difference leads us to introduce the important distinction between discrete and continuous random variables.

Definition 3 (Discrete random variable)**.** *A random variable is discrete if its image (i.e., the set of distinct values taken by the variable) is a countable set.*

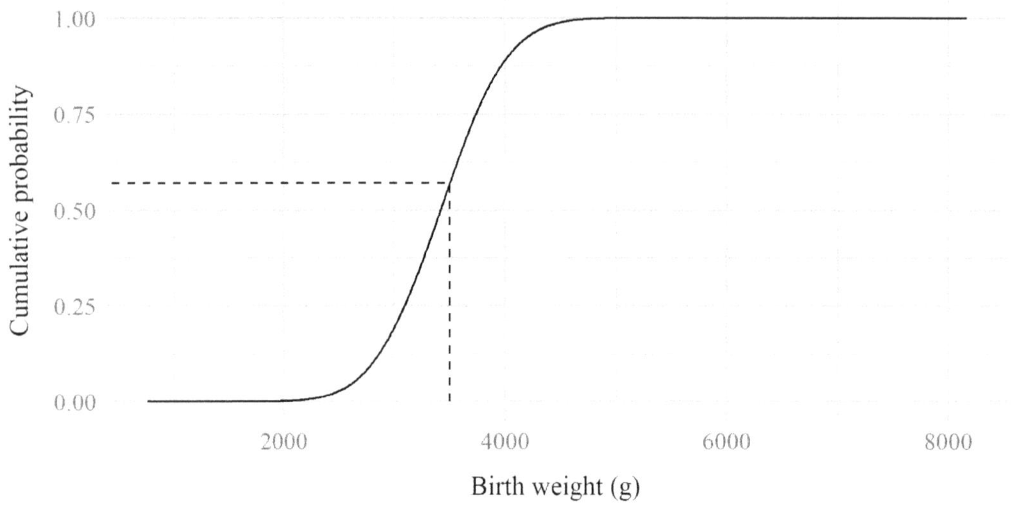

Figure 2.2
Cumulative distribution function estimated for birth weights in the United States population, 2001. Dashed lines mark the coordinates $(y = 3500, F_Y(3500) = 0.57)$.

Rolling a die or testing a patient for diabetes, as in the examples above, are instances of experiments giving rise to discrete random variables. If Y is a discrete variable taking on m distinct values y_j, $j = 1, \ldots, m$, then the function $m_Y(\cdot)$ defined over the real line $\mathbb{R}$ and expressed as

$$m_Y(y) = \begin{cases} \Pr(Y = y_j) & \text{for } y = y_j \\ 0 & \text{for } y \neq y_j \end{cases}$$

is called *probability mass function* (PMF) of Y. It can be noted that this function assigns a probability (mass) equal to $m_Y(y_j)$ to the points y_j, therefore $0 \leq m_Y(y) \leq 1$. Also, we have that $\sum_j^m m_Y(y_j) = 1$. The relationship between the CDF and the PMF of a discrete variable is given by the following equation:

$$F_Y(y) = \sum_{j:y_j \leq y} m_Y(y_j).$$

For example, if Y is the outcome of testing a patient for diabetes we may define the PMF as

$$m_Y(y) = \begin{cases} 1 - u & \text{for } y = 0 \\ u & \text{for } y = 1 \\ 0 & \text{otherwise,} \end{cases}$$

where $u \in (0, 1)$ may be regarded as the prevalence of diabetes in the population. Correspondingly, the CDF is given by

$$F_Y(y) = \begin{cases} 0 & \text{for } y < 0 \\ 1 - u & \text{for } 0 \leq y < 1 \\ 1 & \text{for } y \geq 1, \end{cases}$$

which is a step function with jumps at $y = 0$ and $y = 1$.

By contrast, a continuous random variable can be defined as one whose image is an uncountable set. For example, birth weight is a continuous random variable, as it can take any value in a continuum. The distinction between discrete and continuous random variables, however, can be better appreciated with regard to their probability functions.

Definition 4 (Continuous random variable)**.** *A random variable Y is continuous if and only if there exists a function $f_Y(\cdot)$ such that $F_Y(y) = \int_{-\infty}^{y} f_Y(t)\,\mathrm{d}t$ for every $y \in \mathbb{R}$.*

The function $f_Y(\cdot)$ is called *probability density function* (PDF). We have encountered a particular one in Figure 1.3. The domain and codomain of a PDF are the real line $\mathbb{R}$ and the half line $[0, \infty)$, respectively. Also, we have that $\int_{-\infty}^{\infty} f_Y(t)\,\mathrm{d}t = 1$.

A PDF is conceptually different from a PMF. The latter assigns probability to individual values, whereas the former describes the density of probability. For a continuous random variable, the probability of Y taking any specific value is zero; that is, $\Pr(Y = y) = 0$ for all $y \in \mathbb{R}$. In contrast, it is meaningful to consider events defined over intervals. For example, there would be little point in evaluating the probability of observing a birth weight equal to, say, 3457.26 g, since there are infinitely many possible values. Instead, one may consider the probability of an infant weighing between 3000 and 4000 g. To compute the probability of an event A, we integrate the PDF over A, that is,

$$\Pr(Y \in A) = \int_A f_Y(t)\,\mathrm{d}t.$$

2.1.2 Multivariate

In the previous sections, we discussed univariate distributions. The joint distribution of multiple random variables, say $Y_1, \ldots, Y_k$, carries information about the simultaneous occurrence of events. We refer the reader to Appendix B.1.1 for formal definitions. Figure 2.3 shows the density contours of a bivariate normal distribution. The contours in the figure are *isodensity* contours, that is, lines of constant density $f_{Y_1,Y_2}(y_1, y_2)$. Examples of bivariate variables of interest include height and weight in medical applications, unemployment rate and inflation in macroeconomics, and stock market indices in finance (e.g., see Kotz et al., 2000).

Knowledge of the joint distribution implies knowledge of the *marginal* distributions of each variable Y_j, $j = 1, 2, \ldots, k$. The converse, however, does not generally hold: except in specific cases, the joint distribution cannot be determined from the marginals alone. The relationship between marginal and joint distributions can be studied further through copulas; see Appendix B.2.3.

2.1.3 Conditional

A large number of statistical approaches are based on the idea of describing the probability of an event (conditioned event) *given*, or *conditionally on*, the occurrence of another event (conditioning event). For instance, the probability that an infant is low birth weight (< 2500 g) given that she was born at 37 weeks of gestation; the probability of a housing market crash in a country given the country's withdrawal from the European Union; or the probability of a tornado in Texas given the flap of a butterfly's wings in Brazil. In these examples, the relationships between the conditioning and conditioned events may arise from underlying biological, economic, or natural mechanisms. In statistical applications, such relationships are investigated in terms of stochastic dependence (or lack thereof). Formal definitions are deferred to Appendix B.1.2.

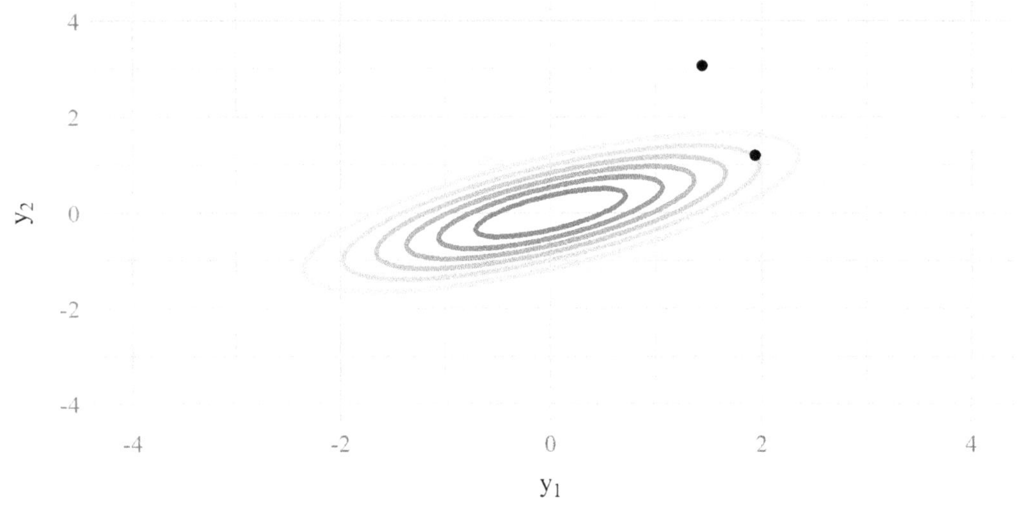

Figure 2.3
Density contours of a bivariate random variable where darker grey corresponds to higher density. The points marked by large dots are discussed further on in Section 2.2.5.

2.2 Quantiles and related functions

2.2.1 Definition

In Chapter 1, we have seen that quantiles characterise the distribution of a random variable by describing how probability is allocated below, above, and between values of the variable.

Definition 5 (Quantile). *The pth quantile, or quantile with index p, of a random variable Y, denoted by ξ_p, is defined as*

$$\xi_p \equiv \inf\{y \in \mathbb{R} : F_Y(y) \geq p\}$$

for any $0 < p < 1$.

It is useful to examine Definition 5 more closely. The expression $F_Y(y) \geq p$ means that the probability that Y is less or equal to y must be *at least* p.[1] Since there may be many

[1]Some authors include the endpoints $p = 0$ and $p = 1$ when defining the pth quantile of a random variable. We prefer to exclude these two values of p (unless noted otherwise) as to avoid any ambiguity or confusion, and for other practical reasons. By Definition 5, the 0-quantile of a distribution is always $-\infty$. This may seem counterintuitive when the distribution $F_Y(y) = 0$ for some finite real value $y \in \mathbb{R}$. For example, in the case of the discrete distribution represented in Figure 2.1 it would be tempting to change the definition of the 0-quantile to $\xi_0 \equiv \sup\{y \in \mathbb{R} : F_Y(y) = 0\}$. However, there are good reasons for not using a different definition of ξ_0 (Embrechts and Hofert, 2013). As for $p = 1$, the CDF reaches this value in the limit when the support of the random variable is unbounded; thus, in that case $\{y \in \mathbb{R} : F_Y(y) \geq 1\} = \emptyset$ (though, by convention, one could take $\inf \emptyset = \infty$). We are not going to get embroiled in these technical aspects. We stress that Definition 5 is appropriate for most of the applications considered in this book, while a discussion on *extremal* quantiles can be found in Section 6.3.3. On that occasion, we will use the typical definition of left and right endpoints of a distribution F, that is, $\alpha(F) = \inf\{x : F(x) > 0\}$ and $\omega(F) = \sup\{x : F(x) < 1\}$, respectively, which may be finite or infinite.

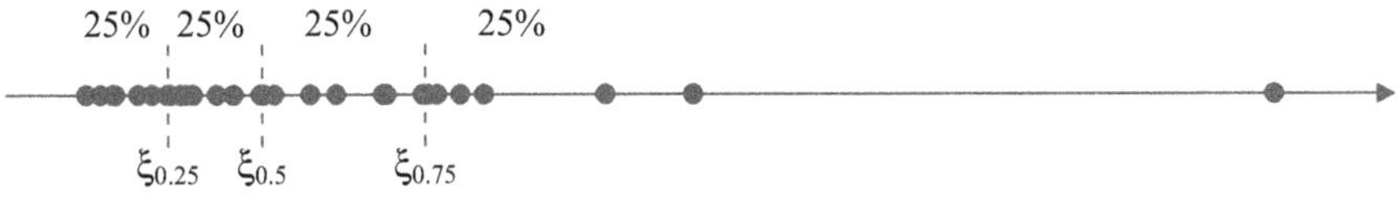

Figure 2.4
Quartiles divide the distribution in four parts weighting 25% each.

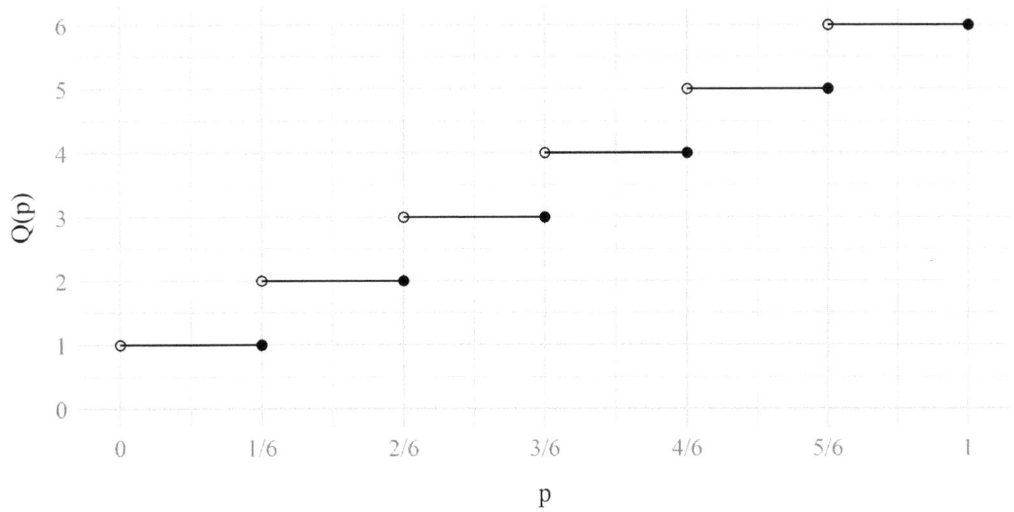

Figure 2.5
Quantile function for the die-rolling experiment.

of such values and following the right-continuousness of F, we take the *infimum*[2] of all the values y such that $F_Y(y) \geq p$.

Quantiles are often called differently as a result of dividing the distribution in a certain number of parts of equal probability. For example, the *deciles* are the quantiles $\xi_{0.1}, \xi_{0.2}, \ldots, \xi_{0.9}$, which partition the distribution in ten parts weighting 10% each. The *quartiles* $\xi_{0.25}, \xi_{0.50}, \xi_{0.75}$ partition the distribution in four parts weighting 25% each (Figure 2.4). Other partitions produce *quintiles* (five parts), *vigintiles* (20 parts) and centiles (100 parts), which are also called percentiles when multiplied by 100. The median is the well-known quantile $\xi_{0.50}$, which splits the distribution into two parts of equal weight.

We may define the *quantile function* (QF) of the random variable Y, $Q_Y(p)$, $0 < p < 1$, as the generalised inverse[3] of the CDF of Y, that is $Q_Y(p) \equiv F_Y^{-1}(p)$. Since, in general, the CDF is not injective, the QF is not the ordinary inverse of the CDF. For example, consider the CDF in Figure 2.1. Its inverse is plotted in Figure 2.5 for $0 < p \leq 1$. The median is $Q_Y(3/6) = 3$. Indeed, $Q_Y(p) = 3$ for any $2/6 < p \leq 3/6$. It can be noted that the discontinuities of the CDF translate into flat segments of the QF and vice versa.

In contrast, if F_Y is continuous and strictly increasing, then the QF reduces to the ordinary inverse, that is, $F_Y^{-1}(p) = y$ when $p = F_Y(y)$. The QF of a continuous variable is

---

[2] In typical situations encountered in the present book, the infimum coincides with the minimum of the set and therefore the quantile is defined as the smallest of the values y such that $F_Y(y) \geq p$.

[3] Often the generalised inverse of a function g is denoted by g^- to distinguish it from the ordinary inverse g^{-1}. We will not use this notational distinction.

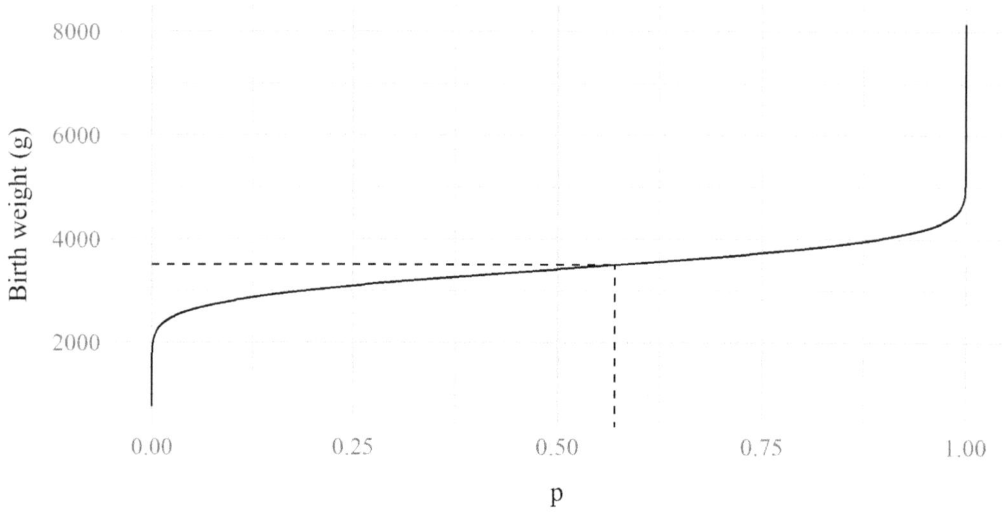

Figure 2.6
Quantile function as inverse of the cumulative distribution function plotted in Figure 2.2 for birth weights in the United States population, 2001. Dashed lines mark the coordinates $(p = 0.57, Q_Y(0.57) = 3500)$.

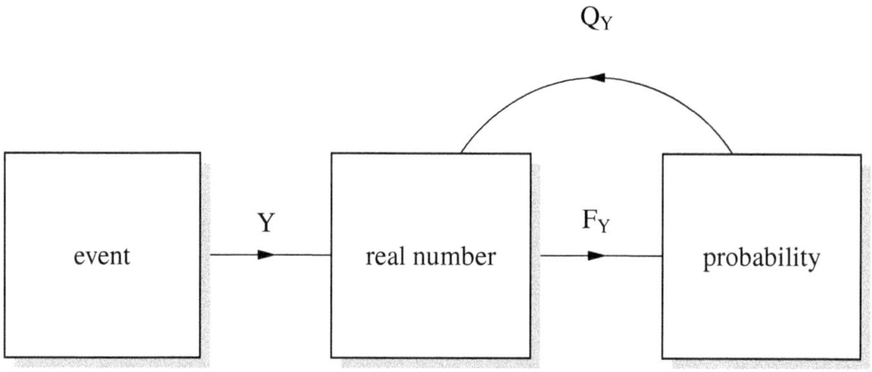

Figure 2.7
Diagram depicting the functional relationships between events, numbers and probabilities.

also easier to plot. Its graph is the rotation of the mirror image of $p = F_Y(y)$ (Figure 2.6). This suggests that the CDF can be regained from its QF with

$$F_Y(y) = \sup\{p \in (0, 1) : Q_Y(p) \leq y\},$$

for $y \in \mathbb{R}$.

Figure 2.7 provides a diagram that conceptually summarises the most important functional relationships considered so far. We move from events to numbers by means of random variables, and from the latter to probabilities by means of CDFs. When the domain of the problem is stated in terms of probability, we can easily work out the corresponding values of the random variable by means of QFs.

2.2.2 Properties

Quantiles enjoy a number of properties, which provide useful rules in the simulation of random variables and the construction of statistical models.

The QF is nondecreasing and left-continuous, and, for $-\infty < y < \infty$ and $0 < p < 1$, satisfies

i. $Q_Y(F_Y(y)) \leq y$;

ii. $F_Y(Q_Y(p)) \geq p$;

iii. $F_Y(y) \geq p$ if and only if $y \geq Q_Y(p)$.

From the last property it follows a result that is particularly useful to generate random variables with a given probability distribution. In the literature, it has been called uniform transformation rule (Gilchrist, 2000) or quantile transformation (Embrechts et al., 1997), but it is more widely known as (inverse) probability integral transformation. The uniform transformation rule for the bivariate case is given in Appendix B.2.1 (Rosenblatt representation).

Proposition 1 (Uniform transformation rule). *Let Y be a random variable with CDF F_Y and let U be a standard uniform distribution. Then the following results hold:*

(a) Y is identically distributed as $F_Y^{-1}(U)$, and we write $Y \stackrel{d}{=} F_Y^{-1}(U)$.

(b) $F_Y(Y) \stackrel{d}{=} U$ if and only if F_Y is continuous.

Proof. Result (a) follows immediately from the identity (Parzen, 1979)

$$\Pr(F_Y^{-1}(U) \leq y) = \Pr(U \leq F_Y(y)) = F_Y(y).$$

This result applies to both continuous and discrete random variables (Parzen, 2004).

To prove (b), define $Z = F_Y(Y)$ for any continuous random variable Y. Then

$$\begin{aligned} F_Z(z) &= \Pr(Z \leq z) = \Pr(F_Y(Y) \leq z) \\ &= \Pr(Y \leq F_Y^{-1}(z)) = F_Y(F_Y^{-1}(z)) = z. \end{aligned}$$

If F_Y is discrete, the last equality in the expression above may not hold. For example, one can easily verify that $F(F^{-1}(p)) > p$ if $p \neq j/6$, $j = 1, 2, \ldots, 6$, for the distribution depicted in Figure 2.5.

□

Remark 1. *One can obtain the counterpart of part (b) in Proposition 1 for a discrete random variable by interpolating F_Y between successive points (Reiss, 1989, p. 34).*

The following rules (Gilchrist, 2000) show that quantiles have 'algebraic' properties that make them particularly suitable for statistical modelling:

iv. $Q_{a+bY}(p) = a + bQ_Y(p)$, $a \in \mathbb{R}$, $b \in \mathbb{R}_{\geq 0}$ (**location–scale-shift rule**);

v. if $Q_1(p)$ and $Q_2(p)$ are two QFs, then their pointwise sum is also a valid QF (**addition rule**);

vi. if $Q_1(p)$ and $Q_2(p)$ are two QFs, then $\pi Q_1(p) + (1 - \pi)Q_2(p)$, $0 \leq \pi \leq 1$, defines a QF lying pointwise between $Q_1(p)$ and $Q_2(p)$ (**intermediate rule**);

vii. if $Q_1(p)$ and $Q_2(p)$ are two positive QFs, then their pointwise product is also a valid QF (**multiplication rule**);

viii. if $Q_Y(p)$ is the QF of Y, the reflection $-Q_Y(1-p)$ is the QF of $-Y$, that is, $-Q_Y(1-p) = Q_{-Y}(p)$ (**reflection rule**);

ix. if $Q_Y(p)$ is the QF of a strictly positive Y, the reciprocal $1/Q_Y(1-p)$ is the QF of $1/Y$, that is, $1/Q_Y(1-p) = Q_{1/Y}(p)$ (**reciprocal rule**);

x. if $h(\cdot)$ is a nondecreasing function[4] on $\mathbb{R}$, then $Q_{h(Y)}(p) = h\{Q_Y(p)\}$. Hence $Q_Y(p) = h^{-1}\{Q_{h(Y)}(p)\}$ (**Q-transformation rule** or equivariance to monotone transformations);

xi. if $h(\cdot)$ is a nondecreasing function on $(0,1)$ such that $\lim_{t\downarrow 0} h(t) = 0$ and $\lim_{t\uparrow 1} h(t) = 1$, then $Q_Y(h(p))$ is a valid QF (**P-transformation rule**).

In general, the expected value is not equivariant to monotone transformations. For example, if h is convex, then $h\{\mathrm{E}(Y)\} \leq \mathrm{E}\{h(Y)\}$ (Jensen's inequality), with equality holding if and only if h is linear. This may seem a minor issue in some applications, but the following simple numerical example shows that by ignoring this rule one may let gross approximations creep into the analysis. Consider the set of numbers $\{1, 2, 3\}$. The median and the mean are both equal to 2. Now let us apply the natural logarithm, thus $h(\cdot) = \log(\cdot)$ and $h^{-1}(\cdot) = \exp(\cdot)$. We obtain $\{0, 0.693, 1.099\}$, whose median and mean are 0.693 and 0.597, respectively. We then apply the inverse transformation to the median of the log-transformed numbers and obtain $\exp(\log 2) = 2$, which is exactly the median of the original set. However, $\exp(0.597) = 1.817$, which is clearly different from the mean of the original set; if used as an approximation, it would be a nonnegligible 9% underestimate of the true mean. In research studies, it is not uncommon to find this kind of approximations, which therefore introduce unnecessary numerical errors whose direction and magnitude depend on the distribution of Y.

2.2.3 Related functions

We now introduce a number of related functions that will recur in other parts of the book. These are particularly useful for characterising continuous random variables, and less so in the discrete case. We start with the *sparsity function* of a continuous random variable (Parzen, 1979).

Definition 6 (Sparsity function). *The* sparsity function *of a random variable Y, denoted by q, is defined as*

$$\frac{\partial Q(p)}{\partial p} \equiv q(p), \tag{2.1}$$

for any $0 < p < 1$, where $Q(p)$ is the QF of Y.

The reason for calling 'sparsity' the quantity in (2.1) (Tukey, 1965) can be easily seen from the following identities:

$$\begin{aligned} F\left(Q(p)\right) &= p, \\ \frac{\partial F\left(Q(p)\right)}{\partial p} &= \frac{\partial p}{\partial p}, \\ f\left(Q(p)\right) q(p) &= 1. \end{aligned}$$

[4]The function $h(\cdot)$ need not be continuous.

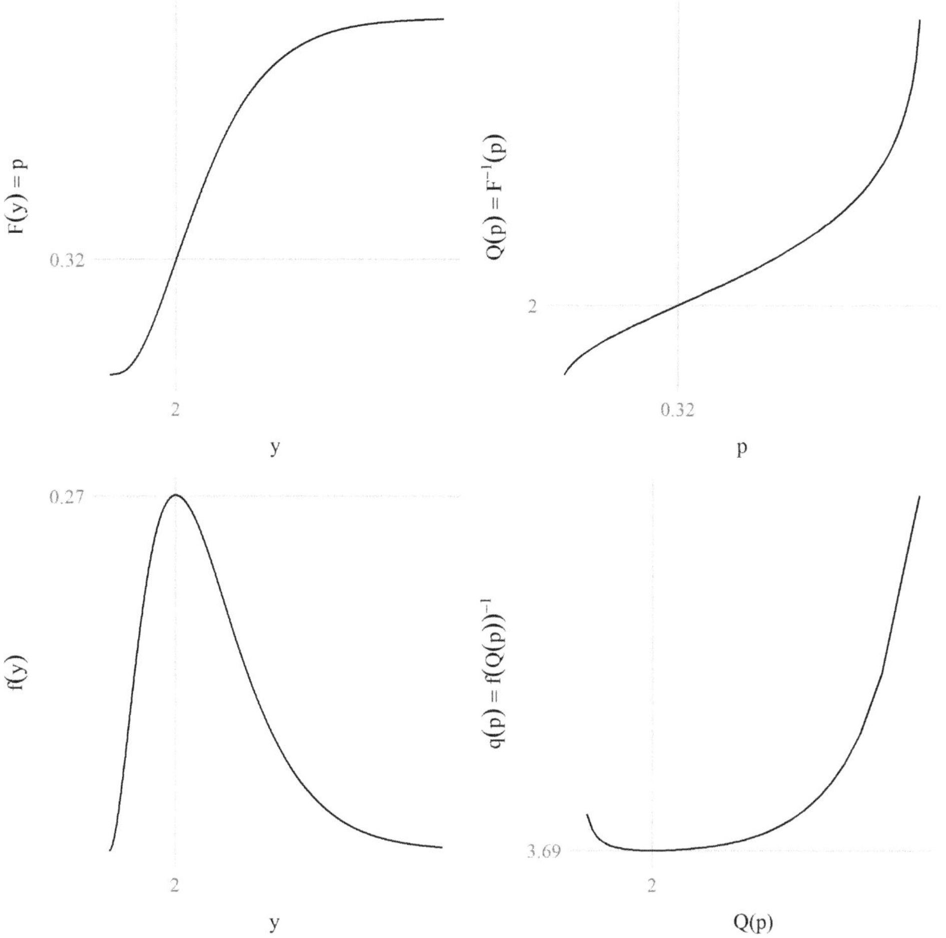

Figure 2.8
Density (lower left), cumulative distribution (upper left), sparsity (lower right), and quantile (upper right) functions of a gamma distribution (with shape parameter 3 and rate parameter 1). Grid lines mark the modal quantile $Q(p^*)$.

Hence

$$q(p) = [f(Q(p))]^{-1},$$

for $f(Q(p)) > 0$. The sparsity at p is the reciprocal of the density function evaluated at the quantile of order p.[5] We can use (2.1) for different purposes. For example, if Y is unimodal,

$$p^* = \underset{p}{\operatorname{argmax}} \frac{1}{q(p)},$$

indexes the *modal* quantile $Q(p^*)$. A graphical summary for a skewed distribution is given in Figure 2.8.

[5]The sparsity function $q(p)$ has been referred to as *quantile density function* by numerous authors (e.g., see Parzen, 1979; Falk, 1986; Jones, 1992), while $f\left(F^{-1}(p)\right)$ has been called *density quantile function* (Parzen, 1979) or p-PDF (Gilchrist, 2000) to stress that it is the PDF expressed in terms of p.

The composition $F\left(F^{-1}(p)\right)$ introduced above is a special case of the comparison distribution function. In general, we may have two distribution functions and we are interested in their composition.

Definition 7 (Comparison distribution functions). *Let F and G denote two CDFs for a common random variable Y. The comparison cumulative distribution function for the distributions F and G is defined as*

$$\upsilon(p; F, G) = G\left(F^{-1}(p)\right), \tag{2.2}$$

for any $0 < p < 1$. Conversely, we define the comparison quantile function for the distributions F and G as

$$\kappa(y; F, G) = G^{-1}\left(F(y)\right), \tag{2.3}$$

for any y such that $0 < F(y) < 1$.

The comparison distribution functions have many applications. For example, they are used in the analysis of a distribution's shape (van Zwet, 1964; Oja, 1981) and are the basis of the P–P and Q–Q plots, which are diagnostic tools to assess departures from theoretical probability models (Parzen, 1993, 2004). More generally, they provide means for comparing distributions. Clearly, $\upsilon(p; F, G) = p$ and $\kappa(y; F, G) = y$ if F and G are monotone increasing and $F \equiv G$.

If F and G are continuous, we obtain the *comparison density function*, say d, by differentiating υ

$$\frac{\partial \upsilon(p; F, G)}{\partial p} \equiv d(p; F, G) = \frac{g\left(F^{-1}(p)\right)}{f\left(F^{-1}(p)\right)}. \tag{2.4}$$

If F and G are discrete, we define d in terms of the PMFs

$$d(p; F, G) = \frac{m_G\left(F^{-1}(p)\right)}{m_F\left(F^{-1}(p)\right)}. \tag{2.5}$$

The comparison distribution functions can be used to define first-order stochastic dominance, a concept that is strictly related to the quantile treatment effect which will be discussed in Section 3.2.2.

Definition 8 (First-order stochastic dominance). *Let F and G be CDFs. We say that G first-order stochastically dominates F if and only if one (and hence all) of the following equivalent conditions holds:*

- $G(y) \leq F(y)$ *for all* $y \in \mathbb{R}$;
- $\upsilon(p; F, G) = G\big(F^{-1}(p)\big) \leq p$ *for all* $p \in (0, 1)$;
- $\kappa(y; F, G) = G^{-1}(F(y)) \geq y$ *for all* $y \in \mathbb{R}$;

with strict inequality for at least one $y \in \mathbb{R}$ (equivalently, for at least one $p \in (0, 1)$).

We have now a fairly complete picture of the main functions that can be used to define a distribution and how they relate one to each other. A summary of these relationships in the continuous case is provided in Table 2.1.

Table 2.1
Relationships between distribution functions of a continuous random variable.

	F	f	Q	q
F		F'	F^{-1}	$1/F'(F^{-1})$
f	$\int f$		$\int 1/f$	$1/f$
Q	Q^{-1}	$1/Q'(Q^{-1})$		Q'
q	$\int 1/q$	$1/q$	$\int q$	

2.2.4 Conditional

Conditional quantiles play a central role in statistical modelling throughout this book. We begin with the simplest setting involving a single conditioning variable.

Let Y and X be two random variables defined on the same probability space. The *conditional quantile function* of Y given $X = x$ is defined as

$$Q_{Y|X=x}(p) \equiv F^{-1}_{Y|X=x}(p), \quad 0 < p < 1,$$

where $F_{Y|X=x}$ denotes the conditional distribution function of Y given $X = x$, and $F^{-1}_{Y|X=x}$ its generalised inverse.

For each fixed value of x, the function $Q_{Y|X=x}(p)$ inherits all the properties of a QF discussed in Section 2.2.2. In particular, it is nondecreasing and left-continuous in p. Thus, conditional quantiles provide a complete description of the distribution of Y at each value of the conditioning variable.

Unlike classical approaches that focus on conditional means, conditional quantiles allow us to study how different parts of the distribution of Y vary with X. This makes them particularly suitable for investigating heterogeneity, tail behaviour, and distributional effects, which are often of primary interest in applications.

Further theoretical results linking conditional and marginal quantiles, as well as their probabilistic representations, are given in Appendix B.2. These results provide useful insights but are not required for the main development of the methods presented in this book.

In many applications, the conditioning variable is multivariate. Let $\boldsymbol{X} = (X_1, \ldots, X_k)$ denote a vector of covariates (which may be all continuous, all discrete, or a mix of continuous and discrete variables). The conditional QF of Y given $\boldsymbol{X} = \boldsymbol{x}$ is defined as

$$Q_{Y|\boldsymbol{X}=\boldsymbol{x}}(p) \equiv F^{-1}_{Y|\boldsymbol{X}=\boldsymbol{x}}(p), \quad 0 < p < 1.$$

For simplicity, we will often write $Q_Y(p \mid \boldsymbol{x})$, or $Q_{Y|\boldsymbol{X}}(p)$ when no ambiguity arises.

Conditional quantile functions and their estimation will be studied extensively in Parts III and IV, where they form the basis of regression models and inferential procedures.

2.2.5 Multivariate

Quantiles are straightforward to define in the univariate case because the real line is naturally ordered. In multivariate settings, however, the lack of a natural ordering makes the definition of quantiles considerably more challenging. Since multivariate quantiles are not pursued further in this book, we limit ourselves to a brief conceptual overview.

As in the univariate case, the joint CDF returns a probability. However, the lack of a natural ordering in multidimensional space means that there may be multiple vectors $\boldsymbol{y} \in \mathbb{R}^k$, $k > 1$, such that $F_{\boldsymbol{Y}}(\boldsymbol{y}) = p$, $0 < p < 1$ (indeed, there is an infinite number of such vectors when $\boldsymbol{Y}$ is continuous). For example, the large dots in Figure 2.3 mark the coordinates of two vectors $\boldsymbol{y}_1 = (y_{11}, y_{12})^\top$ and $\boldsymbol{y}_2 = (y_{21}, y_{22})^\top$ such that $F_{\boldsymbol{Y}}(\boldsymbol{y}_1) = F_{\boldsymbol{Y}}(\boldsymbol{y}_2) = 0.84$. Hence, according to the univariate definition, both $\boldsymbol{y}_1$ and $\boldsymbol{y}_2$ would be quantiles of order 0.84. The definition of quantiles for multivariate data is therefore problematic. Nevertheless, it is possible to consider alternatives to the definition based on the linear ordering of univariate random variables.

Several approaches have been proposed to overcome the lack of ordering (e.g., see Koenker et al., 2018, for a review), most of which replace the idea of a single point-valued quantile with that of a region or contour.

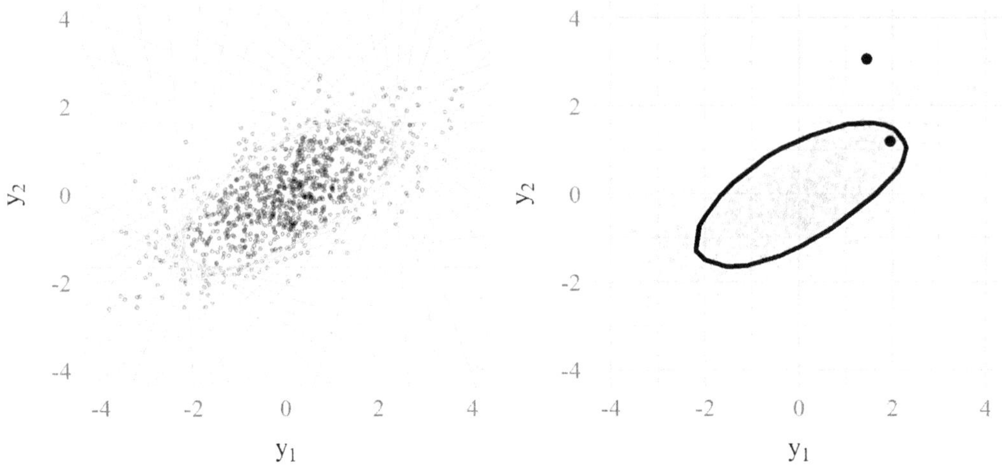

Figure 2.9
Bivariate normal sample with directional quantile hyperplanes (left, grey lines) and their envelope (right, black polygon), corresponding to a projection-based quantile of order $p = 0.05$. The envelope is obtained as the intersection of halfspaces defined by the directional quantiles. Marked points correspond to those in Figure 2.3.

A first class of definitions relies on statistical depth. In particular, halfspace (Tukey) depth assigns to each point $\boldsymbol{y} \in \mathbb{R}^k$ the smallest probability mass contained in any closed halfspace having $\boldsymbol{y}$ on its boundary. Quantile regions of order p are then defined as the sets of points with depth at least p. These regions are convex, nested, and provide a centre-outward ordering of the space. In this sense, multivariate quantiles are no longer points but contours, with deeper points corresponding to more central observations. Importantly, these depth contours can be interpreted as multivariate analogues of univariate quantiles and enjoy appealing geometric properties such as affine equivariance (Hallin et al., 2010).

A second, conceptually simpler approach is based on projections. For any direction $\boldsymbol{d} \in \mathbb{R}^k$, consider the scalar projection $\boldsymbol{d}^\top \boldsymbol{Y}$ and define its univariate pth quantile, $\xi_p = Q_{\boldsymbol{d}^\top \boldsymbol{Y}}(p)$. Collecting these quantiles over all directions yields a family of supporting hyperplanes whose envelope defines a multivariate quantile region. This construction, called the directional quantile envelope (DQE), has a clear probabilistic interpretation, as it reduces the problem to familiar univariate quantiles along all possible directions. Moreover, under suitable definitions, these projection-based envelopes coincide with halfspace depth contours, thereby linking the two approaches (Kong and Mizera, 2012). An illustration of the DQE for $p = 0.05$ is provided in Figure 2.9 using a sample from the bivariate normal distribution depicted in Figure 2.3. For a given direction $\boldsymbol{d} \in \mathbb{R}^2$, the supporting halfspace (left plot, grey lines) is

$$H(\boldsymbol{d}, \xi_{0.05}) = \{\boldsymbol{y} : \boldsymbol{d}^\top \boldsymbol{y} \geq \xi_{0.05}\}.$$

The intersection of many such halfspaces yields the DQE (right plot, black polygon). The two points $\boldsymbol{y}_1$ and $\boldsymbol{y}_2$, although sharing the same cumulative probability, lie one inside and one outside the DQE. Thus, one point is more extreme than the other. This may be informative in applications such as risk assessment (Geraci et al., 2019).

Although conceptually useful, these multivariate notions are included only for completeness. In the remainder of the book, the focus is on univariate quantiles.

2.3 Moments of a random variable

2.3.1 Classical definition

A moment of a random variable (or, equivalently, of its distribution) is the expected value of the variable raised to a given power, provided that the expectation exists. There are different types of moments, defined as follows.

Definition 9 (Moments). *The* r*th raw moment (or moment about the origin)* μ_r *of a random variable* Y *is defined as* $\mu_r \equiv \mathrm{E}(Y^r)$. *The* r*th central moment (or moment about the mean)* $\bar{\mu}_r$ *is defined as* $\bar{\mu}_r \equiv \mathrm{E}\left[(Y - \mu_1)^r\right]$, *where* μ_1 *is the first raw moment of* Y. *The standardised* r*th (central) moment* $\bar{\bar{\mu}}_r$ *is defined as* $\bar{\bar{\mu}}_r \equiv \mathrm{E}\left[(Y - \mu_1)^r / \sigma^r\right]$, *where* $\sigma = \sqrt{\bar{\mu}_2}$.

In Section 1.1, we encountered instances of moments that are commonly used in applied statistics: the mean, μ_1, the variance, $\bar{\mu}_2$, the skewness index, $\bar{\bar{\mu}}_3$, and the kurtosis index, $\bar{\bar{\mu}}_4$. A compact way to represent and derive moments is through the *moment generating function* (MGF) $M_Y(t)$ defined as

$$M_Y(t) \equiv \mathrm{E}\left(e^{tY}\right) = \begin{cases} \sum_{y \in S_Y} e^{ty} m_Y(y) & \text{if } Y \text{ is discrete,} \\ \int_{S_Y} e^{ty} f_Y(y)\, \mathrm{d}y & \text{if } Y \text{ is continuous.} \end{cases}$$

It is easy to show that

$$\mu_r = \left.\frac{\mathrm{d}^r}{\mathrm{d}t^r} M_Y(t)\right|_{t=0}.$$

There are some potential drawbacks in using moments to characterise a probability distribution. The first important question is whether an MGF uniquely characterise a distribution. Under mild conditions, the answer is positive since the MGF completely determines the distribution of a random variable. However, two very different distributions can have almost numerically identical MGFs (McCullagh, 1994). A second possible drawback is related to whether moments always exist. The answer is negative and has practical consequences. For example, nonexistence of moments can occur in heavy-tailed distributions, such as the Cauchy distribution, which casts serious concerns on the fourth standardised moment as measure of tail heaviness. Finally, moments are sensitive to outliers. This is a well-known problem and has important consequences in practical applications.

Finally, we can define the joint (or product or mixed) moments of a bivariate random variable $\boldsymbol{Y} = (Y_1, Y_2)$. The most popular of such moments is the *covariance*, which is a central product moment. If $\boldsymbol{Y}$ is continuous, then the covariance is defined as

$$\begin{aligned} \operatorname{cov}(Y_1, Y_2) &= \mathrm{E}\{(Y_1 - \mathrm{E}(Y_1))(Y_2 - \mathrm{E}(Y_2))\} \\ &= \int_{-\infty}^{\infty}\int_{-\infty}^{\infty} (y_1 - \mathrm{E}(Y_1))(y_2 - \mathrm{E}(Y_2)) f_{Y_1,Y_2}(y_1, y_2)\, \mathrm{d}y_1\, \mathrm{d}y_2. \end{aligned}$$

2.3.2 Quantile-based definition

Let Y be a random variable with distribution function F, and let m and f denote, respectively, the PMF of Y (if discrete) or the PDF of Y (if continuous). In classical introductory textbooks, the expected value of Y is typically defined as a weighted average given by

$$\mathrm{E}(Y) = \begin{cases} \sum_{j: y_j \in S_Y} y_j\, m(y_j) & \text{if } Y \text{ is discrete,} \\ \int_{S_Y} y f(y)\, \mathrm{d}y & \text{if } Y \text{ is continuous,} \end{cases}$$

where S_Y denotes the support of Y. The summation and Riemann integral for, respectively, discrete and continuous distributions can be unified under the Riemann–Stieltjes integral

$$\mathrm{E}(Y) = \int_{-\infty}^{\infty} y \, \mathrm{d}F(y), \tag{2.6}$$

provided that this integral exists. It can be very instructive to change variable in the integral of Equation (2.6). Let $p = F(y)$, so $dp = dF(y)$ and $y = Q(p)$. Then

$$\mathrm{E}(Y) = \int_0^1 Q(p) \, \mathrm{d}p. \tag{2.7}$$

It is argued (Parzen, 2004, p. 654) that '[q]uantile thinking defines statistics as summation done by sorting (ranking) data before adding', a Galtonian-inspired principle which we have discussed in Chapter 1.

Let $g(\cdot)$ be a real function. Equation (2.7) can be generalised to $\mathrm{E}(g(Y)) = \int_0^1 g(Q(p)) \, \mathrm{d}p$ (Parzen, 1979), which motivates the following:

$$\mu_r = \int_0^1 \{Q(p)\}^r \, \mathrm{d}p, \tag{2.8}$$

$$\bar{\mu}_r = \int_0^1 \{Q(p) - \mu_1\}^r \, \mathrm{d}p, \tag{2.9}$$

$$\bar{\bar{\mu}}_r = \frac{\bar{\mu}_r}{(\bar{\mu}_2)^{r/2}}. \tag{2.10}$$

Thus, for example, $\mathrm{var}(Y) = \int_0^1 \{Q(p) - \mu_1\}^2 \, \mathrm{d}p$. In contrast to univariate moments, joint moments depend on the dependence structure, which is captured by a copula C (see Definition (23)). For example,

$$\mathrm{cov}(Y_1, Y_2) = \int_0^1 \int_0^1 \left\{ Q_{Y_1}(u) - \int_0^1 Q_{Y_1}(p) \, \mathrm{d}p \right\} \left\{ Q_{Y_2}(v) - \int_0^1 Q_{Y_2}(p) \, \mathrm{d}p \right\} \mathrm{d}C(u, v),$$

so that both marginal and joint features are expressed entirely in terms of quantile functions.

A particularly important class of functionals arises when the integrand is linear in the quantile function. Specifically, let $w : [0,1] \to \mathbb{R}$ be a weight function such that $\int_0^1 |w(p)| \, dp < \infty$. Then

$$L(F) = \int_0^1 w(p) \, Q(p) \, \mathrm{d}p \tag{2.11}$$

defines an *L-functional.*[6] A discrete version of (2.11) is obtained by replacing the integral with a finite weighted sum,

$$L = \sum_{j=1}^{m} w_j \, Q(p_j), \tag{2.12}$$

where $0 < p_1 < \cdots < p_m < 1$ and the weights w_j approximate $w(p)$. Replacing $Q(p_j)$ with sample order statistics (Section 4.1) yields the usual definition of *L-statistics.*

[6] Although defined in terms of the quantile function, L-functionals admit the equivalent representation

$$L(F) = \int w\{F(y)\} \, y \, \mathrm{d}F(y),$$

which highlights that the weights depend on the rank $F(y)$ rather than on y itself.

Equation (2.7) shows that the mean is the special case of (2.11) with constant weight $w(p) \equiv 1$. More generally, L-functionals correspond to weighted averages of the QF, and hence to weighted averages of the ordered sample values. A notable example is the trimmed mean, obtained by taking $w(p) \propto I(\alpha \leq p \leq 1-\alpha)$ for some $\alpha \in (0, 1/2)$,

$$\mu_\alpha = \frac{1}{1-2\alpha}\int_0^1 I(\alpha \leq p \leq 1-\alpha) Q_Y(p)\,\mathrm{d}p, \tag{2.13}$$

which assigns zero weight to the lower and upper $(100 \cdot \alpha)\%$ tails and constant weight elsewhere.

2.4 Statistical modelling and inference

2.4.1 Objectives and approaches

Statistical inference concerns the process of drawing conclusions about a population from observations obtained on a sample of statistical units belonging to the population. This is an inductive process consisting of several stages, typical of scientific experiments, simplified as follows:

1. identification of the population;
2. sampling – that is, selection by means of *random* procedures of a subset of the population and collection of data;
3. modelling and inference – that is, choice of a family $\mathcal{M}$ of probability models;
4. validation.

Commonly, the target of the inference concerns a parameter of interest, say θ, representing an unknown feature of the population, such as a population mean or a probability function. Parameters that are not of immediate interest, yet accounted for in the analysis, are called *nuisance* parameters.

A population may be finite or infinite, real or hypothetical; in any case, it needs to be well defined. The sampling procedure is ideally designed *before* the data collection takes place so that the random mechanism of selection of the sample units is known exactly. A random sample is formally defined as follows.

Definition 10 (Random sample). *A collection of random variables $Y_1, \ldots, Y_n$ is a random sample of size n from a population with CDF F_Y if the Y_is are mutually independent, $Y_i \perp\!\!\!\perp Y_j$, $i \neq j$, and have common distribution $F_{Y_i} = F_Y(y)$, $i = 1, \ldots, n$.*

This is the definition of the random sample *par excellence*. The acronym IID, which stands for *independent and identically distributed*, is conventionally adopted to indicate that the probability distribution for one unit in the sample is the same for all units, that is, $F_{Y_1}(y) = \ldots = F_{Y_n}(y)$, for all $y \in \mathbb{R}$, and that the sample observations have been collected in such a way that the probability F_{Y_i}, $i = 1, \ldots, n$, is not affected by the value taken by any other variable Y_j, $i \neq j$. However, everyday research shows that Definition 10 often clashes with empirical evidence since either or both of the two IID attributes cannot be assumed. While observations may still be independent, they are often not identically distributed (NID). In this situation, the F_{Y_i}s may differ from one another in terms of location, or scale, or shape (see Chapter 3), or any combinations of these, and modelling typically requires more effort than when there is simply lack of independence.

Modelling and inference (on which we will focus during the remainder of the book) take on specific forms depending on whether the approach is parametric, nonparametric or semiparametric.

In the classical *parametric* framework, the family $\mathcal{M}$ is completely specified except for a low-dimensional vector of parameters $\boldsymbol{\theta}$. Cox and Hinkley (1997) identify two main questions related to the stage of modelling and inference, namely (a) whether the sampled observations are consistent with the family $\mathcal{M}$; and (b) what can be said about the unknown parameter $\boldsymbol{\theta}$, or the prediction of observations not included in the sample. For example, in Section 1.2 we considered the distribution of birth weights. Let Y represent the birth weight of a singleton born at term in the US population (Figure 1.3). (For the moment, we ignore the fact that population birth weight data do not come from a random sampling procedure.) We have seen that a reasonable family $\mathcal{M}$ of models for Y is the normal distribution[7] with mean μ and variance σ^2. Thus, in the modelling step, we have graphically assessed the conformity of the data with $\mathcal{M}$ (Figure 1.3). Under the assumption $Y_1, \ldots, Y_n \overset{\text{IID}}{\sim} \mathcal{N}\left(\mu, \sigma^2\right)$, our inference would then focus on $\boldsymbol{\theta} = (\mu, \sigma)$, a two-dimensional parameter whose knowledge would completely determine the birth weight distribution.

In a *nonparametric* approach, the parameter of interest is often an entire function and therefore belongs to an infinite-dimensional space. For instance, a family $\mathcal{M}$ of nonparametric models for the distribution of birth weights could be the class of all symmetric and continuous densities. In this case, the parameter of interest is the density f. In general, a nonparametric method may rely on parametric distributions. For example, consider the model

$$Y_i = g\left(\boldsymbol{x}_i\right) + \varepsilon_i,$$

where $\boldsymbol{x}_i$ is a vector of known birth weight determinants, $\varepsilon_i \overset{\text{IID}}{\sim} \mathcal{N}\left(0, \sigma^2\right)$ and the distribution of ε_i is independent of $\boldsymbol{x}_i$. In a typical setting, a nonparametric estimator of the unknown function g would be one that can be used for a broad class of functions, for example, the class of smooth functions.

Distribution-free methods are often labelled as nonparametric, although they involve different ideas (Cox and Hinkley, 1997; Stuart et al., 1999). For example, the model

$$Y_i = \boldsymbol{x}_i^\top \boldsymbol{\beta} + \varepsilon_i,$$

where $\varepsilon_i \overset{\text{NID}}{\sim} F_{\varepsilon_i}$ and the form of F is not specified, could be regarded as distribution-free. A distinction has to be made also between robust, nonparametric and distribution-free procedures. *Robustness* is a notion that will recur often throughout the present book since it represents a salient property of quantiles. Robustness can be defined as the insensitivity of the procedure to small departures from the assumptions of the theoretical model (Huber, 1981). Nonparametric procedures may or may not be robust. This is also true for distribution-free methods, although most of these have also good robustness properties (Huber, 1981). Moreover, *resistant* procedures (i.e., a procedure insensitive to small changes in the underlying sample) can be classified as robust, although, again, there are conceptual differences (Huber, 1981).

Finally, semiparametric models may be considered as an hybrid of parametric and nonparametric components, whereby the parameter of interest is finite-dimensional and the nuisance parameter is infinite-dimensional (Begun et al., 1983). This would be the case of

[7]Contrary to this statement, we shall see in Section 6.3.3 that the normal distribution is not reasonable to model extreme birth weights.

the distribution-free model introduced above, where $\boldsymbol{\beta}$ is the parameter of interest and f is the nuisance (see, e.g., Powell, 1994, for an alternative definition of semiparametric models).

There is clearly no universally accepted 'best' approach. There are arguments in favour of one or the other, but often experimental, analytical and computational issues will in fact naturally lead to the best, available alternative. Traditionally, parametric models have had priority over nonparametric ones, for their ability to describe important properties of the system under study by means of parameters or constants that have immediate interpretation. But also because the statistical properties of parametric models are well studied and theoretically more advantageous (consider, e.g., the Cramér–Rao bound on the variance of an estimator).

2.4.2 Estimation, prediction, and hypothesis testing

Loosely speaking, statistical models rely on two components, one stochastic and one deterministic. The former defines the probabilistic framework, the conceptual laboratory where events take place and are measured. The latter attempts to explain aspects that are not subject to probability laws but are assumed to be fixed and unknown. It is important to stress that the deterministic component is intrinsically tied to the probabilistic one and that the goal of statistical modelling is to provide an optimal, sensible description of both. Achieving this goal relies on estimation, prediction, and tests of hypotheses, which represent important problems of the inductive process. Clearly, the different types of procedures, as discussed in the previous section, have a considerable weight in this process.

Estimation concerns obtaining an *estimate* of the parameter of interest θ or of a function of the parameter $h(\theta)$ using the information provided by the data. An estimate is either a single value (*point estimation*) or an interval of values (*interval estimation*). The statistics that produce these values are called *estimators*, which are both random variables and functions of the sample. For example, in Chapter 1 we encountered the sample mean $\hat{\theta} = 1/n \sum_{i=1}^{n} Y_i$, which is a random variable with its own distribution and a function of $Y_1, \ldots, Y_n$.

Estimators can be derived using different methods. The classical method of *maximum likelihood* finds its basis in the likelihood function.

Definition 11 (Likelihood function). *The likelihood function of n random variables $\boldsymbol{Y} = (Y_1, \ldots, Y_n)$ is defined to be the joint probability mass or density function of $\boldsymbol{Y}$, which is considered to be a function of θ.*

The likelihood function will be denoted as $L(\theta; \boldsymbol{y})$, where $\boldsymbol{y} = (y_1, \ldots, y_n)$. In the simplest case of a random sample from a population $F_Y(y)$, $L(\theta; y_1, \ldots, y_n) = \prod_{i=1}^{n} f(y_i)$ if Y is continuous or $L(\theta; y_1, \ldots, y_n) = \prod_{i=1}^{n} m_Y(y_i)$ if Y is discrete.

So, for example, the vector $\hat{\boldsymbol{\theta}} = (3416, 477)^{\top}$ contains the point estimates of the location and scale parameters of the normal distribution fitted to the US birth weight data in Figure 1.3. In this case, the estimates are given by the sample mean and standard deviation,[8] which coincide with the maximum likelihood estimators under the normal model. When a parametric model for F_Y is not available or the corresponding likelihood is analytically too complex to handle, one can resort to alternative approaches. As we shall see later, these include matching methods such as the method of moments (Mood et al., 1974) and the method of quantiles (Gilchrist, 2000), estimating equations, and optimisation of loss functions.

Hypothesis testing concerns the assessment of the consistency of the data with a particular parameter value θ_0. The hypothesis that $\theta = \theta_0$ is called *null hypothesis* and

[8]Sample moments are *nonparametric* estimates of population moments.

conventionally denoted by H_0. In this book, we will often consider null hypotheses of the type $H_0 : G_Y^{-1}(p) = F_Y^{-1}(p)$ and $F_{Y|\mathbf{X}=\boldsymbol{x}_1}^{-1}(p) = F_{Y|\mathbf{X}=\boldsymbol{x}_2}^{-1}(p)$, for specific values of p, $0 < p < 1$, or for $p \in \mathcal{P}$, $\mathcal{P} \subseteq (0, 1)$. The former type of hypotheses arises, for example, in goodness-of-fit problems $H_0 : \kappa(y; F, G) = y$ where F is a theoretical model and G is estimated from the data.

Another important inferential goal is that of prediction of as yet unobserved random variables from the same random system that generated the data (Cox, 2006). In quantile inference, prediction is a natural consequence of estimation. Suppose a random sample $Y_1, \ldots, Y_n$ of size n is taken from a population F_Y. It might be of interest to predict the outcome of a new observation, say Y_{n+1}, based on the data. As in the case of estimation, we can distinguish between a single value (*point prediction*) or an interval of values (*interval prediction*). Individual sample quantiles may be taken as optimal point predictions (Gneiting, 2011). Estimating two quantiles, say ξ_p and ξ_{1-p}, $0 < p < 0.5$, provides the lower and upper bounds of a prediction interval where we expect that $100 \cdot (1 - 2p)\%$ of future observations will fall, for the data-generating distribution F_Y.

3

Features of a distribution and statistical modelling

3.1 Location, scale, and shape of a distribution

The distinction between location, scale and shape of a distribution can be at times artificial, since these features may be intrinsically interrelated (e.g., they cannot be controlled separately). For example, consider the Poisson distribution (see definition in Appendix A.2), which depends on one parameter only, say λ. In statistical applications, the Poisson distribution is used to model count data such as the number of car accidents during a week or the number of visits of a patient to their doctor during a year. Often, the focus of the analysis is on estimating the expected number of events, which is given by λ. However, the variance of a Poisson distribution also is equal to λ. Indeed, moments of order higher than 2, which are often used to characterise aspects of the shape of the distribution, depend on the same parameter.[1] The histograms of three simulated Poisson samples with $\lambda = 3$, $\lambda = 7$, and $\lambda = 1000$ are depicted in Figure 3.1. For small values of λ, the mean is close to zero, the bulk of the observations falls within a relatively narrow range and the distribution is skewed to the right. As λ increases, the distribution shifts towards larger values, the spread increases, and it becomes more and more symmetric. Every component of the LSS in a Poisson model is therefore controlled by λ.

Since the cumulative distribution and quantile functions are two sides of the same coin, the LSS of a distribution can be examined with one or the other. There are a number of quantile-based measures of LSS; these will be discussed in some detail in Section 3.1.4 further below.

3.1.1 Location

The location of a distribution is often associated with the 'central tendency' or 'typical value' of the distribution. Mean, median, mode, for example, can be all considered measures of central tendency. However, they have different properties and, importantly, they also differ because of their interpretation. In Section 1.2, we have seen that there are practical consequences if we choose the sample mean or the sample median as measures of central tendency when the distribution has heavy tails or if it is skewed. Therefore it is important to carefully consider which measure is most appropriate to summarise the data, especially because many studies in applied research often focus mainly on location.

Imagine a world populated by a number of Bart Simpson, depicted in Figure 3.2, who go around on skateboards yelling 'eat my shorts!', and suppose that, despite their identical look, they differ by height. If we were to describe as briefly as possible such peculiar population,

[1] This fact can be more clearly understood by examining the moment generating function (see definition in Section 2.3.1) of $Y \sim \text{Pois}(\lambda)$, which is given by

$$M_Y(t) = \exp\{\lambda(\exp(t) - 1)\}$$

and whose derivatives at $t = 0$, which yield the moments of the distribution, are all functions of λ.

DOI: 10.1201/9781003326410-3

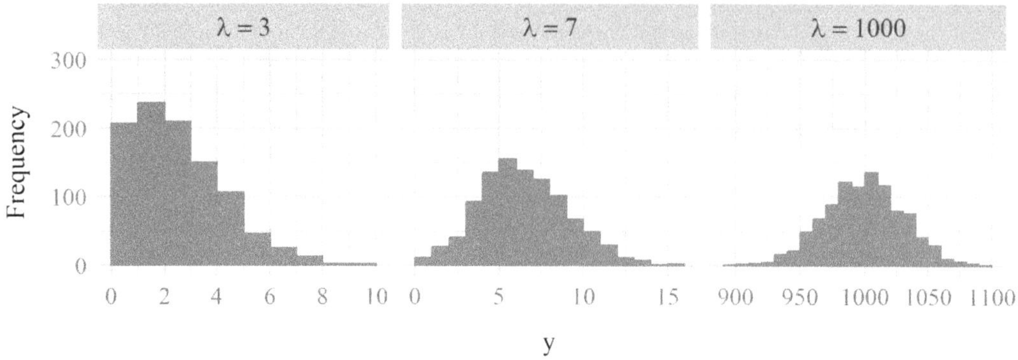

Figure 3.1
Frequency diagrams of Poisson distribution for different values of λ.

Figure 3.2
Ordered Bart population illustrating a location-shift distribution.

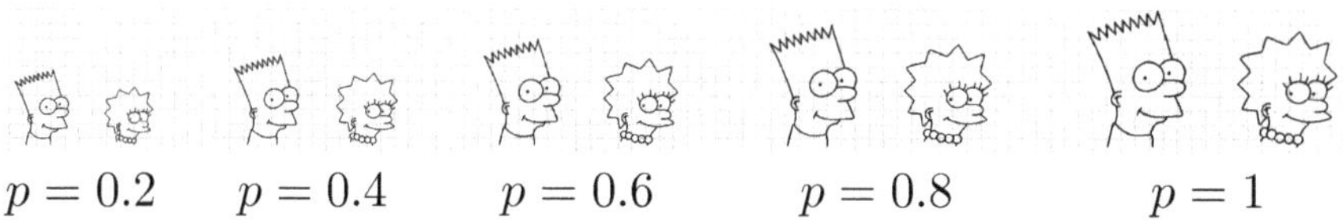

Figure 3.3
Paired Bart's and Lisa's sorted heights differ by a constant shift $\mu_{\text{B}} - \mu_{\text{L}}$.

we could pick the middle Bart as representative of all the others. If in addition Bart's height distribution follows the density $f_0(y - \mu_{\text{B}})$, where μ_{B} is middle Bart, then any Bart from this population is simply a deviation from middle Bart, that is, $Y = \mu_{\text{B}} + U$ where $f_0(u)$ describes the distribution of deviations. Such model is called *location family*. The parameter μ_{B} represents a shift which determines the location of $f_Y(y)$ and, thus, is called *location parameter*. Each quantile of Y is a shifted version of the corresponding quantile of U, that is $Q_Y(p; \mu_{\text{B}}) = \mu_{\text{B}} + Q_0(p)$.

Imagine now that there are two worlds, one inhabited by copies of Bart and one by as many copies of his sister Lisa. Suppose also that their height distributions have densities $f_0(y - \mu_{\text{B}})$ and $f_0(y - \mu_{\text{L}})$, respectively, with $\mu_{\text{B}} > \mu_{\text{L}}$. These models imply that the height distributions in the two populations are exactly the same, except for a *constant* shift. In other words, if we sort the two populations according to their heights and set them side by side (Figure 3.3), then the difference between sorted heights

$$\{\mu_{\text{B}} + Q_0(p)\} - \{\mu_{\text{L}} + Q_0(p)\} = \mu_{\text{B}} - \mu_{\text{L}}$$

does not depend on the particular rank p.

3.1.2 Scale

The scale of a distribution is related to its dispersion, that is, the amount by which a set of observations deviate from their 'location' (e.g., mean) or from each other. It provides information about the variability of observations: the larger the dispersion, the more diverse the observations, the 'less predictable' the outcome. Used along with location, the scale increases our knowledge of the distribution since it tells us how 'far' one can expect to go. Variance (standard deviation), range, and inter-quartile range (IQR) are all measures of dispersion but, again, their interpretations differ.

For a continuous random variable Y, a *scale family* of distributions can be expressed in terms of the PDF $f_0(y/\sigma)/\sigma$, where $\sigma > 0$ is called the *scale parameter*. Suppose that, in a given population and on any given day, individuals experience fluctuations in (systolic) blood pressure around their own average level. We may then model the deviation of the ith individual from their mean as a random variable $Y_i = \sigma_i Y_0$, where Y_0 is centred at 0 and has PDF f_0. In this case, σ_i captures the extent of within-day variability for individual i. This could be a realistic situation in which sources of heterogeneity such as age, sex, and lifestyle factors determine different responses to similar stimuli. Some of these factors may be observed, others may not be observed or even be observable. Heterogeneity of this kind is a classic example of NID observations.

3.1.3 Shape

The shape of a distribution refers to every other feature of a distribution that is distinct from location and scale. This definition is purposely vague as there are a number of aspects one can consider and which can be formalised in different ways. Some distributions are more peaked than others, have heavier tails, longer tails, or have more than one mode. Differences in distributional shape concern how the probability mass is distributed once location and scale differences have been accounted for. For example, distributions that belong to the same *location–scale family* $f_0\left((y-\mu)/\sigma\right)/\sigma$ have the same shape.

We begin with skewness and kurtosis, which are perhaps the most common shape attributes reported in basic statistics textbooks. Their definitions are traditionally operational and have been tied to specific moments of the distribution. The former is often defined as the lack of symmetry and is associated with the standardised third central moment $\bar{\bar{\mu}}_3$ (Definition 9), while the latter as the degree of tail heaviness and is associated with the standardised fourth central moment $\bar{\bar{\mu}}_4$. While the concept of skewness is intuitive, kurtosis is not. The literature abounds with articles debating the nature of kurtosis as something related to the weight of the tails (thus propensity to generate outliers) or peakedness or both. If one ties the notion of kurtosis to the fourth moment (Pearson, 1905), then one must invariably conclude that kurtosis has little, if anything, to do with the centre (peak) of the distribution (Westfall, 2014). But defining what is being measured by the tool that is measuring it is putting the cart before the horse and leaves little room for alternative measures of kurtosis. Most importantly (and ironically!), the fourth moment of heavy-tailed distributions may be infinite or fail to exist (Jones et al., 2011). In other words, the moments do not necessarily measure what they are supposed to (Mood et al., 1974). This is due to the fact that moments are basically averages that 'hide' the local features of a distribution. This is true for the first moment as discussed in the first chapter, as well as for moments of higher order. Two distributions may share the same value of $\bar{\bar{\mu}}_3$ or $\bar{\bar{\mu}}_4$, but have very different distributional shapes. In general, there is widespread agreement that moments are not a good measure of shape and that they may not be able to discriminate between different distributions (Balanda and MacGillivray, 1988; Groeneveld, 1998). Balanda and

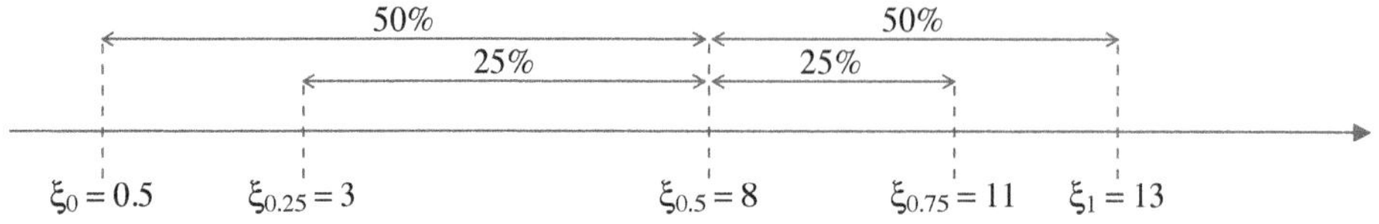

Figure 3.4
Five-number summary of recovery time in diseased patients in a hypothetical example.

MacGillivray (1988), for example, examine the concept of kurtosis in great detail. They conclude that

> [...] it is best to define kurtosis vaguely as the location- and scale-free movement of probability mass from the shoulders of a distribution into its centre and tails and to recognise that it can be formalised in many ways.

In this book, we entertain this view of kurtosis since all parts of the distribution (centre, shoulders and tails) come together while location and scale are kept separate. After all, describing the shape of a distribution as platykurtic (flatter than normal), mesokurtic (normal) or leptokurtic (more peaked than normal), as originally intended (Pearson, 1905), does involve the idea that movements of the probability mass affect both the tails and the centre of the distribution (Jones et al., 2011). It is just unfortunate that $\bar{\mu}_4$ is unable to capture both aspects at once and that this index is tied to a word that literally means 'curved' (from the Greek *kurtos*).

Suppose a hypothetical scenario where a patient has been diagnosed with a certain debilitating disease and she receives information on the prognosis from her doctor. She is told that, on average, patients with this disease recover about 9 years after the start of treatment. However, about 50% of the patients recover before 8 years and 50% after. These two numbers clearly help the patient put things in perspective. She is also told that, based on a large number of reported cases, recovery time is at least 6 months and 13 years at most, about 50% of the patients recover between 3 and 11 years, while the remaining 50% is equally split before 3 years and after 11 years.

We can put these numbers together into the five-number summary (Figure 3.4) to understand the shape of the distribution. We have FNS $= \{0.5, 3, 8, 11, 13\}$. Since the difference $\xi_{0.75} - \xi_{0.5} = 3$ is smaller than $\xi_{0.5} - \xi_{0.25} = 5$, the lower half of the distribution is more spread out than the upper half, suggesting left skewness. This is confirmed when using more extreme quantiles, since $\xi_1 - \xi_{0.5} = 5 < 7.5 = \xi_{0.5} - \xi_0$. If a new treatment was to be discovered and this treatment produced a distribution with median and range similar to the old one, but opposite sign of skewness, then this would be good news for the patients, as it would indicate that extremely long recovery times become less common and that a larger proportion of patients recover relatively early.

We will see in the next section and, more extensively, in Chapter 11, that other concepts of shape of a distribution such as kurtosis, peakedness, or tail weight may elicit (besides purely academic interest) practical and useful information when applied as criteria for comparisons.

3.1.4 Quantile-based measures of LSS

In this section, we discuss quantile-based measures of LSS and occasionally compare them to moment-based measures to better delineate their advantages and disadvantages.

Table 3.1
Median, mean and cumulative probability at the mean of the equivalised disposable income (euros) in European countries, 2015.

Country	Country code	Median	Mean	$F(\mu)$
Croatia	HR	5,017	5,729	0.59
Czechia	CZ	6,903	7,901	0.63
Denmark	DK	31,357	36,198	0.64
France	FR	21,590	25,206	0.64
Germany	DE	20,193	22,906	0.60
Hungary	HU	4,366	4,900	0.60
Ireland	IE	20,172	23,926	0.61
Italy	IT	17,149	19,515	0.60
Netherlands	NL	23,432	26,342	0.61
Norway	NO	43,315	46,574	0.58
Poland	PL	5,307	6,120	0.62
Portugal	PT	8,036	9,776	0.63
Romania	RO	2,322	2,776	0.63
Serbia	RS	2,265	2,695	0.59
Slovakia	SK	6,689	7,208	0.58
Sweden	SE	26,822	28,982	0.57
Switzerland	CH	41,396	47,351	0.62
United Kingdom	GB	20,000	23,779	0.62

A well-known quantile-based measure of location is the median while its moment-based counterpart is the mean. Common arguments in favour of the former are its robustness against outliers and transformation equivariance. On the other hand, common arguments *against* the median are lack of a simple formula for its calculation and lack of additivity. In practice, counterarguments can be found for either sets of reasons, and those counterarguments then rebutted in a spiral loop, depending on which axe one wants to grind. In our view, when choosing one or the other, interpretation has priority over many other considerations.

A classical example concerns income and wealth statistics. The European Union (EU) Statistics on Income and Living Conditions (EU-SILC) is a multi-country survey that provides data on the equivalised disposable income (EDI, euros) of thousands of households (in the following example a total of 149,443 households for 18 countries). The estimated median and mean EDI by country for the year 2015 are shown in Table 3.1. The third column, denoted by $F(\mu)$, shows the cumulative probability estimated at the mean by using the country-specific empirical CDF (ECDF).

Consider Switzerland and Norway, the two countries that may contend for the top of the list. The former has greater mean income at 47,351 euros. This value represents the 62th percentile of Switzerland's income distribution. On the other hand, Norway's lower mean income at 46,574 euros represents the 58th percentile of the country's income distribution. If we were asked to compare a summary of the income distributions of two countries by contrasting the quantile ξ_{p_1} for one country and ξ_{p_2} for the other, with $p_1 \neq p_2$, we would immediately raise our eyebrows. But this is exactly what we are doing when comparing the means in Table 3.1. On the other hand, the median always carries the same probability level 0.5. If we put Switzerland and Norway on an equal footing, then the latter is better off than the former based on median income (43,315 euros for Norway and 41,396 euros for Switzerland).

To put things in context, consider the following first-order approximation to the quantile function (QF) $Q(p)$ around 0.5:

$$Q(p) = Q(0.5) + q(\tau_p)(p - 0.5) \tag{3.1}$$

for $p \neq 0.5$ and $\tau_p \in (\min\{p, 0.5\}, \max\{p, 0.5\})$. Since the first derivative of the QF is the sparsity function, denoted by q as in Equation (2.1), then, by using the expression in Equation (2.7), the expected value of a continuous variable Y can be written as

$$\mathrm{E}(Y) = \int_{\mathcal{I}} Q(p)\,\mathrm{d}p = Q(0.5) + \int_{\mathcal{I}} q(\tau_p)(p - 0.5)\,\mathrm{d}p, \tag{3.2}$$

where $\mathcal{I} = (0, 0.5) \cup (0.5, 1)$. That is, the mean is given by the median plus a weighted average of the probability area below and above the median, with weights given by the sparsity function (i.e., inversely proportional to the density at the corresponding quantiles). If the density is symmetric, then the rightmost integral in (3.2) will be zero. Otherwise, the 'weights' will skew the mean to one side of the median or the other. Now, if we take the difference between two expectations, we have

$$\mathrm{E}(Y_1) - \mathrm{E}(Y_2) = \{Q_1(0.5) - Q_2(0.5)\} + \left\{\int_{\mathcal{I}} [q(\tau_1) - q(\tau_2)](p - 0.5)\,\mathrm{d}p\right\}.$$

The second addendum in the equation above is what muddies the waters of an otherwise 'fair' comparison between medians. Of course, an analogous conclusion can be made when comparing quantiles of any order $a \in (0, 1)$ by using $Q(p) = Q(a) + q(\tau_p)(p - a)$.

A well-known quantile-based measure of scale is the inter-quartile range $IQR = Q(0.75) - Q(0.25)$. This can be generalised to the inter-quantile or inter-percentile range $IPR(p) = Q(1 - p) - Q(p)$ at level $p < 0.5$, with the extreme case $p = 0$ leading to the range. Arguments similar to those mentioned above can be made in favour of (or against) a quantile-based measure of scale relative to, say, the classical standard deviation.

Finally, there are a number of quantile-based measures for shape, particularly for skewness and kurtosis (Groeneveld and Meeden, 1984; Groeneveld, 1998; Jones et al., 2011). At the very beginning of Chapter 1, we came across the Bowley skewness index. Here, we define the general form of the quantile-based skewness as

$$\lambda_p = \frac{Q(1 - p) + Q(p) - 2Q(0.5)}{IPR(p)}$$

for $p < 0.5$ (hence, $\lambda_{0.25}$ is the Bowley index). Such an index not only is location and scale invariant, but is also bounded between -1 and 1 (as opposed to the Pearson's third standardised moment which can be infinite or even undefined). When λ_p is near the bounds -1 or 1, then $Q(1 - p) \approx Q(0.5)$ or $Q(p) \approx Q(0.5)$, respectively, which is a form of extreme skewness (at a given quantile level p, that is). Hinkley (1975) used the symmetry condition $\lambda_p = 0$ to estimate the parameter of the Box–Cox transformation.

As for kurtosis, define the 'tails' of the distribution as those lying outside the quantiles $Q(p)$ and $Q(1 - p)$. Building on the results by Horn (1983) and Ruppert (1987), Staudte (2017) considered the following identity:

$$\underbrace{\frac{IPR(p)}{IPR(r)}}_{\text{kurtosis}} = \underbrace{\frac{IPR(p)}{IPR(q)}}_{\text{tail weight}} \cdot \underbrace{\frac{IPR(q)}{IPR(r)}}_{\text{peakedness}}, \tag{3.3}$$

where $0 < p < q < r < 0.5$. The kurtosis measure $IPR(p)/IPR(r)$ is therefore the product of the tail weight index $IPR(p)/IPR(q)$ and the peakedness index $IPR(q)/IPR(r)$. These quantile-based measures of shape are sign, location and scale invariant. The formal justifications for their names are detailed in Staudte (2017) but they can be appreciated intuitively. For practical purposes, say $p = 0.1$, $q = 0.25$ (hence $IPR(0.25) = IQR$) and $r = 0.4$. In this case, the tail weight index contrasts the distance between the two most extreme deciles to the central 50% of the distribution. Thus a distribution has heavier tails than another if its tail index has a numerator that is proportionally larger than the denominator. This is consistent with the 'practical tail index' of Morgenthaler and Tukey (2000). Analogously, by contrasting the IQR to the distance between $Q(0.4)$ and $Q(0.6)$, two quantiles sufficiently (but not too) close to the median, a more pointed distribution will have the denominator of its peakedness index proportionally larger than the denominator. This is consistent with the peakedness measure of Horn (1983).

One clear disadvantage of quantile-based measures of LSS is the difficulty of choosing the levels p or q or r in the definitions above. While some degree of arbitrariness is unavoidable, the choice should be guided by a number of considerations. For example, for a more informative (discriminating) kurtosis index based on a variety of distributional shapes, Staudte (2017) recommends $r \in [0.3, 0.4]$. As for p, Staudte (2017) considers setting its value as a function of r so that the kurtosis index takes the value 3 when the distribution is normal (as it does $\bar{\bar{\mu}}_4$).

One can establish a relationship between the location (median), scale (IQR) and shape of a distribution within the following identity:

$$Q(p) = \underbrace{Q(0.5)}_{\text{median}} + \frac{1}{2}\underbrace{IPR(0.25)}_{\text{IQR}} \times \underbrace{\frac{IPR(p)}{IPR(0.25)}}_{\text{shape index}} \cdot \left(\underbrace{\frac{Q(1-p) + Q(p) - 2Q(0.5)}{IPR(p)}}_{\text{skewness index}} - 1\right), \tag{3.4}$$

for $p < 0.5$.[2] The quantity $IPR(p)/IPR(0.25)$ in Equation (3.4) is loosely defined as the 'shape index' (Gilchrist, 2000, p. 72), although it can be seen as the tail-weight measure given in Equation (3.3) when $p < 0.25$. For symmetric distributions, the contribution of the skewness index vanishes. The interested reader is referred to van Zwet (1964) and Oja (1981) for a theoretical discussion on related issues about ordering of distributions based on LSS.

The function `qlss` from the `Qtools` package (Geraci, 2026, 2016b) provides a quantile-based LSS summary with the indices defined in (3.4)) of either a *theoretical* or an *empirical* distribution. It returns an object of class `qlss`, which is a list containing measures of location (median), scale (IQR and IPR), and shape (skewness and shape indices) for each of the probabilities specified in the argument `probs` (by default, `probs` $= 0.1$). The quantile-based LSS summaries of the standard normal distribution is given in R Code 3.1 for $p = 0.1$ and $p = 0.4$.

Following on from the EU-SILC data example introduced earlier, the LSS summaries of the countries' income distributions with $p = 0.1$, $q = 0.25$ and $r = 0.4$ are shown in Figure 3.5. Favourable traits of a more equal income distribution are lower (right) skewness and higher peakedness. Two countries like Sweden and Ireland stand out as opposites. The lower inequality of the former country relative to the latter is confirmed by the Gini coefficient calculated on the same data (0.26 and 0.31, respectively). The calculation of the

[2]This equation appears in Gilchrist (2000, p. 74) with a '+1' inside the brackets, which is appropriate to the case $p > 0.5$. For the present formulation, which is written for $p < 0.5$, the correct sign is negative. See also Benjamini and Krieger (1996, eq. 1).

R Code 3.1: Quantile-based location, scale, and shape for the normal distribution with the R package `Qtools`.

```
qlss(fun = "qnorm", probs = c(0.1, 0.4))
> call:
> qlss.default(fun = "qnorm", probs = c(0.1, 0.4))
>
> Unconditional Quantile-Based Location, Scale, and Shape
>
> ** Location **
> Median
> [1] 0
> ** Scale **
> Inter-quartile range (IQR)
> [1] 1.34898
> Inter-quantile range (IPR)
>       0.1       0.4
> 2.5631031 0.5066942
> ** Shape **
> Skewness index
> 0.1 0.4
>   0   0
> Shape index
>      0.1      0.4
> 1.900031 0.375613
```

Gini coefficient for all countries in the dataset is shown in R Code 3.2 using the `reldist` package (Handcock and Morris, 1999; Handcock, 2023).

3.2 Unconditional and conditional modelling

Statistical modelling is a delicate and central step of data analysis. The choice of the model should align with the goal of the analysis and it typically requires an iterative selection process (Cox and Hinkley, 1997). Traditionally, parametric modelling has been favoured over nonparametric modelling for reasons in part explained in Section 2.4. If the prescribed model is too simplistic or inadequate, there are three alternative strategies one can consider:

1. force the data to follow the prescribed model;
2. find a more complex model;
3. relax (some) parametric assumptions.

The first solution is often achieved by means of transformations to remove (partially or completely) features of the data that do not conform to the assumed model. This is a potentially hazardous approach as it may actually lead to undesirable side effects. First of all, even a technically well-executed transformation may produce a result that does not

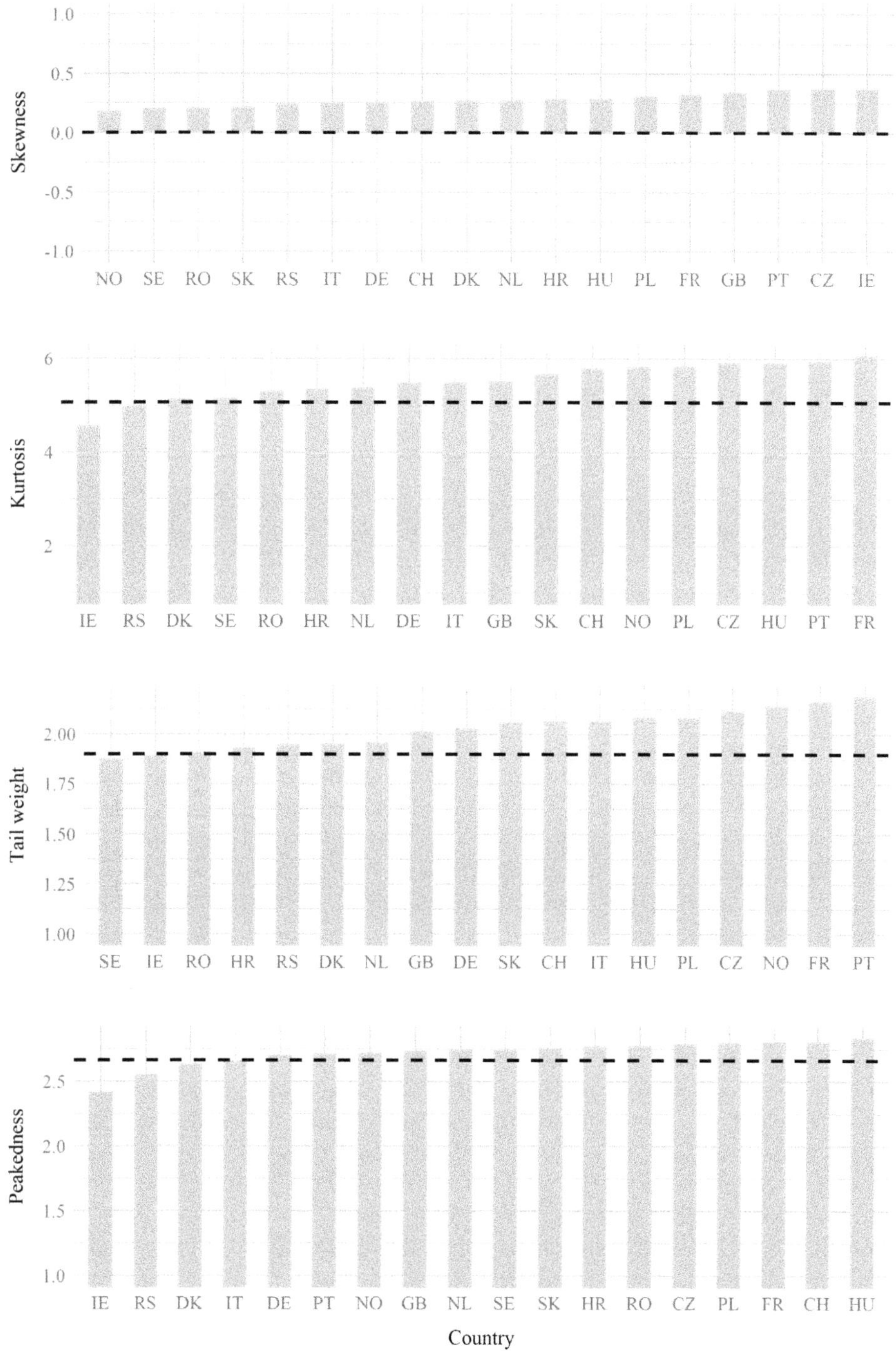

Figure 3.5
Quantile-based measures of skewness, kurtosis, tail weight and peakedness for the distribution of equivalised disposable income in European countries, 2015. The values of the indices for the standard normal distribution (see R Code 3.1) are marked by dashed lines for reference.

R Code 3.2: Gini coefficient for the distribution of equivalised disposable income in European countries, 2015.

```
library(reldist)
tapply(eusilc2015$EDI, eusilc2015$Country, gini)
>          Croatia          Czechia          Denmark           France
>        0.3311472        0.2439403        0.2794872        0.2900673
>          Germany          Hungary          Ireland            Italy
>        0.3038050        0.2622058        0.3138245        0.3239069
>      Netherlands           Norway           Poland         Portugal
>        0.2556184        0.2427791        0.3057232        0.3437969
>          Romania           Serbia         Slovakia           Sweden
>        0.3745405        0.3980310        0.2253468        0.2587271
>      Switzerland United Kingdom
>        0.2968295        0.3197087
```

address the goal of the analysis. An example would be applying a log transformation to reduce skewness and then taking the mean. However, absolute changes on the log-scale then become relative changes on the original scale. Second, it may introduce substantial numerical approximations. One can verify this by first calculating the mean (or any other moment) of a log-transformed variable and then by transforming back (e.g., see Section 2.2.2). While transformations can be a powerful mathematical device for statistical modelling, they typically have a limited scope and cannot be a substitute for a thoughtful search of the appropriate model.

Rather than forcing the data to adapt to the model, a more logical approach is to search for a model that fits the data better. This requires making adjustments to the current model or introducing a different model, typically at the expense of the model's complexity. Parametric modelling, however, may prove to be particularly arduous. Non- and semiparametric alternatives provide an excellent solution which may also increase the robustness of the results.

In the next two sections, we introduce a distinction between two types of modelling which motivates Part II and III of this book. *Unconditional modelling*[3] refers to the case in which the focus of the analysis is on the distribution F_Y (or on a related statistic) of a univariate random variable Y. In contrast, the target of the analysis in *conditional modelling* is to observe how the distribution of Y changes conditionally on different values of another variable X. The former type of modelling does not exclude the case in which the variable Y is actually observed in a subpopulation, or the case in which a variable X is changed from one value to another and the goal is to assess related changes in the *marginal* (i.e., not conditional on X) distribution of Y.

The rationale for discussing unconditional and conditional modelling separately is that, typically, theoretical and methodological results for univariate distributions are more abundant than, or temporally precede, those for conditional distributions. For example, asymptotic results for unconditional sample quantiles were available at least since the 1940s (Kendall, 1940). Those for conditional (regression) quantiles appeared some 40 years later

[3]In the statistical literature, the term 'marginal' is often used instead. However, this term implies that some form of integration has taken place as in the case of a marginal distribution obtained from a multivariate one (Appendix B.1.1). However, we may be interested in situations where neither a joint distribution nor a multivariate dataset are available. Moreover, in regression analysis, 'marginal' may also refer to the *marginal effect* of a variable on the outcome distribution. To avoid confusion, we prefer the term 'unconditional'.

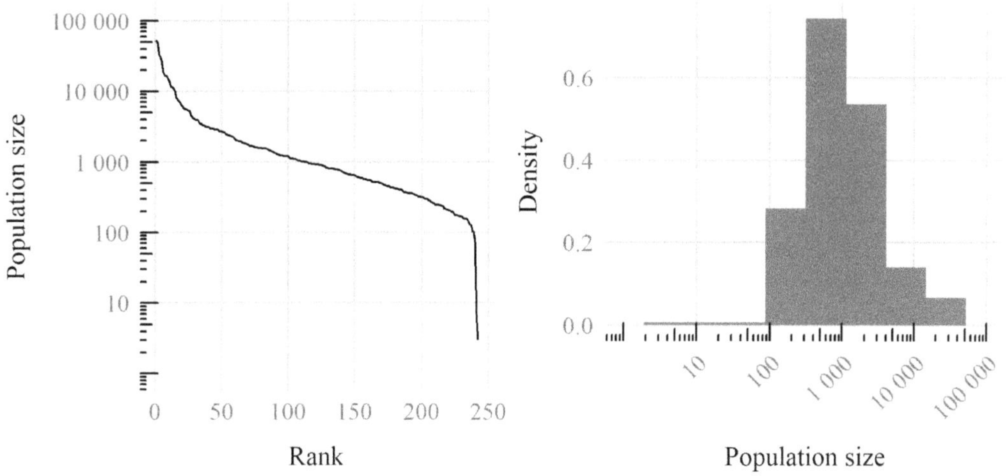

Figure 3.6
Rank-size plot (left) and histogram (right) of population size for incorporated places in West Virginia in 2000. Population size is shown on the $\log_{10}$-scale.

(Koenker and Bassett, 1978). This is naturally a consequence of the different complexity of the problems.

3.2.1 Unconditional modelling

The study of the distribution of a univariate random variable Y, either discrete or continuous, may be required for reporting purposes or may be of interest due the regularities exhibited by the population F_Y over time and/or space, regularities that are eventually used as a base for forecasts about a future event. For example, reports from vital statistics agencies, censuses, cancer registries, and other population-based data collection systems provide detailed statistics about individual variables, often with the purpose of looking at trends over time. Two such instances include prediction of the future evolution of common diseases like cancer (International Agency for Research on Cancer, 2014) or the demand of energy (U.S. Energy Information Administration, 2016). Such reports do not typically try to establish associations between distributions.

To illustrate the basic idea, we consider the rank-size distribution of 242 city populations in West Virginia as estimated by the 2000 US Census (Figure 3.6). A rank-size distribution is the distribution of size by rank, in decreasing order of size, or, using more familiar terms, a QF with inverted domain. A number of phenomena, including the size of human settlements, show a remarkable regularity in the shape of their rank-distributions – a fact also known as rank-size rule and Zipf's law (Gabaix, 1999). In his paper, Reed (2002) proposed a stochastic model to describe the foundation of settlements and their subsequent evolution after foundation. That study is an example of unconditional modelling ('macro view') in which differences between settlements, potentially related to known explanatory factors, are treated as random. In the words of Reed (2002, p. 2):

> The explanation for the observed regularity in the size distribution is essentially mathematical in that it is a consequence of stochastic processes with certain characteristics. This does not mean that geographic, economic, and other factors are

> not important in determining the growth and eventual size of any city. Rather it means that when looking at the distribution of settlement sizes over a whole region, the effects of the variation in these factors can be modeled effectively by stochastic processes [...].

By looking at the rank-size plot in Figure 3.6, we observe that the great majority of cities have a small population: approximately only 10% is above 5000 and 5% is above 10,000. This distribution is characterised by a steep slope at larger sizes and a slow rate of decrease as the size decreases resulting in a rather long tail. Note that the rank is in decreasing order, hence the QF is actually skewed to the left. It is of interest to segment the distribution into 'pieces' to study such regularities, including known threshold effects such as the *King effect*, that are observed across countries and over time. In other words, it is of interest to model the shape of the distribution by using quantiles.

We conclude this section by briefly introducing an approach to unconditional modelling in which the goal of the analysis is to assess how a change in a variable X affects the unconditional distribution of Y. Suppose X is binary and $\Pr(X = 1) = \pi$. What would be the effect of a change in the proportion π on the population average $\mu = \mathrm{E}(Y)$? In a conditional regression model

$$\mathrm{E}(Y \mid X = x) = \underbrace{\mathrm{E}(Y \mid X = 0)}_{\beta_0} + \underbrace{\{\mathrm{E}(Y \mid X = 1) - \mathrm{E}(Y \mid X = 0)\}}_{\beta_1} x$$

the parameter β_1 not only measures the impact of X on the conditional mean, but it can be interpreted also as the effect of X on the unconditional mean since

$$\mu = (1 - \pi)\,\mathrm{E}(Y \mid X = 0) + \pi\,\mathrm{E}(Y \mid X = 1).$$

and $\partial\mu/\partial\pi = \mathrm{E}(Y \mid X = 1) - \mathrm{E}(Y \mid X = 0) = \beta_1$.

A related question concerns the effect of X on the quantiles of Y itself, rather than on the conditional quantiles of $Y \mid X$. This is relevant, for example, in poverty and inequality studies, where the distribution of wealth is examined at the population level. Conditional quantiles do not generally provide this information directly, although specific methods have been developed for this purpose (Firpo et al., 2009). Unconditional quantile regression will be briefly covered in Chapter 12.

3.2.2 Conditional modelling

Arguably, the majority of the statistical methods deal with the comparison of probability distributions in different populations. Comparison is the primary means by which differences are assessed and associations between variables are investigated, with the ultimate goal of building a case for establishing causal relationships. Contrasting the cholesterol level in patients who received a placebo with that of patients who received the treatment will help us understand whether a drug is effective at reducing high cholesterol. Likewise, comparing mortality rates in socioeconomically advantaged and disadvantaged populations will inform policymakers about existing health disparities. The typical statement 'we study the association between the exposure X (e.g., treatment, income) and the response Y (e.g., cholesterol or mortality)' reflects a comparison of features of Y given X. What are the features we are comparing, then?

Although distributions can (and should) be compared in terms of their LSS, the analysis focuses most often on location, occasionally on scale, and rarely on shape. Consider a model where Y is a normally distributed random variable whose mean is a linear function of X, that is, $\mu_x = a + bx$, and its variance is σ^2, for all values of X. This is the standard setup of a mean regression model, in which the following distributions are being compared:

$$\begin{array}{ll} \mathcal{N}\left(a+b,\sigma^2\right) & \text{when } X=1 \\ \mathcal{N}\left(a+2b,\sigma^2\right) & \text{when } X=2 \\ \mathcal{N}\left(a+3b,\sigma^2\right) & \text{when } X=3 \\ \cdots & \cdots \end{array}$$

Conditional modelling can be easily illustrated starting from the simple case in which two populations are contrasted in relation to some characteristic Y. Let $F_0(y)$ and $F_1(y)$ denote the CDFs of Y in the two populations, and let $\mu_0 = \int_{-\infty}^{\infty} y \,\mathrm{d}F_0(y)$ and $\mu_1 = \int_{-\infty}^{\infty} y \,\mathrm{d}F_1(y)$ be their respective means. The two CDFs are plotted in Figure 3.7. The question of whether and how F_1 differs from F_0 can be approached by comparing, for any fixed value y, the distance between the CDFs along the probability (vertical) axis. Therefore

$$\delta(y) = F_1(y) - F_0(y).$$

A discrepancy between the two distributions at some value y would result in a positive or negative $\delta(y)$. The largest absolute difference $|\delta(y)|$ over y is used to calculate the Kolmogorov–Smirnov test statistic (Mood et al., 1974). Figure 3.7 shows the case in which F_1 stochastically dominates F_0, so that $\delta(y) < 0$ for all $y \in \mathbb{R}$. A complementary approach is to compare, for any fixed value p, the distance between the CDFs along the quantile (horizontal) axis. Therefore

$$\Delta(p) = F_1^{-1}(p) - F_0^{-1}(p)$$

gives the contrast between the pth quantiles of the two distributions. This is also called the 'quantile treatment effect' (QTE)[4] (Doksum, 1974; Lehmann, 1975; Koenker and Xiao, 2002). An illustration is given in Figure 3.7.

Now, define the indicator variable

$$X = \begin{cases} 1 & \text{if } Y \sim F_1, \\ 0 & \text{if } Y \sim F_0, \end{cases}$$

and the conditional distribution of Y given X

$$F_{Y|X=x}(y) = \begin{cases} F_1(y) & \text{if } x = 1, \\ F_0(y) & \text{if } x = 0. \end{cases}$$

Then the mean regression model

$$\mathrm{E}(Y \mid X = x) = \underbrace{\mu_0}_{\text{intercept}} + \underbrace{(\mu_1 - \mu_0)}_{\text{slope}}\, x \tag{3.5}$$

arises from the contrast between F_0 and F_1. Indeed, regression analysis revolves around the modelling of differences between populations. The regression slope can be written in terms

[4]Rather than taking differences, another approach consists in taking 'ratios'. This leads to the comparison distribution, that is, $v(p; F_0, F_1) = F_1\left(F_0^{-1}(p)\right)$, which has been introduced in Section 2.2.3. In particular, its derivative yields the comparison density d, which takes the form of a ratio of densities (or PMFs); see (2.4) and (2.5) (see also Section 11.3). This connection underlies Q–Q plots (Section 7.1.6).

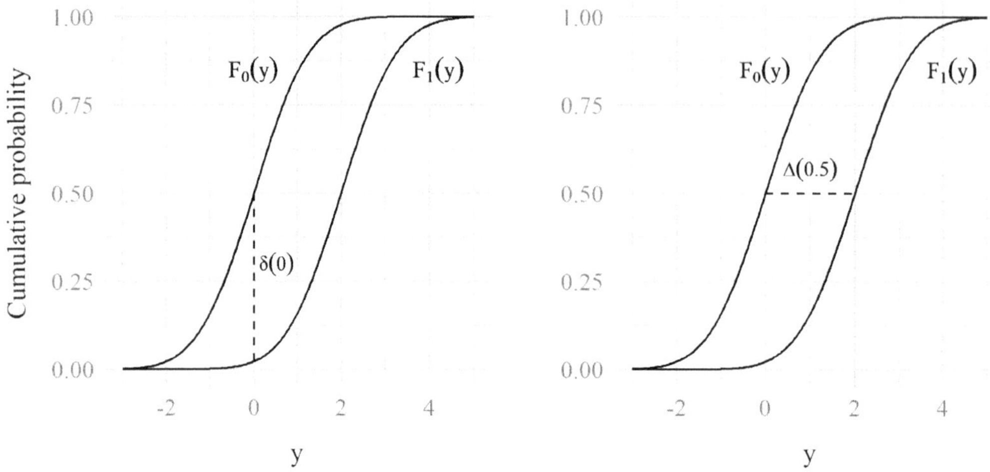

Figure 3.7
Left: Probability difference (à la Kolmogorov-Smirnov). Right: Quantile difference (à la Lehmann-Doksum).

of either the conditional distribution function $F_{Y|X}$ or the *conditional quantile function* $Q_{Y|X} \equiv F^{-1}_{Y|X}$, that is

$$\mu_1 - \mu_0 \equiv \int_{-\infty}^{\infty} y \left\{ \mathrm{d}F_{Y|X=1} - \mathrm{d}F_{Y|X=0} \right\}$$
$$= \int_0^1 \left\{ Q_{Y|X=1}(p) - Q_{Y|X=0}(p) \right\} \mathrm{d}p = \int_0^1 \Delta(p)\, \mathrm{d}p. \tag{3.6}$$

The classical regression model for the mean thus takes a weighted average (uniform in p) of the quantile differences $\Delta(p)$. Often, the question of interest is whether the quantile differences are constant over p (i.e., the shift of the CDF is uniform along the horizontal axis) and therefore whether the mean provides an exhaustive summary of $\Delta(p)$, $0 < p < 1$.

We now illustrate an empirical example using the `faithful` dataset, which contains 272 observations on waiting time (minutes) between eruptions and the duration (minutes) of the eruption for the Old Faithful geyser in Yellowstone National Park, United States.

Waiting times between eruptions are plotted against eruption durations in Figure 3.8. Two clusters of observations can be defined for durations below three minutes ($x = 0$) and above three minutes ($x = 1$) (see also Azzalini and Bowman, 1990). Figure 3.8 also shows that the empirical CDF of waiting time for $x = 1$ appears shifted to the right compared to that for $x = 0$. Aside from a shift in location, there seem to be no other differences between the two CDFs.

R Code 3.3 shows the quantile differences $\Delta(p)$ for $p \in \{0.1, 0.25, 0.5, 0.75, 0.9\}$. The distribution of waiting times is shifted by an approximately constant amount of 26 min at all considered values of p. The difference in means is 25.5 min, while the ratio of standard deviations is close to one, indicating similar dispersion in the two distributions. Moreover, according to the Kolmogorov–Smirnov test (R Code 3.4), the data do not provide evidence against the assumption of normality for waiting times for durations below or above three

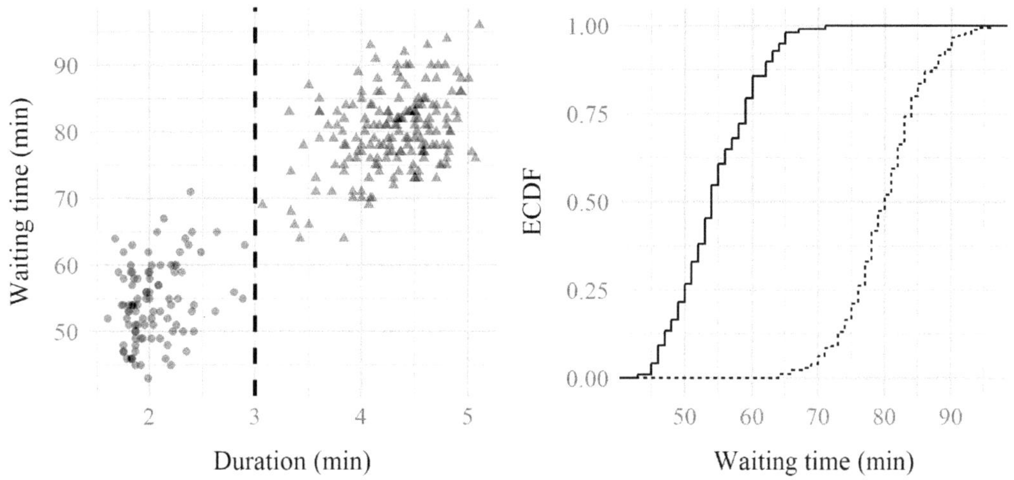

Figure 3.8
Left: Waiting times between eruptions against durations of eruptions (dashed vertical line drawn at 3 minutes) in the Old Faithful Geyser dataset. Right: Empirical cumulative distribution function of waiting time by duration of eruption (solid, shorter than 3 minutes; dashed, longer than 3 minutes).

R Code 3.3: Comparison of waiting times between eruptions by eruption duration in the Old Faithful Geyser dataset.

```
data(faithful)
x <- as.numeric(faithful$eruptions >= cutoff)
y <- faithful$waiting
p <- c(0.1, 0.25, 0.5, 0.75, 0.9)
qhat <- tapply(y, x, quantile, probs = p)
muhat <- tapply(y, x, mean)
sdhat <- tapply(y, x, sd)
mat <- rbind(qhat[[1]], qhat[[2]], qhat[[2]] - qhat[[1]])
mat <- cbind(mat, mean = c(muhat, diff(muhat)), sd =
c(sdhat, sdhat[2]/sdhat[1]))
rownames(mat) <- c("x = 0", "x = 1", "Delta")
round(mat,1)
>         10% 25% 50% 75%  90% mean  sd
> x = 0   47  50  54  59 62.4 54.5 5.8
> x = 1   73  76  80  84 88.0 80.0 6.0
> Delta   26  26  26  25 25.6 25.5 1.0
```

R Code 3.4: Kolmogorov–Smirnov normality test for waiting times between eruptions by eruption duration in the Old Faithful Geyser dataset.

```
ks.test(scale(y[x == 0]), "pnorm")

> Warning in ks.test.default(scale(y[x == 0]), "pnorm"): ties should not be
present for the one-sample Kolmogorov-Smirnov test

>
>  Asymptotic one-sample Kolmogorov-Smirnov test
>
> data:  scale(y[x == 0])
> D = 0.080155, p-value = 0.5614
> alternative hypothesis: two-sided
ks.test(scale(y[x == 1]), "pnorm")

> Warning in ks.test.default(scale(y[x == 1]), "pnorm"): ties should not be
present for the one-sample Kolmogorov-Smirnov test

>
>  Asymptotic one-sample Kolmogorov-Smirnov test
>
> data:  scale(y[x == 1])
> D = 0.051679, p-value = 0.7382
> alternative hypothesis: two-sided
```

minutes.[5] A model such as the one in Equation (3.5) would therefore provide an adequate summary of the relationship between X (duration of eruption below or above 3 minutes) and Y (waiting time between eruptions).

[5]The warning `ties should not be present` reminds us that the classical one-sample Kolmogorov–Smirnov test assumes a continuous distribution, so ties technically invalidate the exact null distribution of the statistic. A simple workaround is to add jitter to restore continuity artificially, for example, `jitter(scale(y[x == 0]))`. See also R Code 8.17.

Part II

Unconditional quantiles

4

Sample quantiles

4.1 Order statistics and ranks

Order statistics and ranks play an important role in data analysis and can be used to conduct inference on quantiles. For a thorough treatment of order statistics, the reader is referred to Reiss (1989) and David and Nagaraja (2003), among others.

Definition 12 (Order Statistics). *Let $Y_1, \ldots, Y_n$ be a sample of n random variables. If $Y_1, \ldots, Y_n$ are arranged in increasing order of magnitude and then written as*

$$Y_{(1)} \leq Y_{(2)} \leq \ldots \leq Y_{(n)},$$

then we call $Y_{(i)}$ the ith order statistic, $i = 1, \ldots, n$.

Two particular order statistics are the minimum, $Y_{(1)}$, and the maximum, $Y_{(n)}$.

The order statistic $Y_{(i)}$ is sometimes denoted as $Y_{i:n}$ to give emphasis to the sample size. Correspondingly, the CDF of the ith order statistic is denoted by $G_{(i)}(y)$ or by $G_{i:n}(y)$. For a sample $Y_1, \ldots, Y_n \overset{\text{IID}}{\sim} F_Y$, the CDF of the ith order statistic is given by

$$G_{(i)}(y) = \sum_{r=i}^{n} \binom{n}{r} \{F_Y(y)\}^r \{1 - F_Y(y)\}^{n-r}. \tag{4.1}$$

One can immediately recognise that the CDF in (4.1) is the sum of binomial PMFs with parameters n and $F_Y(y)$. Owing to the relation between binomial sums and the incomplete beta function (see Definition 15 in Appendix A.1), we can write $G_{(i)}(y)$ as

$$G_{(i)}(y) = B(i, n - i + 1, F(y)). \tag{4.2}$$

It follows from (4.1) that the minimum and maximum have CDF

$$G_{(1)}(y) = 1 - \{1 - F_Y(y)\}^n \tag{4.3}$$

and

$$G_{(n)}(y) = \{F_Y(y)\}^n, \tag{4.4}$$

respectively. The expressions above hold for any population, continuous or discrete. However, if the Y_is are not identically distributed, then order statistics have a rather more complicated distribution (Bapat and Beg, 1989).

Order statistics are necessarily dependent on one another because of the inequality relations among them (David and Nagaraja, 2003). The expression of the joint distribution of two or more order statistics is conceptually similar to (4.1) (David and Nagaraja, 2003, p. 11), although more cumbersome.

DOI: 10.1201/9781003326410-4

As we will see in Section 4.2, order statistics are instrumental for the estimation of quantiles. Similarly, the *ranks* $R_1, \dots, R_n$, which determine the position of each Y_i, $i = 1, \dots, n$, relative to all the others, play the same ordering role of the CDF (Lehmann, 1975). Ranks are used to define rank statistics which may involve observations from two or more variables (e.g., the Kendall rank correlation coefficient) or observations from the same variable for two or more groups (e.g., the Wilcoxon–Mann–Whitney rank-sum test). Ranks, too, play a role in quantile inference (Chapter 7).

Remark 2. *The ranks R_i, $i = 1, \dots, n$, are themselves random variables. Indeed, they have a discrete uniform distribution.*

For example, consider the sample $\boldsymbol{y} = (7, 5, 9, 3, 1)$ of size $n = 5$. Then the order statistics take the values $y_{(1)} = 1$, $y_{(2)} = 3$, $y_{(3)} = 5$, $y_{(4)} = 7$, and $y_{(5)} = 9$, and the ranks are $r_1 = 4$, $r_2 = 3$, $r_3 = 5$, $r_4 = 2$, and $r_5 = 1$. Thus, $y_{(j)} = y_i$ if $r_i = j$, $i, j = 1, \dots, n$.

4.2 Empirical distribution

Let $Y_1, \dots, Y_n \overset{\text{IID}}{\sim} F_Y$. The sample CDF or empirical CDF (ECDF) of $Y_1, \dots, Y_n$ is defined as

$$\hat{F}_Y(y) = \frac{1}{n} \sum_{i=1}^{n} I\left(Y_i \leq y\right), \qquad y \in \mathbb{R}, \tag{4.5}$$

where I denotes the indicator function. The random variable $I\left(Y_i \leq y\right)$ is Bernoulli with probability $p = F_Y(y)$. Therefore, $\mathrm{E}\left(\hat{F}_Y(y)\right) = F_Y(y)$ and $\mathrm{var}\left(\hat{F}_Y(y)\right) = n^{-1}F_Y(y)(1 - F_Y(y))$ for every fixed value of y. The estimator $\hat{F}_Y(y)$ is unbiased for $F_Y(y)$ and its variance converges to zero when the sample size goes to infinity. By the Glivenko–Cantelli theorem, $\hat{F}_Y$ converges uniformly almost surely to F_Y. The pointwise (for fixed y) asymptotic distribution of $\hat{F}_Y(y)$ is found by applying the CLT

$$\sqrt{n}\left\{\hat{F}_n(y) - F_Y(y)\right\} \xrightarrow{d} \mathcal{N}\left(0, F_Y(y)\{1 - F_Y(y)\}\right), \tag{4.6}$$

when $n \to \infty$. This result is generalised to a Gaussian process when the above empirical process is seen as a function of y (e.g., see Pollard, 1984).

The sample CDF (4.5) is a piecewise-constant function (or step function), regardless of whether the random variable Y is continuous or discrete. At small sample sizes, the function behaves irregularly, with adverse effects on estimation and inference in general. However, when Y is continuous, the discrete behaviour of this function becomes less jerky as the sample size increases. This is shown in Figure 4.1 which depicts the ECDFs of three samples of different sizes drawn from a gamma distribution. This notwithstanding, a large sample size cannot obviate the discreteness resulting from rounding continuous measurements. An example is given in Figure 4.2 with the ECDF of almost three and a half million observations of gestational age in the US population: despite the considerably large sample size, there are only 23 unique values. While in practice rounding is inevitable, excessive rounding results in a large proportion of ties. To exemplify with numbers, the ECDF in Figure 4.2 yields $\hat{F}(38) = 0.38$ and $\hat{F}(39) = 0.62$, a rather dramatic jump between two 'contiguous' observations of gestational age.

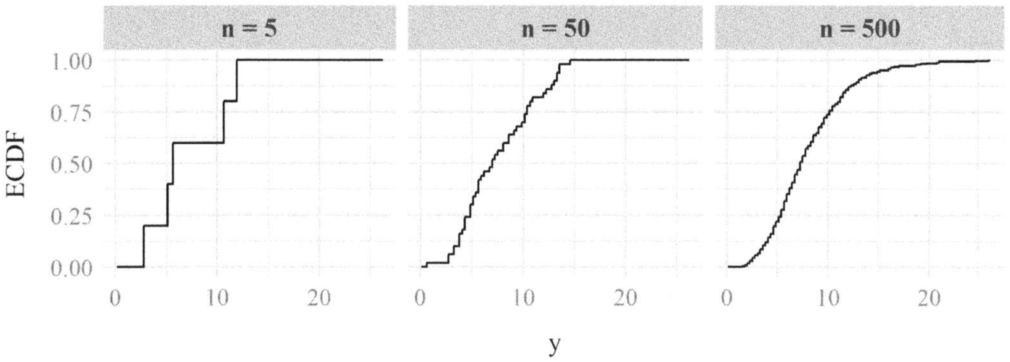

Figure 4.1
Empirical cumulative distribution functions of samples from a gamma distribution for different sample sizes n.

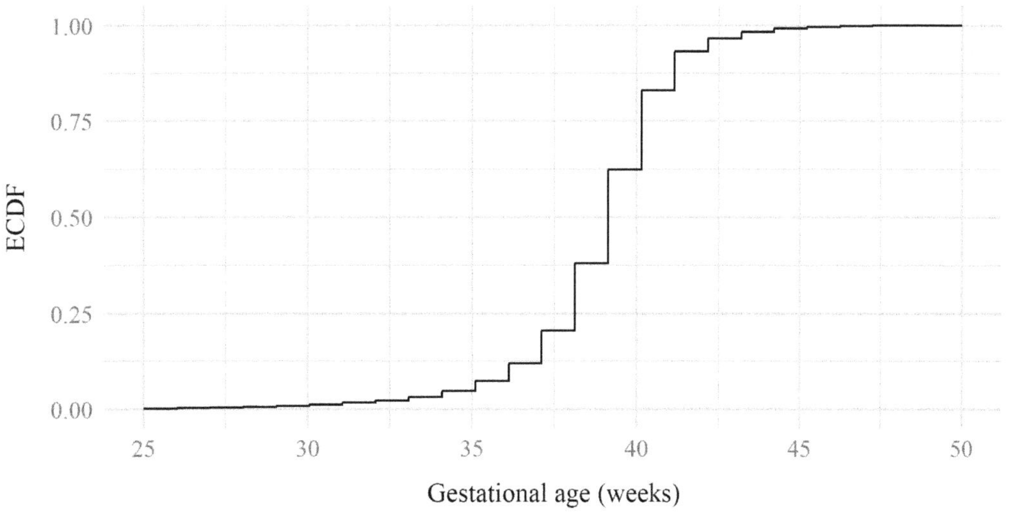

Figure 4.2
Empirical cumulative distribution function of gestational age, United States, 2001.

4.3 Sample quantiles

The (classical) sample quantile function (QF), defined as the generalised inverse of the sample CDF in (4.5), is such that

$$\hat{Q}_Y(p) = Y_{(j)}, \qquad \text{for } (j-1)/n < p \leq j/n. \tag{4.7}$$

This function, too, is discrete. However, this is *one* of several possible definitions of sample quantiles. Statistical packages often provide alternative estimators, typically based on one or two order statistics. Many of these estimators can be written as (Hyndman and Fan, 1996)

$$\tilde{Q}(p) = (1-\gamma)Y_{(j)} + \gamma Y_{(j+1)}, \tag{4.8}$$

where $(j-m)/n \leq p < (j-m+1)/n$, for some $m \in \mathbb{R}$ and $0 \leq \gamma \leq 1$. The parameter γ depends on $h = np + m - j$, where $j = \lfloor pn + m \rfloor$. For example, one obtains the classical definition (4.7) by setting $m = 0$ and

$$\gamma = \begin{cases} 0 & \text{if } h = 0 \\ 1 & \text{if } h > 0, \end{cases}$$

which is implemented as `type = 1` in the `R` function `quantile`; see Table 4.4 for the complete list of quantile estimators available in this `R` function.

Table 4.1 summarises properties desirable in quantile estimators (Hyndman and Fan, 1996). In particular, since property P1 does not hold for sample quantile functions based on discrete data, in the next section we consider *mid* functions. Note also that property P6 does not apply to the sample median obtained with Equation (4.7).

When $F_Y(y)$ is continuous, sample quantiles are consistent estimators of population quantiles (see Box 4.1 for technical details). This result can also be understood by viewing the sample quantile as a functional of the empirical distribution function $\hat{F}_Y$, which converges uniformly almost surely to F_Y.

The asymptotic distribution of sample quantiles can be derived by exploiting this connection and applying the delta method (Tanner, 1996). In Section 4.2, we have seen that, for fixed y, the estimator (4.5) is an average of Bernoulli random variables with parameter

Table 4.1
Six properties for sample quantiles adapted from Hyndman and Fan (1996, Table 1 p. 361).

P1:	$\tilde{Q}(p)$ is a continuous function of p.
P2:	$\frac{1}{n}\sum_{i=1}^{n} I\left(y_i \leq \tilde{Q}(p)\right) \geq p$.
P3:	$\sum_{i=1}^{n} I\left(y_i \leq \tilde{Q}(p)\right) = \sum_{i=1}^{n} I\left(y_i \geq \tilde{Q}(1-p)\right)$.
P4:	Where $\tilde{Q}^{-1}(y)$ is uniquely defined, $\tilde{Q}^{-1}(y_{(j)}) + \tilde{Q}^{-1}(y_{(n-j+1)}) = 1$ for $j = 1, \ldots, n$.
P5:	Where $\tilde{Q}^{-1}(y)$ is uniquely defined, $\tilde{Q}^{-1}(y_{(1)}) > 0$ and $\tilde{Q}^{-1}(y_{(n)}) < 1$
P6:	$\tilde{Q}(0.5)$ is equal to the sample median defined by $(y_{(k)} + y_{(k+1)})/2$ if $n = 2k$, $y_{(k+1)}$ if $n = 2k+1$.

Box 4.1: Consistency of sample quantiles

Classical results on the consistency of sample quantiles rely on the uniqueness of the corresponding population quantiles.

Theorem 1 (!t). *Let* $\xi_p = F_Y^{-1}(p)$ *be the unique pth quantile of* F_Y *and consider the sample* $Y_1, \ldots, Y_n \overset{\text{IID}}{\sim} F_Y$. *Then the order statistic* $Y_{(k_n)}$ *such that* $k_n/n \to p$ *as* $n \to \infty$ *is a strongly consistent estimator of* ξ_p, *that is*

$$\Pr\left\{\sup_{m \geq n} |Y_{(k_n)} - \xi_p| > \epsilon\right\} \to 0 \qquad \text{as } n \to \infty,$$

for every $\epsilon > 0$.

For a proof of Theorem 1, see Serfling (2002).

Remark 3. *As stated, Theorem 1 does not ensure uniform convergence in the family of all distributions* F *with uniquely defined* ξ_p. *Zieliński (1998) provides a necessary and sufficient condition for the uniform strong consistency of* $Y_{(k_n)}$.

Remark 4. *Sample quantiles are strongly consistent also if* Y *is discrete, unless both* $F_Y(\xi_p) = p$ *and* F_Y *is flat in a right-neighbourhood of* ξ_p *(Serfling, 2002).*

$F_Y(y)$, and is therefore asymptotically normal. This pointwise binomial structure provides intuition for the behaviour of $\hat{F}_Y$, which, viewed as a stochastic process, converges to a Gaussian process. We now sketch how this result leads to the asymptotic normality of sample quantiles. A rigorous proof is given by Walker (1968).

In addition to being continuous, we assume that F_Y is strictly increasing on its support. Fix y such that $p = F_Y(y)$ and let $\xi_p = F_Y^{-1}(p)$. Applying the delta method to (4.6) yields

$$\sqrt{n}\left(F_Y^{-1}\{\hat{F}_n(y)\} - \xi_p\right) \xrightarrow{d} \mathcal{N}\left(0, \frac{p(1-p)}{\{f_Y(\xi_p)\}^2}\right),$$

where we used the transformation $h(u) = F_Y^{-1}(u)$, $0 < u < 1$, and its derivative

$$\frac{\partial h(u)}{\partial u} = \frac{1}{f_Y(F_Y^{-1}(u))}.$$

To complete the argument, we replace the random variable $F_Y^{-1}\{\hat{F}_n(y)\}$ with the order statistic $Y_{(r_n)}$, where r_n is a sequence of integers such that $n^{-1/2}(r_n - np) \to 0$. Since $|Y_{(r_n)} - F_Y^{-1}\{\hat{F}_n(y)\}| \to 0$ almost surely, we have our final result

$$\sqrt{n}\left(Y_{(r_n)} - \xi_p\right) \xrightarrow{d} \mathcal{N}\left(0, \frac{p(1-p)}{\{f_Y(\xi_p)\}^2}\right). \tag{4.9}$$

Taking $r_n = \lceil np \rceil$, we obtain the asymptotic normality of the sample quantile $\hat{Q}_Y(p)$ defined in (4.7).

The asymptotic variance in (4.9) can be written as $p(1-p)\{q(p)\}^2$, where $q(p)$ is the sparsity function (see Equation (2.1)). Hence, the uncertainty about the true value of the population quantile is inversely proportional to the population density evaluated at the quantile of interest. The density f is unknown in nonparametric settings or is known up to a low-dimensional parameter in parametric modelling. In either case, the estimation of f requires some additional assumptions (Chapter 7).

The above result can be extended to obtain the joint distribution of sample quantiles. Suppose $p_1, \ldots, p_m$ are distinct quantile levels. We are interested in the joint distribution of $\hat{\boldsymbol{Q}} = \left(\hat{Q}_Y(p_1), \ldots, \hat{Q}_Y(p_m)\right)$. Since order statistics are necessarily dependent (Section 4.1), we should expect the elements of $\hat{\boldsymbol{Q}}$ to be dependent as well. We have the following result:

$$\sqrt{n}(\hat{\boldsymbol{Q}} - \boldsymbol{\xi}) \xrightarrow{d} \mathcal{N}(\boldsymbol{0}, \boldsymbol{\Sigma}), \tag{4.10}$$

where $\boldsymbol{\xi} = (\xi_{p_1}, \ldots, \xi_{p_m})^\top$ is the vector of population quantiles and the generic rsth element of $\boldsymbol{\Sigma}$ is

$$\sigma_{rs} = \frac{\min\{p_r, p_s\} - p_r p_s}{f(\xi_{p_r})f(\xi_{p_s})}, \qquad 1 \leq r \leq s \leq m,$$

with $\sigma_{sr} = \sigma_{rs}$ (Walker, 1968).

The asymptotic normality of sample quantiles for continuous distributions can also be proven via Bahadur's representation of quantiles. See Box 4.2 for technical details. In the next section, we introduce a particular definition of sample quantiles that offers inferential advantages when the distribution is discrete.

Box 4.2: Bahadur representation of quantiles

The classical and elegant result of Bahadur (1966) is central to the asymptotic theory for quantile estimators, either conditional or unconditional. There are several refinements and extensions of this result (see Sen, 1968; Ghosh, 1971; Kiefer, 1967; Csörgő, 1983, among others). Here, we provide the (weaker) version given in DasGupta (2008, p. 95) which suffices to establish the asymptotic normality of sample quantiles via the CLT.

Theorem 2. *Let $Y_1, \ldots, Y_n \stackrel{\text{IID}}{\sim} F_Y$ and suppose F_Y is differentiable at $\xi_p = F^{-1}(p)$ with $f(\xi_p) = F'(\xi_p)$. Then the sample quantile estimator $\hat{\xi}_{p,n}$ admits the representation*

$$\hat{\xi}_{p,n} = \xi_p + \frac{p - \hat{F}_n(\xi_p)}{f(\xi_p)} + R_n,$$

which can be rewritten as

$$\hat{\xi}_{p,n} - \xi_p = \frac{1}{n}\sum_{i=1}^{n} Z_i + R_n,$$

where $Z_i = \frac{p - I(Y_i \leq \xi_p)}{f(\xi_p)}$ and $R_n = o_{\mathbb{P}}(n^{-1/2})$.

The first term on the right-hand side of the equation above is the sample mean of the IID sequence $\{Z_i\}$ while the remainder becomes negligible in probability as $n \to \infty$ (the exact order of R_n is given in Kiefer, 1967). Theorem 2 thus links the asymptotic normality of $\hat{\xi}_{p,n}$ to that of the empirical distribution function $\hat{F}_n$ (Serfling, 2002). See Bahadur (1966) and Ghosh (1971) for alternative proofs.

4.4 Mid functions

In general, sample quantiles as defined in (4.7) may not be consistent for the population quantiles when the underlying distribution is discrete. For example, not only does the sample median $\hat{Q}_Y(0.5) = Y_{(\lceil n/2 \rceil)}$ lack asymptotic normality if Y is discrete, but its limiting distribution appears to be discrete in the presence of ties, even if Y is continuous (Genton et al., 2006). An example of a sample for continuous measurements with a substantial occurrence of ties has been given in Figure 4.2.

An illustration of the lack of consistency of a sample quantile is given in R Code 4.1 with data simulated from a Poisson distribution with parameter $\lambda = 5$. The QF of this distribution is shown later in Figure 4.3. The population median is $\xi_{0.5} = 5$, where the distribution function takes the value $F_Y(\xi_{0.5}) = 0.6160$. According to the remark in Box 4.1, $\hat{Q}_Y(0.6160)$ is not consistent for $\xi_{0.6160} = 5$ since $F_Y(\xi_{0.6160}) = 0.6160$ and F_Y is flat to the right of 5. In other words, $Q_Y(p)$ has a jump at $p = 0.6160$. However, the sample median $\hat{Q}_Y(0.5)$ is consistent for $\xi_{0.5}$ since $F_Y(\xi_{0.5}) \neq 0.5$. That is, $\xi_{0.5}$ lies on a flat segment of the QF. Table 4.2 shows that, for selected values of n going from 10 to 10,000, the relative frequency (estimated probability) with which $\hat{Q}_n(0.6160)$ equals $\xi_{0.6160}$ bounces around 0.5 (e.g., see example in Jentsch and Leucht, 2016), while that for the sample median goes to 1.

The discreteness of the observations, whether it be genuine or the result of aggressive rounding, may lead to inaccurate inferences due to the discontinuities of the ECDF. For instance, certain discrete test statistics are affected by a loss of power due to lack of continuity (e.g., see Agresti, 2013). To correct for this problem, a few approaches have been

R Code 4.1: Simulation study on the consistency of sample quantiles.

```
n <- c(10, 50, 100, 500, 1000, 5000, 10000)
R <- 5000
lambda <- 5

# CDF at jumps
round(ppois(0:5, lambda = lambda), 4)
> [1] 0.0067 0.0404 0.1247 0.2650 0.4405 0.6160
# population median
print(xi <- qpois(0.5, lambda = lambda))
> [1] 5
# CDF at the median
print(p <- ppois(xi, lambda = lambda))
> [1] 0.6159607
# simulation for consistency
set.seed(456)
pn_jump <- rep(NA, length(n))
pn_step <- rep(NA, length(n))
for(j in 1:length(n)){
x <- replicate(R, rpois(n[j], lambda = lambda))
pn_jump[j] <- mean(apply(x, 2, quantile, probs = p, type = 1) == xi)
pn_step[j] <- mean(apply(x, 2, quantile, probs = 0.5, type = 1) == xi)
}
```

Table 4.2
Consistency of the pth sample quantile when p is at the jump or on the step.

n	10	50	100	500	1000	5000	10,000
Jump	0.34	0.52	0.50	0.51	0.50	0.51	0.50
Step	0.38	0.71	0.86	1.00	1.00	1.00	1.00

proposed, including the mid-p-value (Lancaster, 1961) which is related to the subject of this section.

We now introduce the mid-cumulative distribution function (mid-CDF) (Parzen, 1993, 2004), a modification of the standard CDF that plays an important role in discrete modelling and in samples with ties. The definition of the mid-CDF is given below in its general form. It represents a bridge between discrete and continuous variables.

Definition 13 (Mid-distribution function). *For a random variable Y with CDF $F_Y(y)$, either continuous or discrete, the function*

$$\mathcal{F}_Y(y) \equiv \Pr(Y \leq y) - 0.5 \cdot \Pr(Y = y)$$

is called mid-distribution function (mid-CDF). If Y is continuous, then $\mathcal{F}_Y(y) = F_Y(y)$ since $\Pr(Y = y) = 0$.

Remark 5. *Suppose that the random variable T is a discrete test statistic for some null hypothesis H_0. Then it is easy to find that*

$$\begin{aligned} p_m &= \Pr(T > t \mid H_0) + 0.5 \cdot \Pr(T = t \mid H_0) \\ &= 1 - \{\Pr(T \leq t \mid H_0) - 0.5 \cdot \Pr(T = t \mid H_0)\} \\ &= 1 - \mathcal{F}_T(t \mid H_0) \end{aligned}$$

where p_m is the one-sided mid-p-value.

We are interested in the quantile analogue of the mid-CDF. Since the mid-CDF of a continuous random variable reduces to the ordinary CDF, we focus on the discrete case.

Definition 14 (Mid-quantile function). *Let Y be a discrete random variable and let $y_1, \ldots, y_m$ denote the m distinct values that Y can take on with corresponding probabilities $p_1, \ldots, p_m$. We also define the probabilities $\pi_1 = p_1/2$ and $\pi_j = \mathcal{F}(y_j) = \sum_{i=1}^{j-1} p_i + p_j/2$, for $j = 2, \ldots, m$. The following function*

$$\mathcal{Q}_Y(p) \equiv \mathcal{F}_Y^{-1}(p) = \begin{cases} y_1, & \text{if } p < \pi_1, \\ y_j, & \text{if } p = \pi_j,\ j = 1, \ldots, m, \\ (1-\gamma)y_j + \gamma y_{j+1}, & \text{if } p = (1-\gamma)\pi_j + \gamma\pi_{j+1}, \\ & 0 < \gamma < 1,\ j = 1, \ldots, m, \\ y_m, & \text{if } p > \pi_m. \end{cases}$$

is called mid-quantile function (mid-QF) (Ma et al., 2011). If $m = \infty$, then the last category is suppressed.

The mid-QF is piecewise linear and it connects the points (π_j, y_j). One can verify that $\mathcal{Q}(\mathcal{F}(y_j)) = y_j$. Importantly, mid-quantiles cannot be obtained by inverting $\mathcal{F}_Y(y)$, as defined above, at points that are not in the support of Y. However, we can consider $\mathcal{Q}(p)$

as the inverse of a continuous function, say $\mathcal{F}_Y^c(y)$, that interpolates the points (y_j, π_j), $j = 1, \ldots, m$.

The corresponding sample mid-CDF is $\hat{\mathcal{F}}_Y(y) = \hat{F}_Y(y) - 0.5 \cdot \hat{p}(y)$, where $\hat{F}_Y(y)$ is given in (4.5) and

$$\hat{p}(y) = \frac{1}{n} \sum_{i=1}^{n} I(Y_i = y) \tag{4.11}$$

is the sample relative frequency of y.

To define the sample mid-quantiles, we need to introduce some more notation. Let z_j, $j = 1, \ldots, s$, be s distinct values that occur in the sample and let $\hat{p}_j = \hat{p}(z_j)$, $j = 1, \ldots, s$, be the corresponding relative frequencies. Then

$$\hat{\mathcal{Q}}_Y(p) = (1 - \gamma) z_j + \gamma z_{j+1}, \tag{4.12}$$

where j is such that $p = (1 - \gamma)\hat{\mathcal{F}}(z_j) + \gamma \hat{\mathcal{F}}(z_{j+1})$, $0 < \gamma \leq 1$, $z_0 = z_1$, $z_{s+1} = z_s$, and $\hat{\mathcal{F}}(z_j) = \hat{F}_Y(z_j) - 0.5 \cdot \hat{p}_j$.

Ma et al. (2011) showed that if the underlying distribution F_Y is absolutely continuous, then the sample mid-quantiles have the same asymptotic properties as the classical sample quantiles. More importantly, if F_Y is discrete with finite support, then the sample mid-quantiles are consistent estimators of the population mid-quantiles and are asymptotically normal. The precise form of the asymptotic variance depends on the position of p: whether it lies below the first mid-probability or above the last, strictly between two consecutive mid-probabilities, or exactly at a mid-probability. In the first case, the estimator converges in probability to the corresponding boundary value; in the other two cases, it satisfies an asymptotic normal law of the form

$$\sqrt{n} \left\{ \hat{\mathcal{Q}}_n(p) - \mathcal{Q}_Y(p) \right\} \xrightarrow{d} \mathcal{N}(0, \omega), \tag{4.13}$$

where the expression of ω depends on the local configuration of the support points and probability masses around p. We refer the reader to Ma et al. (2011) for the exact variance formulae.

To illustrate an application of the sample mid-QF to discrete data, we simulated data from a Poisson distribution (R Code 4.2) and applied the `midquantile` function[1] from the `Qtools` package. Table 4.3 shows the estimated classical sample quantile (4.7) and mid-quantile (4.12) functions evaluated at $p = \hat{F}_Y(z_j)$. These are values of p where the sample QF jumps. As one can gather from Figure 4.3, the graph of $\hat{\mathcal{Q}}_Y(p)$ nicely mimics the QF of a continuous variable by linearly interpolating through the steps of the discrete QF.[2] This is especially desirable when the underlying distribution is continuous but the sample contains many ties, as in the gestational age data discussed in Section 4.2. The mid-QF for those data is plotted in Figure 4.4.

Another advantage of mid-quantiles is that the ordinary quantiles can be recovered from the mid-quantiles. The goal is to map $\hat{\mathcal{Q}}_Y(p)$ back to an estimate of $Q_Y(p) \equiv \inf\{y \in \mathbb{R} : F_Y(y) \geq p\}$. Let us denote such an estimate by $\tilde{Q}_Y(p)$. This can be achieved in a few simple steps. The estimates $\hat{F}_Y(z_j)$, $j = 1, \ldots, s$, which are necessary to estimate the mid-quantiles, are used to obtain the mid-probabilities $\hat{\mathcal{F}}_Y(y)$ as described above. Successively, we identify the index j such that $\hat{\mathcal{F}}_Y(z_j) \leq p < \hat{\mathcal{F}}_Y(z_{j+1})$. If Y is a variable whose values are

[1] The sample mid-probabilities are available via `Qtools::midecdf`. The estimator of the *continuous* mid-distribution function $\mathcal{F}^c$ is returned as `Qtools::midecdf()$fn()`.

[2] Still, the mid-QF is not continuously differentiable.

R Code 4.2: Sample quantiles and mid-quantiles for Poisson data.

```
library(Qtools)
set.seed(3874)
z <- rpois(1000, lambda = 5)
p <- ecdf(z)(sort(unique(z)))
Qhat <- quantile(z, probs = p, type = 1)
Qmid <- midquantile(z, probs = p)
tab <- data.frame(p, Qhat, Qmid$y)
colnames(tab) <- c("p","Q(p)","Qmid(p)")
all.equal(as.numeric(midq2q(Qmid)), as.numeric(Qhat))
> [1] TRUE
```

Table 4.3
Quantiles and mid-quantiles of the Poisson data generated with R Code 4.2.

p	0.008	0.043	0.131	0.265	0.448	0.611	0.757
$\hat{Q}_Y(p)$	0	1	2	3	4	5	6
$\hat{\mathcal{Q}}_Y(p)$	0.186	1.285	2.396	3.423	4.529	5.528	6.566
p	0.869	0.935	0.974	0.988	0.997	1.000	
$\hat{Q}_Y(p)$	7	8	9	10	11	12	
$\hat{\mathcal{Q}}_Y(p)$	7.629	8.629	9.736	10.609	11.750	12.000	

irregularly spaced, then one would naturally take $\tilde{Q}_Y(p) = z_{j+1}$ if $p > \hat{F}_Y(z_j)$ or $\tilde{Q}_Y(p) = z_j$ otherwise, with the understanding that $\tilde{Q}_Y(p) = z_1$ if $p < \hat{F}_Y(z_1)$. If, on the other hand, Y is a variable with equally spaced values (e.g., a count), then we can take $\tilde{Q}_Y(p) = \lceil \hat{\mathcal{Q}}_Y(p) \rceil$ if $p > \hat{F}_Y(z_j)$ or $\tilde{Q}_Y(p) = \lfloor \hat{\mathcal{Q}}_Y(p) \rfloor$ otherwise, with the understanding that $\tilde{Q}_Y(p) = \lceil \hat{\mathcal{Q}}_Y(p) \rceil$ if $p < \hat{F}_Y(z_1)$.

A numerical test is given in the last line of R Code 4.2, which confirms the equality of $\tilde{Q}_Y(p)$, recovered using the function `Qtools::midq2q`, and the sample quantiles $\hat{Q}(p)$. This is an important property that sets mid-quantiles apart from so-called 'quantile-like' alternatives like expectiles. The latter have a non-trivial relationship with quantiles (Jones, 1994) and require non-trivial relabelling procedures that may lead to poor approximations on the tails (Koenker, 2013). M-quantiles, which represent a generalisation of expectiles, suffer from a similar shortcoming.

Sample mid-quantiles are not free from limitations. Consider a Bernoulli random variable Y. For $y \in \{0, 1\}$, the mid-CDF is then given by $\mathcal{F}_Y(y) = (y+1-\pi^*)/2$, where $\pi^* = \Pr(Y = 1)$. Therefore, $\pi_1 = \mathcal{F}_Y(0) = 0.5 - 0.5\pi^*$ and $\pi_2 = \mathcal{F}_Y(1) = 1 - 0.5\pi^*$. By definition, mid-quantiles are equal to 0 if $p < \pi_1$ and to 1 if $p > \pi_2$. Otherwise, $\mathcal{Q}_Y(p) = (1-\gamma)y_1 + \gamma y_2 = \gamma$ for $p = (1-\gamma)\pi_1 + \gamma\pi_2$. From the latter expression we get $\gamma = (p - \pi_1)/(\pi_2 - \pi_1) = 2p - 1 + \pi^*$. In summary, we obtain the following conditional mid-QF:

$$\mathcal{Q}_Y(p) = \begin{cases} 0 & \text{if } p < \pi_1 \\ 2p - 1 + \pi^* & \text{if } \pi_1 \leq p \leq \pi_2 \\ 1 & \text{if } p > \pi_2. \end{cases}$$

On the interval $[\pi_1, \pi_2]$, this function is linear with slope determined by π^*, and therefore characterises the Bernoulli distribution. However, it cannot distinguish between distribu-

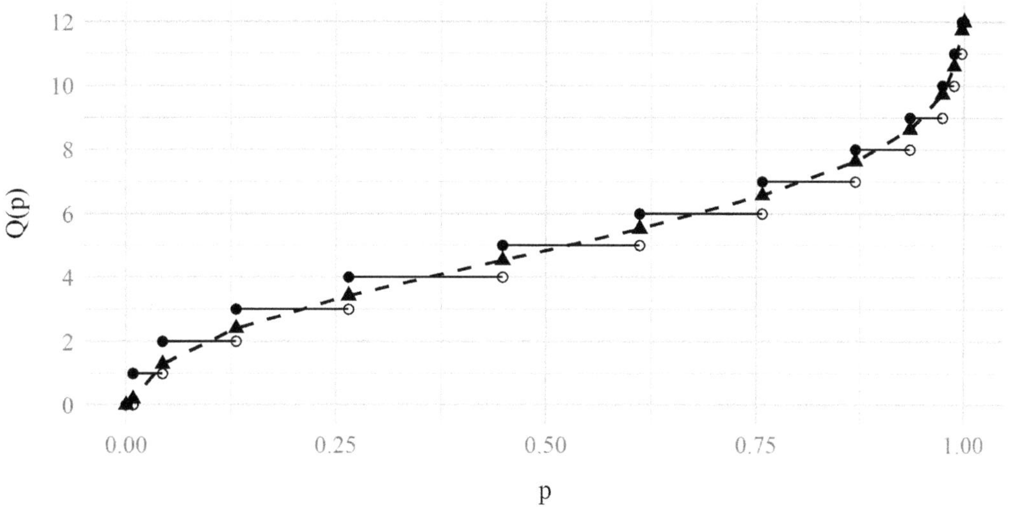

Figure 4.3
Quantile (grey solid line) and mid-quantile (black dashed line) functions for the Poisson data generated with R Code 4.2. The triangles mark points with coordinates $(\hat{F}_Y(z_j), \hat{\mathcal{Q}}_Y(\hat{F}_Y(z_j)))$.

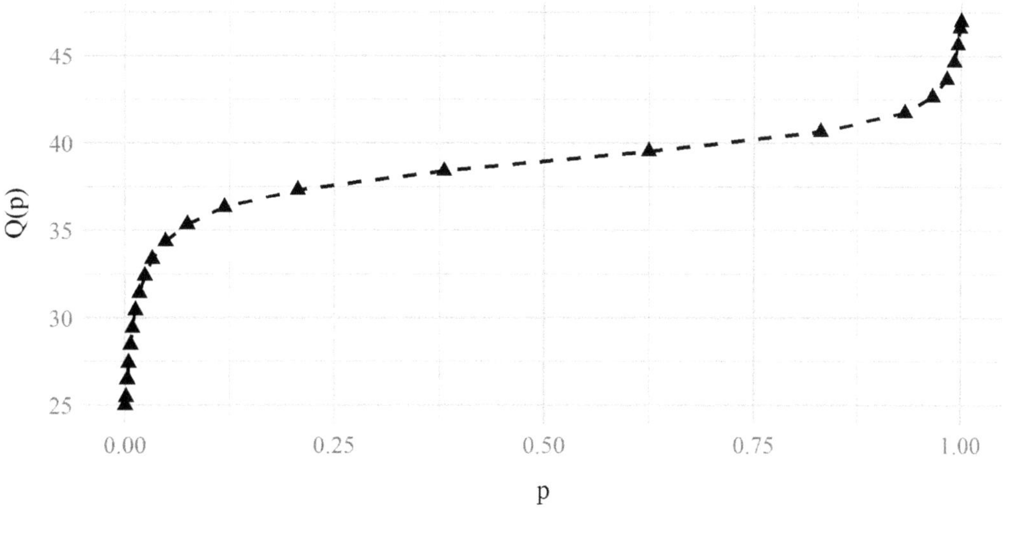

Figure 4.4
Mid-quantile (black dashed line) function for gestational age, United States, 2001. The triangles mark points with coordinates $(\hat{F}_Y(z_j), \hat{\mathcal{Q}}_Y(\hat{F}_Y(z_j)))$.

tions with overlapping tails where it is constant. An alternative is to build a *smooth* substitute using fractional order statistics (Stigler, 1977), which are conceptually similar to mid-quantiles but are otherwise based on stochastic, rather than linear, interpolation (e.g., see Wang and Hutson, 2011).

4.5 Simulation study

The reader might be wondering whether, among the multitude of quantile estimators, one should be preferred over the others. Some guidelines about desirable properties of sample quantiles have been given in Section 4.3. However, it is legitimate to consider practical differences of existing implementations in software packages, especially at different sample sizes.

We therefore run a small simulation study to assess the comparative performance of sample quantiles in terms of bias and mean squared error (MSE). We use the `R` function `quantile` from the `stats` package, although a similar experiment could be done by using a different statistical software. As summarised in Table 4.4, there are 9 types of sample quantiles available from the function `quantile`. They can all be defined as special cases of (4.8). Type 1 is the classical sample quantile. Types 4 to 9 are continuous functions of p and can be defined as interpolating functions of the points $(u_k, y_{(k)})$, where

$$u_k = \frac{k - \alpha}{n - \alpha - \beta + 1}$$

and α and β are constants. This is equivalent to using the representation (4.8) with $m = \alpha + p(1 - \alpha - \beta)$ and $\gamma = h$. For example, type 4 is the interpolating function proposed by Parzen (1979). According to Hyndman and Fan (1996), only type 5 possesses all the properties in Table 4.1. These authors, however, recommend type 8 as it is approximately median-unbiased for any underlying distribution F_Y.

We generate data from two distributions, a standard normal and a Poisson with parameter $\lambda = 1000$; the latter value is large enough for the distribution to be approximately symmetric. We subsequently estimate nine deciles, $p \in \{0.1, 0.2, \ldots, 0.9\}$, using the nine sample quantile estimators mentioned above, as well as the sample mid-quantile (4.12), which we label 'type 10'. We consider four different sample sizes $n \in \{5, 10, 25, 50\}$ and 10,000 replications of each sample. At smaller sample sizes ($n = 5, 10$), most quantile estimators show considerable bias (Figure 4.5), especially types 1, 3, and 4. At larger sample sizes, the magnitude of the bias becomes smaller and close to 0 for all estimators and deciles. Similarly, the MSE puts types 1, 3, and 4 at a disadvantage at smaller n, but it decreases

Table 4.4
There are 9 types of quantile estimators available in the `R` function `quantile` and these are of the form $\tilde{Q}(p) = (1-\gamma)Y_{(j)} + \gamma Y_{(j+1)}$ as defined in (4.8).

Type	m	γ	Continuous function of p
1	0	$\begin{cases} 0 & \text{if } h = 0 \\ 1 & \text{otherwise} \end{cases}$	no
2	0	$\begin{cases} 0.5 & \text{if } h = 0 \\ 1 & \text{otherwise} \end{cases}$	no
3	-0.5	$\begin{cases} 0 & \text{if } h = 0 \text{ and } j \text{ even} \\ 1 & \text{otherwise} \end{cases}$	no
4	0	h	yes
5	0.5	h	yes
6	p	h	yes
7	$1-p$	h	yes
8	$(p+1)/3$	h	yes
9	$p/4+3/8$	h	yes

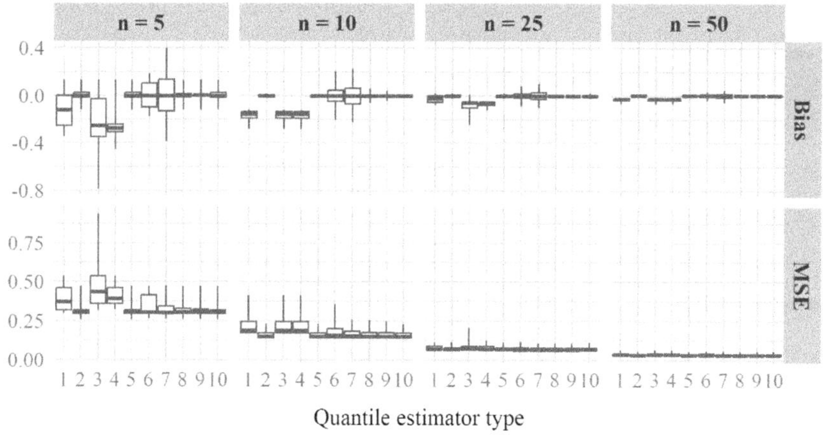

Figure 4.5
Data generated from a standard normal distribution. Each boxplot shows the distribution across deciles of the bias and mean squared error (MSE) for each quantile estimator (types 1–9 as in `stats::quantile(..., type)`, and type 10 for mid-quantiles), averaged over 10,000 replications.

for all estimators as n increases and levels off at similar values. We see that, in general, type 7 does not perform particularly well despite being the default quantile estimator in the R function `quantile`. Among estimators that have 'jumps', type 2 behaves favourably and similarly to types 8, 9, and 10, which are continuous functions of p. Type 2, indeed, compromises by averaging two successive order statistics when np is a whole number.

For data generated from a Poisson distribution, we observe again that types 1, 3, and 4 are less competitive, either in terms of bias or MSE (Figure 4.6). However, as before, the performance of all estimators is broadly similar at the largest sample size.

In Chapter 7, we study inferential properties, with particular emphasis on the coverage of interval estimators.

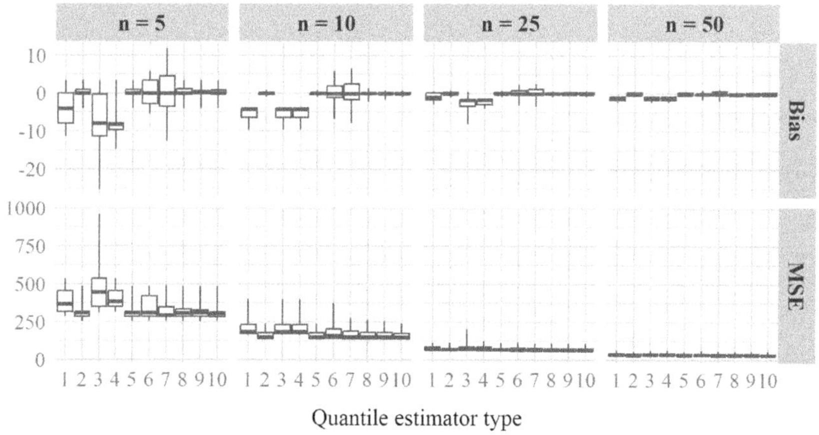

Figure 4.6
Data generated from a Poisson ($\lambda = 1000$) distribution. Each boxplot shows the distribution across deciles of the bias and mean squared error (MSE) for each quantile estimator (types 1–9 as in `stats::quantile(..., type)`, and type 10 for mid-quantiles), averaged over 10,000 replications.

5

Computation of quantiles

5.1 Sorting

As discussed in Chapter 4, sorting a sample of observations is a natural step in the estimation of sample quantiles. Consider the definition of sample quantile based on order statistics given in (4.8). To select the order statistics of interest, we clearly need to sort the observations first.[1] The computational complexity of a sorting algorithm is therefore key to the performance of the estimation procedure. As the sample size n increases, so does the computational time. For example, the `R` function `sort` implements radix sorting, shellsort, and quicksort algorithms (Knuth, 1998). The `R` function `quantile` makes a call to `sort` (which employs a more efficient *partial* sorting).

A good behaviour for a sorting algorithm is $O(n \log n)$ time complexity, although in practice differences in CPU times between commonly available algorithms are negligible when dealing with small problems. R Code 5.1 shows the timing for three different sorting algorithms as implemented in the function `sort`. Given the extremely small execution time for a single evaluation, the function `microbenchmark` (Mersmann, 2024) repeatedly evaluates (by default, 100 times) the expressions to accurately measure CPU times at a sub-millisecond scale. At $n = 1000$, the quicksort algorithm is the fastest, followed by shellsort and radix. With a 100-fold larger sample size, the radix algorithm outperforms the other two algorithms. However, at these sample sizes, times are in the order of milliseconds or microseconds, and differences between algorithms are inconspicuous.[2] In conditional quantile estimation, sorting observations can be very laborious. Therefore, we need to turn to more efficient computational algorithms as those discussed in the next sections of this chapter.

5.2 Linear programming

The problem of estimating the pth quantile of a random variable Y can be formulated as a generalisation of the LAD minimisation problem, that is,

$$\min_{\xi \in \mathbb{R}} \mathrm{E}\left\{\psi_p(Y - \xi)\right\} = \min_{\xi \in \mathbb{R}} \int_{-\infty}^{\infty} \psi_p(y - \xi)\, \mathrm{d}F_Y(y), \tag{5.1}$$

[1]See also the simulation-based approach by Dunn (1991), and the remedian method of Rousseeuw and Bassett (1990) and Lau and Boos (1994), which are both efficient alternatives to sorting algorithms.

[2]For extremely large datasets, exact sorting may become computationally or memory intensive, motivating the use of approximate or streaming quantile algorithms that avoid complete sorting (Greenwald and Khanna, 2001).

DOI: 10.1201/9781003326410-5

R Code 5.1: CPU time benchmarking of sorting algorithms in R.

```
library(microbenchmark)
set.seed(564)
# n = 1000
y <- rnorm(1e3)
microbenchmark('quick' = sort(y, method = "quick"),
    'shell' = sort(y, method = "shell"),
    'radix' = sort(y, method = "radix"))
> Unit: microseconds
>   expr  min    lq    mean median     uq   max neval cld
>  quick 22.9 37.45  63.916  52.55  77.10 216.9   100  a
>  shell 46.2 52.35  97.484  71.35 109.70 722.5   100   b
>  radix 55.2 62.30 105.937  84.50 139.85 417.7   100   b
# n = 100000
y <- rnorm(1e5)
microbenchmark('quick' = sort(y, method = "quick"),
    'shell' = sort(y, method = "shell"),
    'radix' = sort(y, method = "radix"))
> Unit: milliseconds
>   expr    min       lq      mean   median       uq     max neval cld
>  quick 6.5110  7.83375 10.497948  9.87000 12.63075 16.8547   100 a
>  shell 8.8810 10.50560 14.103450 13.45970 17.09800 22.4241   100  b
>  radix 4.0956  6.68225  8.748011  8.85255 10.48440 13.7828   100   c
```

for some $p \in (0, 1)$, where the integral is understood in the Lebesgue–Stieltjes sense and

$$\psi_p(r) = r\{p - I(r < 0)\} = \begin{cases} |r| \cdot (1-p) & \text{if } r < 0, \\ |r| \cdot p & \text{if } r \geq 0, \end{cases} \tag{5.2}$$

with $r \in \mathbb{R}$. The loss function ψ_p (sometimes called 'check' function) takes the absolute value of its argument and assigns the weights $1-p$ or p to, respectively, negative or positive arguments (e.g., see Fox and Rubin, 1964). This function is depicted in Figure 5.1 for $p = 0.5$ and $p = 0.75$, along with the quadratic loss function $\varphi(r) = r^2$ used in ordinary least squares (LS) estimation. One can verify that $\psi_{0.5}$ is simply (half) the absolute value function discussed in Section 1.3 apropos the median. When generalised to other quantiles, we observe that the left and right segments have different slopes, for example, the right segment is steeper than the left when $p = 0.75$.

If Y is continuous with density f_Y, we can write (5.1) as the sum of two integrals, one for values of Y below ξ and one for values above ξ, that is,

$$\begin{aligned} \mathrm{E}\{\psi_p(Y-\xi)\} &= \int_{-\infty}^{\infty} \psi_p(y-\xi) f_Y(y)\,\mathrm{d}y \\ &= (p-1)\int_{-\infty}^{\xi} (y-\xi) f_Y(y)\,\mathrm{d}y + p\int_{\xi}^{\infty} (y-\xi) f_Y(y)\,\mathrm{d}y. \end{aligned} \tag{5.3}$$

An analogous expression can be written as a summation when Y is discrete. In particular, if Y has probability mass function $m_Y(y_j) = \Pr(Y = y_j)$, then

$$\mathrm{E}\{\psi_p(Y-\xi)\} = (p-1)\sum_{j:y_j<\xi} (y_j - \xi)\, m_Y(y_j) + p \sum_{j:y_j \geq \xi} (y_j - \xi)\, m_Y(y_j).$$

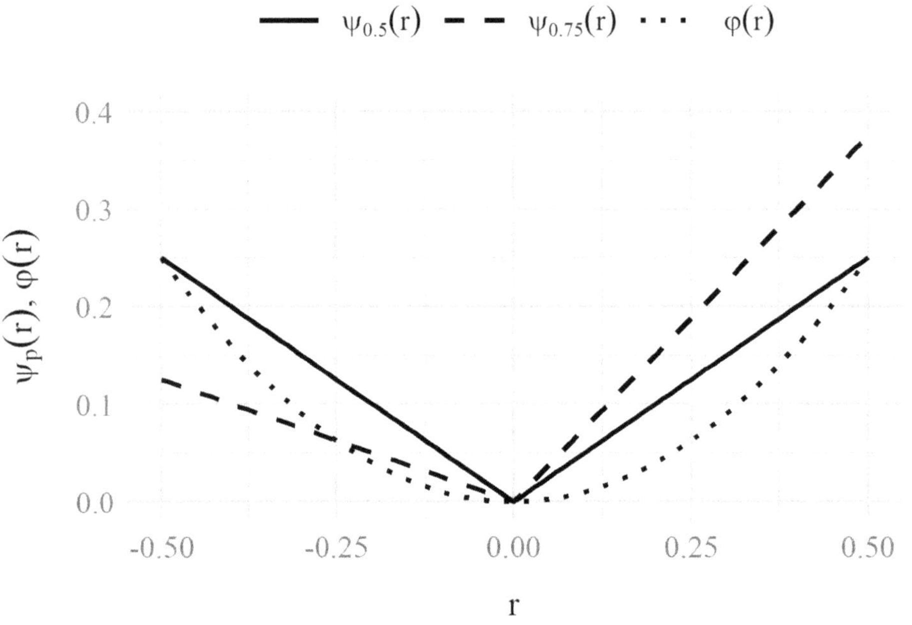

Figure 5.1
Quantile loss function ψ_p for two values of p and the quadratic loss function φ.

By differentiating (5.3) with respect to ξ, we obtain

$$(1-p)\int_{-\infty}^{\xi} f_Y(y)\,\mathrm{d}y - p\int_{\xi}^{\infty} f_Y(y)\,\mathrm{d}y = (1-p)F_Y(\xi) - p\{1 - F_Y(\xi)\}$$
$$= F_Y(\xi) - p. \tag{5.4}$$

Thus, in the continuous case, the first-order condition is

$$F_Y(\xi) - p = 0,$$

whose solution set is $\{\xi^* : F_Y(\xi^*) = p\}$. If F_Y is strictly increasing, the solution is unique and $\xi^* = F_Y^{-1}(p)$. In general, the objective function may exhibit flat regions, and therefore the minimiser need not be unique; any value satisfying $F_Y(\xi^-) \leq p \leq F_Y(\xi)$ is optimal. Non-uniqueness arises, for example, in the discrete case or when F_Y is locally flat.

Consider an IID sample $Y_1, \ldots, Y_n$ with common CDF F_Y and corresponding observations $y_1, \ldots, y_n$. The empirical analogue of (5.1) is

$$\min_{\xi \in \mathbb{R}} \frac{1}{n} \sum_{i=1}^{n} \psi_p(y_i - \xi). \tag{5.5}$$

A well-known case arises for $p = 0.5$, the median, which we have seen in (1.4). Although (5.5) is convex, it does not admit a closed-form solution in general. However, it can be reformulated as a *linear program* (LP), which enables efficient computation. An LP is an optimisation problem of the type

$$\begin{aligned} &\underset{\boldsymbol{z}}{\text{minimise}} && g(\boldsymbol{z}) \\ &\text{subject to} && h_k(\boldsymbol{z}) \leq b_k, \quad k = 1, \ldots, m \end{aligned} \tag{5.6}$$

where the *objective function* $g : \mathbb{R}^q \to \mathbb{R}$ and the *constraint functions* $h_k : \mathbb{R}^q \to \mathbb{R}$ are *linear*. These problems emerge when resources are limited and must be allocated to perform specific actions towards a desired goal. A linear program is a special case of a convex optimisation problem (Boyd and Vandenberghe, 2004) with applications in a large number of fields, from portfolio optimisation in finance to shape design in engineering.

In canonical form (Berman, 1973; Koenker, 2005), an LP problem is given by

$$\begin{aligned} &\underset{\boldsymbol{z}}{\text{minimise}} \quad && \boldsymbol{c}^\top \boldsymbol{z} \\ &\text{subject to} && \boldsymbol{A}\boldsymbol{z} - \boldsymbol{b} \in T, \\ & && \boldsymbol{z} \in S, \end{aligned} \tag{5.7}$$

where S and T are closed convex cones. Along with it, one can formulate the corresponding *dual program* which not only may provide an interesting interpretation in terms of the original (primal) problem, but it can also lead to an efficient computational method for solving it (Boyd and Vandenberghe, 2004), both of which happen to be the case in quantile estimation (Koenker and Bassett, 1978; Koenker, 2005). Roughly speaking, the dual program exchanges the roles of the primal variables and constraints, while reversing the optimisation direction. The dual to (5.7) is

$$\begin{aligned} &\underset{\boldsymbol{d}}{\text{maximise}} \quad && \boldsymbol{b}^\top \boldsymbol{d} \\ &\text{subject to} && \boldsymbol{c} - \boldsymbol{A}^\top \boldsymbol{d} \in S^*, \\ & && \boldsymbol{d} \in T^*, \end{aligned} \tag{5.8}$$

where $\boldsymbol{d}$ is the dual optimisation variable, with dimension equal to that of $\boldsymbol{b}$, while the set $K^* = \{\boldsymbol{x} \mid \boldsymbol{y}^\top \boldsymbol{x} \geq 0 \text{ for all } \boldsymbol{y} \in K\}$ denotes the dual cone of K. Since the optimal value $\boldsymbol{b}^\top \hat{\boldsymbol{d}}$ gives the best lower bound on $\boldsymbol{c}^\top \boldsymbol{z}$, any reduction in the *duality gap* $\boldsymbol{c}^\top \boldsymbol{z} - \boldsymbol{b}^\top \boldsymbol{d}$ is an improvement towards the optimal solution (Koenker, 2005). If the duality gap is zero, then we say that strong duality holds (Boyd and Vandenberghe, 2004).

Quantile estimation can be cast into an LP problem by reformulating (5.5) into the same form as (5.7). We start by rewriting the check function as

$$\psi_p(r) = p \cdot u + (1 - p) \cdot v, \tag{5.9}$$

where $u = \max(r, 0) \equiv (r)_+$ and $v = \max(-r, 0) \equiv (-r)_+$ are nonnegative slack variables representing the positive and negative parts of r, respectively. Then the objective function in (5.5) is proportional to $\sum_{i=1}^{n} p \cdot u_i + (1 - p) \cdot v_i$, where $u_i = (y_i - \xi)_+$, $v_i = (\xi - y_i)_+$, and y_i is the observation of Y_i. By definition, we must have $y_i = \xi + u_i - v_i$, $i = 1, \ldots, n$. Let $\boldsymbol{u} = (u_1, \ldots, u_n)^\top$, $\boldsymbol{v} = (v_1, \ldots, v_n)^\top$, and $\boldsymbol{y} = (y_1, \ldots, y_n)^\top$. Putting all this together we obtain

$$\begin{aligned} &\underset{(\xi, \boldsymbol{u}, \boldsymbol{v})}{\text{minimise}} \quad && 0 \cdot \xi + p\mathbf{1}_n^\top \boldsymbol{u} + (1 - p)\mathbf{1}_n^\top \boldsymbol{v} \\ &\text{subject to} && \xi \mathbf{1}_n + \boldsymbol{u} - \boldsymbol{v} = \boldsymbol{y}, \\ & && (\xi, \boldsymbol{u}, \boldsymbol{v}) \in \mathbb{R} \times \mathbb{R}^{2n}_{\geq 0}. \end{aligned} \tag{5.10}$$

The correspondence between the minimisation problem in (5.10) and that in (5.7) is obtained by letting $\boldsymbol{z} = (\xi, \boldsymbol{u}^\top, \boldsymbol{v}^\top)^\top$, $\boldsymbol{c} = (0, p\mathbf{1}_n^\top, (1 - p)\mathbf{1}_n^\top)^\top$, $\boldsymbol{A} = [\mathbf{1}_n \quad \boldsymbol{I}_n \quad -\boldsymbol{I}_n]$, $\boldsymbol{b} = \boldsymbol{y}$, $S = \mathbb{R} \times \mathbb{R}^{2n}_{\geq 0}$, and $T = \mathbf{0}_n$. The ensuing dual LP is

$$\begin{aligned} &\underset{\boldsymbol{d}}{\text{maximise}} \quad && \boldsymbol{y}^\top \boldsymbol{d} \\ &\text{subject to} && \mathbf{1}_n^\top \boldsymbol{d} = 0, \\ & && \boldsymbol{d} \in [p - 1, p]^n. \end{aligned} \tag{5.11}$$

This follows from the fact that $S^* = 0 \times \mathbb{R}^{2n}_{\geq 0}$ and $T^* = \mathbb{R}^n$, therefore the constraints $\boldsymbol{c} - \boldsymbol{A}^\top \boldsymbol{d} \in S^*$ and $\boldsymbol{d} \in T^*$ essentially reduce to $\mathbf{1}_n^\top \boldsymbol{d} = 0$ and $p - 1 \leq d_i \leq p$, $i = 1, \ldots, n$. Problem (5.11) can be reparameterised as

$$\begin{aligned} \underset{\boldsymbol{d}}{\text{maximise}} \quad & \boldsymbol{y}^\top \boldsymbol{a} \\ \text{subject to} \quad & \mathbf{1}_n^\top \boldsymbol{a} = (1-p)\mathbf{1}_n^\top \mathbf{1}_n, \\ & \boldsymbol{a} \in [0,1]^n, \end{aligned} \tag{5.12}$$

by letting $\boldsymbol{a} = \boldsymbol{d} + (1-p)\mathbf{1}_n$. This formulation shows that quantile estimation amounts to balancing positive and negative residuals with asymmetric weights, a feature reflected in both the primal and dual programs. The optimal values $\hat{a}_i$, $i = 1, \ldots, n$, are called rankscores (Hájek and Šidák, 1967) and have an important role in certain quantile inference methods (see Section 8.3.2). Their values are determined by the ranks of the corresponding observations y_i in the sense that $\hat{a}_i = 0$ if $R_i/n < p$, $\hat{a}_i = 1$ if $(R_i - 1)/n > p$, and $\hat{a}_i = R_i - np$ if $(R_i - 1)/n \leq p \leq R_i/n$.

While (5.10) does not have a simple analytical solution (as the one for, e.g., LS problems), linear programs can be solved fairly easily with effective optimisation algorithms such as Dantzig's simplex method and the more recent interior point methods (Boyd and Vandenberghe, 2004). From a practical point of view, the major difference between these algorithms is their performance depending on the size of the problem to solve. This will be discussed in more detail in Chapter 8, since linear programs have a more important role in estimation of conditional quantiles.

An example is provided in R Code 5.2. For simplicity, we generate a sample of size $n = 5$ from a standard normal distribution (hereafter referred to as the test data) and consider the estimation of the quantile $p = 0.75$. We first formulate the problem in canonical LP form and then solve it using the `lpSolve` package (Berkelaar, 2024). In the order provided in our example, the solution is a vector of length $2n + 1 = 11$ that contains the optimal values of the slack variables $\hat{\boldsymbol{u}}$ and $\hat{\boldsymbol{v}}$, followed by $\hat{\xi}$. The latter is exactly the estimate we obtain from the standard `R` function `quantile`. We can also compute the residuals as $\hat{\boldsymbol{u}} - \hat{\boldsymbol{v}}$. We can verify that $\sum \psi_p(\hat{u}_i - \hat{v}_i)$ is equal to the value of the LP objective function at the optimum. Note also that one of the residuals is exactly 0. In fact, an important property of linear programming is that an optimal solution can be taken to be basic; in the present problem this implies that at least one residual is exactly zero. In the presence of ties, the optimal solution to (5.10) identifies with multiple occurrences of the same y value.

It is worth having a glance at the dual program (5.12). As mentioned earlier, the solution to this problem gives the rankscores. R Code 5.3 shows that the value of the objective function at the solution $\boldsymbol{y}^\top \hat{\boldsymbol{d}} = \boldsymbol{y}^\top\{\hat{\boldsymbol{a}} - (1-p)\mathbf{1}\} = 1.4026$ coincides with that obtained from the primal program (i.e., the duality gap is zero).

The examples presented in this section are merely for didactic purposes since, as soon as n is increased to a few hundred, generating a sparse $3n \times (2n+1)$ matrix and then solving via `lp` becomes rather inefficient. Optimised LP code for quantile estimation is available in the R package `quantreg` (Koenker, 2025).

5.3 Nonsmooth optimisation

The loss function ψ_p is continuous and piecewise linear. Its derivative

$$\frac{\partial \psi_p(r)}{\partial r} = p - I(r < 0) \tag{5.13}$$

R Code 5.2: Estimating quantiles via a simplex algorithm with the R package `lpSolve`.

```
library(lpSolve)
# generate data
set.seed(564)
n <- 5
print(y <- rnorm(n))
> [1]  0.4605501 -0.7750968 -0.7159321  0.6882645 -2.0544591
# define the LP problem
p <- 0.75
obj <- c(rep(p, n), rep(1 - p, n), 0)
S <- cbind(diag(2*n), 0)
A <- cbind(diag(n), -diag(n), 1)
b <- c(rep(0, 2*n), y)
constr <- c(rep(">=", 2*n), rep("=", n))
# solve
fit.lp <- lp("min", obj, rbind(S, A), constr, b)
fit.lp
> Success: the objective function is 1.40257
# compare solution with standard R function
print(xi <- fit.lp$solution[2*n+1])
> [1] 0.4605501
quantile(y, probs = p, names = FALSE)
> [1] 0.4605501
# check linear constraints xi + u - v = y
u <- fit.lp$solution[1:n]
v <- fit.lp$solution[(n+1):(2*n)]
all.equal(xi + u - v, y)
> [1] TRUE
# calculate loss function 'psi'
print(r <- u - v)
> [1]  0.0000000 -1.2356469 -1.1764822  0.2277145 -2.5150091
sum(r * (p - ifelse(r < 0, 1, 0)))
> [1] 1.40257
```

is continuous everywhere except at 0, the point at which ψ_p reaches its minimum. This means that we cannot solve (5.5) by simply setting the ordinary first derivative of ψ_p equal to zero and then solving for ξ (as we would do, for example, to solve an analogous LS problem). However, at 0 the function ψ_p admits directional derivatives in all directions.

Let $r = y - \xi$, then the directional derivative of $\psi_p(r)$ with respect to ξ in direction $d \neq 0$ is given by

$$\begin{aligned}\partial_d \psi_p(r) &= \left.\frac{\partial \psi_p(y - \xi - td)}{\partial t}\right|_{t=0} \\ &= \frac{\partial}{\partial t}(y - \xi - td)\,[p - I(y - \xi - td < 0)]|_{t=0} \\ &= \begin{cases} -d[p - I(y - \xi < 0)] & \text{if } y \neq \xi, \\ -d[p - I(d > 0)] & \text{if } y = \xi. \end{cases} \end{aligned} \tag{5.14}$$

R Code 5.3: Estimating rankscores by solving the dual program corresponding to the primal program in R Code 5.2.

```
constr <- c(rep(">=", n), rep("<=", n), rep("=", 1))
fitd.lp <- lp("max", y, rbind(diag(n),diag(n),
t(matrix(1, nrow = n))), constr, c(rep(0, n), rep(1, n), n*(1-p)))
fitd.lp
> Success: the objective function is 0.8034021
a <- fitd.lp$solution
d <- round(a - (1 - p), 3)
sum(y*d)
> [1] 1.40257
cbind(y, u, v, a, d, rank = rank(y))
>              y         u        v    a     d rank
> [1,]  0.4605501 0.0000000 0.000000 0.25  0.00    4
> [2,] -0.7750968 0.0000000 1.235647 0.00 -0.25    2
> [3,] -0.7159321 0.0000000 1.176482 0.00 -0.25    3
> [4,]  0.6882645 0.2277145 0.000000 1.00  0.75    5
> [5,] -2.0544591 0.0000000 2.515009 0.00 -0.25    1
```

If all the directional derivatives are nonnegative at a point $\hat{\xi}$, then $\hat{\xi}$ is a minimiser of (5.5) (Koenker, 2005, p. 32). Since ψ_p is Lipschitz, then, according to Proposition 2.3.2 in Clarke (1990), $\hat{\xi}$ must satisfy the first-order condition $0 \in \partial^C \psi_p(y - \hat{\xi})$, where

$$\partial^C \psi_p(y - \xi) = \begin{cases} -[p - I(y - \xi < 0)] & \text{if } y \neq \xi, \\ -(\omega + p - \frac{1}{2}) : |\omega| \leq \frac{1}{2} & \text{if } y = \xi, \end{cases} \tag{5.15}$$

is Clarke's (generalised) derivative. This first-order condition is necessary for optimality and, since ψ_p is convex, also sufficient (Rockafellar, 1990). This condition is analogous to the familiar first-order condition in LS estimation, except that the equality symbol '=' is replaced with '∈' since $\partial^C \psi_p$ is set-valued (see also Koenker, 2005, p. 38).

The derivative (5.13) can be readily used in a steepest descent algorithm to find the minimiser of (5.5). For example, Bottai et al. (2015) proposed an iterative algorithm in which the value of ξ at the $(k+1)$th iteration, $\xi^{(k+1)}$, is given by

$$\xi^{(k+1)} = \xi^{(k)} - \delta^{(k)} g_p'(\xi^{(k)}), \tag{5.16}$$

where $g_p(\xi) = \sum_{i=1}^n \psi_p(y_i - \xi)$, $g_p'(\xi) = -\sum_{i=1}^n p - I(y_i < \xi)$, and $\delta^{(k)}$ is the step size at iteration k. Assuming that $y_i \neq \xi^{(k)}$ for all $i = 1, \ldots, n$ and all k, the sequence $\{\xi^{(k)} : k \in \mathbb{N}\}$ so determined converges efficiently to a minimiser. The assumption of non-zero residuals presupposes the continuity of Y, for which the algorithm was developed. In this case, the probability $\Pr(Y_i = \xi^{(k)}) = 0$, $i = 1, \ldots, n$, for any value $\xi^{(k)}$. Even if a zero residual does occur, the algorithm moves off of it in the direction $g'(\xi^{(k)})$.

An instance of the gradient descent algorithm (Bottai et al., 2015) is shown in R Code 5.4 using the test data from R Code 5.2 and the function `lqm` in the R package `lqmm` (Geraci, 2014). The sequence of decreasing values of the objective function g terminates after 25

R Code 5.4: Estimating quantiles via gradient-based optimisation using the R package `lqmm` on the test data from R Code 5.2.

```
library(lqmm)
print(y)
> [1]  0.4605501 -0.7750968 -0.7159321  0.6882645 -2.0544591
# optimise - set verbose = TRUE to print out tracing information
p <- 0.75
fit.lqm <- lqm(y ~ 1, tau = p, control = lqmControl(verbose = FALSE))
#  (1) logLik = 2.107483988920
#  (2) logLik = 1.486746584515
#  Decreasing step...
#  (3) logLik = 1.486746584515
#  Decreasing step...
#  (4) logLik = 1.486746584515
#  (5) logLik = 1.439171853438
# [omitted]
#  Decreasing step...
#  (23) logLik = 1.402585249608
#  (24) logLik = 1.402575213032
#  Decreasing step...
#  (25) logLik = 1.402575213032
coef(fit.lqm)
> (Intercept)
>   0.4605441
```

iterations.[3] The value of g at convergence and the resulting estimate of $\xi_{0.75}$ are within 10^{-5} of those found in the solution of the LP problem.

Derivative-free optimisation techniques (e.g., pattern search, modified Nelder–Mead, implicit filtering) can be considered for general problems, though typically at the cost of slower convergence rates (e.g., Conn et al., 2009).

5.4 Smooth approximations

Smooth approximations of the absolute value function have often been used to obtain estimators that inherit some of the robustness of the L_1-norm and, at the same time, can be subjected to algorithms for smooth functions. A case in point is the M-estimator of a location parameter. It hinges upon a local smoothing of the absolute residuals in an interval around 0, with the width of the interval determined by a 'tuning' constant.

Consider the Huber's loss function (Huber, 1964)

$$\kappa_c(r) = \begin{cases} \frac{1}{2}r^2 & \text{if } |r| < c, \\ c|r| - \frac{1}{2}c^2 & \text{if } |r| \geq c, \end{cases} \tag{5.17}$$

[3]The `logLik` reported in the `lqm` output corresponds to the asymmetric Laplace likelihood, whose maximisation is equivalent to minimising the quantile loss function ψ_p. This connection will be discussed in Section 6.3.1.

where $c > 0$ is a tuning constant. This function is continuously differentiable and has first derivative

$$\frac{\partial \kappa_c(r)}{\partial r} = \begin{cases} r & \text{if } |r| < c, \\ c \cdot \operatorname{sgn}(r) & \text{if } |r| \geq c. \end{cases} \tag{5.18}$$

When applied to the estimation of the location parameter μ, that is,

$$\min_{\mu \in \mathbb{R}} \frac{1}{n} \sum_{i=1}^{n} \kappa_c(Y_i - \mu),$$

κ_c robustifies the residuals with magnitude larger than c, that is, $|Y_i - \mu| > c$. The smaller the tuning parameter c, the closer κ_c gets to the median loss function $\psi_{0.5}$. On the other hand, when $c \to \infty$, κ_c becomes the quadratic loss.

The following asymmetric generalisation of (5.17) was proposed by Chen (2007):

$$\kappa_{c,p}(r) = \begin{cases} -qr - \frac{1}{2}q^2 c & \text{if } r \leq -qc, \\ \frac{1}{2c}r^2 & \text{if } -qc \leq r \leq pc, \\ pr - \frac{1}{2}p^2 c & \text{if } r \geq pc, \end{cases} \tag{5.19}$$

where $q = 1 - p$. This function, too, is continuously differentiable and has first derivative

$$\frac{\partial \kappa_{c,p}(r)}{\partial r} = \begin{cases} -q & \text{if } r \leq -qc, \\ \frac{1}{c}r & \text{if } -qc \leq r \leq pc, \\ p & \text{if } r \geq pc. \end{cases} \tag{5.20}$$

The slope of $\kappa_{c,p}$ is the same as that of ψ_p for values outside the interval $(-qc, pc)$. On the other hand, the cusp at 0 is replaced by a smooth (quadratic) curve. When $c \to 0^+$, the interval $(-qc, pc)$ shrinks and the absolute function takes over the smooth approximation, that is, $\kappa_{c,p} \to \psi_p$. An illustration is given in Figure 5.2, where $\kappa_{c,p}$ is depicted for two values of c, with $p = 0.75$. See also Oh et al. (2011) for a similar approximation.

Following these observations, it is natural to consider an iterative algorithm whereby $\kappa_{c,p}$ approaches ψ_p as $c_k \to 0$ through a decreasing sequence $\{c_k\}$. Figure 5.3 shows an instance of the sequence of minimisers

$$\hat{\mu}_k = \operatorname*{argmin}_{\mu \in \mathbb{R}} g_{k,0.75}(\mu),$$

where k indexes the sequence $\{c_k\}$ (not iterations of a single algorithm), $g_{k,p}(\mu) = \sum_{i=1}^{n} \kappa_{c_k,p}(y_i - \mu)$, and $c_k \in \{1, 0.9, \ldots, 0.1, 0.05, 10^{-3}, 10^{-4}\}$. The data are those generated in R Code 5.2, allowing comparison with the LP solution. As c goes to 0, the estimates $\hat{\mu}_k$ and the corresponding values $g_k(\hat{\mu}_k)$ approach, respectively, the LP solution and optimal value obtained for $\xi_{0.75}$. In this example, each $\hat{\mu}_k$ is estimated via the descent algorithm discussed in Section 5.3, opportunely modified by using $g_{k,p}$ instead of g_p (see R Code 5.5 for details).

The algorithms developed by Clark and Osborne (1986) and Madsen and Nielsen (1993) for median estimation, and subsequently extended by Chen (2007) to any quantile $p \in (0, 1)$, are based on the same mechanism illustrated in Figure 5.3. In practice, these algorithms are more sophisticated than the simple procedure used in our example. In particular, the methods of Madsen and Nielsen (1993) and Chen (2007) start from a 'warm' estimate of μ and a suitable initial value of c, which is then decreased at each iteration according to a prescribed strategy. Stopping criteria are also provided. Moreover, the algorithm converges

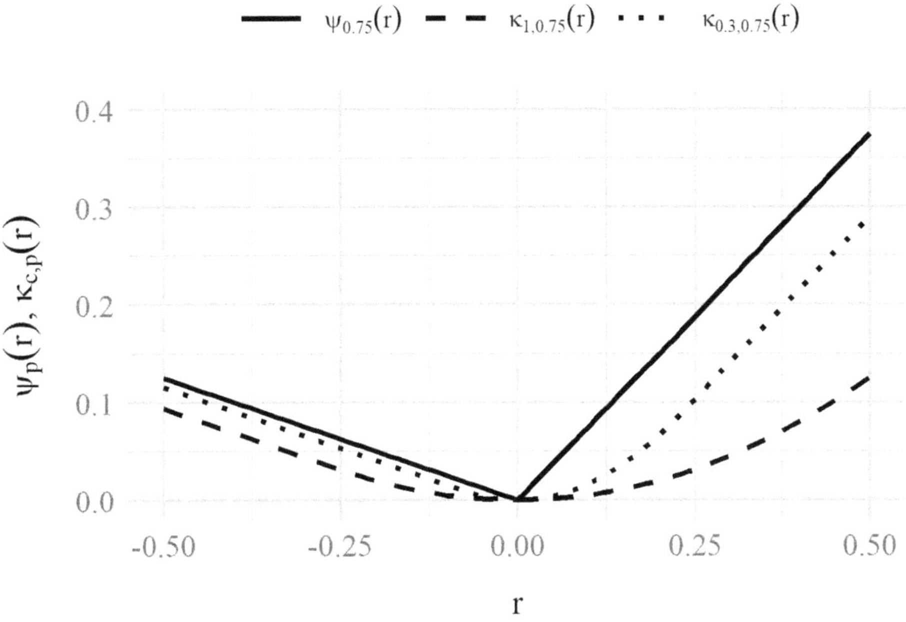

Figure 5.2
Loss function ψ_p for $p = 0.75$ and its smooth approximation $\kappa_{c,p}$ for $c = 1$ and $c = 0.3$.

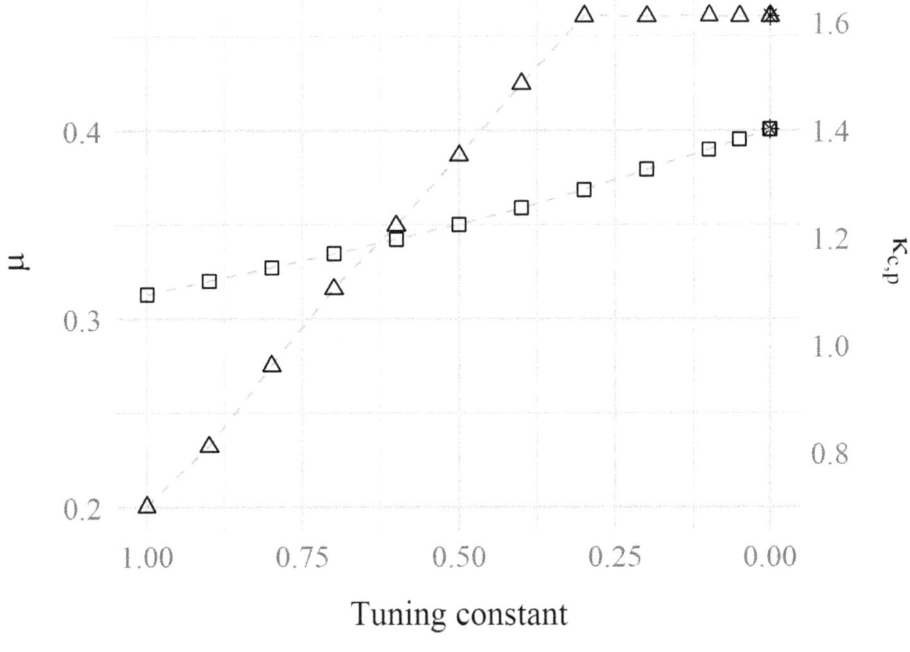

Figure 5.3
Estimates (squares) and optimal values of the loss function $\kappa_{c,0.75}$ (triangles) for a decreasing sequence of values of the tuning constant c using the test data from R Code 5.2. The asterisks mark the optimal value of the LP objective function and the quantile estimate.

R Code 5.5: Estimating quantiles via (modified) gradient-based optimisation using the R package `lqmm` on the test data from R Code 5.2. Note the argument `smooth = TRUE` in `lqmControl`.

```
library(lqmm)
print(y)
> [1]  0.4605501 -0.7750968 -0.7159321  0.6882645 -2.0544591
p <- 0.75
# define decreasing sequence of tuning constants
omicron <- c(sort(seq(0.1, 1, by = 0.1), decreasing = TRUE),
 0.05, 0.001, 0.0001)
# optimise for each value of the sequence
K <- length(omicron)
mu <- g <- rep(NA, K)
for(k in 1:K){
    fit.lqm <- lqm(y ~ 1, tau = p, control = lqmControl(method =
    "gs2", smooth = TRUE, omicron = omicron[k]))
    mu[k] <- coef(fit.lqm)
    # obtain quantile loss from asymmetric Laplace log-likelihood
    g[k] <- -(fit.lqm$logLik - n*log(p*(1 - p)/fit.lqm$scale))*
    fit.lqm$scale
}
c(estimate = mu[K], objective = g[K])
>  estimate objective
> 0.4605291 1.4025351
```

in a finite number of steps, so that it is not necessary for c to approach zero to obtain an exact minimiser of (5.5).

The function in (5.19) is one of several that can be used to approximate the check function. Surveying all such approximations is beyond the scope of this section. We provide just one additional example of a function that has nice properties and has shown a good performance in conditional quantile estimation. The definition of ψ_p given in (5.9) based on the plus (or 'ramp') function suggests using the integral of a sigmoid (i.e., logistic) function (Chen and Mangasarian, 1996):

$$(x)_+ \approx x + c\log\{1 + \exp(-x/c)\}, \tag{5.21}$$

where $c > 0$. The approximation (5.21) improves when $c \to 0^+$. It follows that a smooth approximation of the check function is given by

$$\begin{aligned}\chi_{c,p}(r) &= p[r + c\log\{1 + \exp(-r/c)\}] + q[-r + c\log\{1 + \exp(r/c)\}] \\ &= pr + c\log\{1 + \exp(-r/c)\},\end{aligned} \tag{5.22}$$

where $q = 1 - p$ as before. Both $\chi_{c,p}$ and its first derivative

$$\frac{\partial \chi_{c,p}(r)}{\partial r} = p - \frac{1}{1 + \exp(r/c)} \tag{5.23}$$

are continuously differentiable. Indeed, $\chi_{c,p}$ is smooth, that is, differentiable of all orders, whereas (5.19) is only once continuously differentiable. Schmidt et al. (2007) applied the symmetric equivalent of (5.22) (i.e., $p = 0.5$) to smooth the L_1 penalty in regularised models,

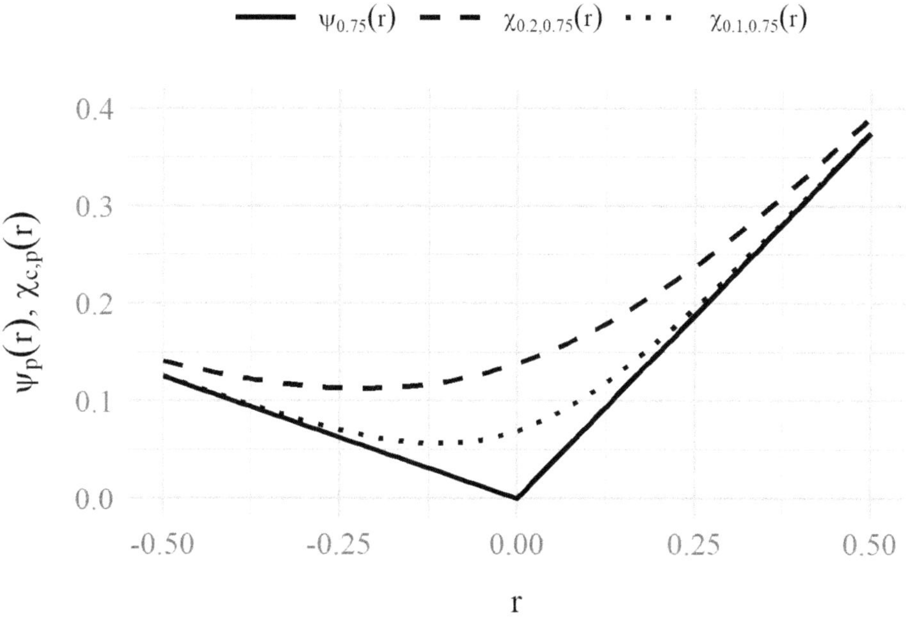

Figure 5.4
Loss function ψ_p for $p = 0.75$ and its smooth approximation $\chi_{c,p}$ for $c = 0.2$ and $c = 0.1$.

while Zheng (2011) proposed a gradient descent algorithm based on (5.22). An illustration of the behaviour of $\chi_{c,p}(r)$ for two values of c and for $p = 0.75$ is given in Figure 5.4.

We conclude this section with a brief discussion of methods that exploit LS estimation in different ways. We start with the proposal of Soliman et al. (1988), in which the LS residual with the smallest magnitude is used to identify the observation that *should* coincide with the median. This idea is motivated by the relationship between the L_1 estimate and the observation (or observations) with zero residual as discussed at the end of Section 5.2. However, Bassett and Koenker (1992) provided counterexamples where Soliman et al.'s (1988) approach fails.

The majorise-minimise (MM) algorithm of Hunter and Lange (2000) (see also Hunter and Lange, 2004) is rooted in the older idea of estimating the median via an iterative reweighted LS algorithm (e.g., see Mosteller and Tukey, 1977). Given the current estimate $\mu^{(k)}$, the $(k+1)$th iterate is

$$\mu^{(k+1)} = \frac{\sum_{i=1}^{n} w_i^{(k)} y_i}{\sum_{i=1}^{n} w_i^{(k)}}, \tag{5.24}$$

where $w_i^{(k)} = |y_i - \mu^{(k)}|^{-1}$. Hunter and Lange (2000) sketches the main idea by extending (5.24) to an arbitrary quantile $p \in (0, 1)$, giving

$$\mu_p^{(k+1)} = \frac{n(2p-1) + \sum_{i=1}^{n} w_i^{(k)} y_i}{\sum_{i=1}^{n} w_i^{(k)}}, \tag{5.25}$$

where $w_i^{(k)} = |y_i - \mu_p^{(k)}|^{-1}$. A crude implementation is given in R Code 5.6 to estimate $\xi_{0.75}$ using the test data from R Code 5.2. In this example, the algorithm starts from the sample mean and stops when $|\mu_{0.75}^{(k+1)} - \mu_{0.75}^{(k)}| < 10^{-8}$. The algorithm converges to the same

R Code 5.6: Estimating quantiles via iterative reweighted least squares on the test data from R Code 5.2.

```
print(y)
> [1]  0.4605501 -0.7750968 -0.7159321  0.6882645 -2.0544591
# IRLS algorithm
p <- 0.75
STOP <- FALSE; mu_old <- mean(y); k <- 1; eps <- .Machine$double.eps^(2/3)
while(!STOP & k < 100){
    # uncomment the line below to print out tracing information
    # cat("iteration", k, "; mu = ", mu_old, "\n")
    w <- 1/abs(y - mu_old + eps)
    mu_new <- (n*(2*p - 1) + sum(w*y))/sum(w)
    if(abs(mu_new - mu_old) < 1e-8) STOP <- TRUE
     else {mu_old <- mu_new; k <- k + 1}
}
# iteration 1 ; mu =  -0.4793347
# iteration 2 ; mu =  -0.3317401
# iteration 3 ; mu =  -0.1364635
# iteration 4 ; mu =  0.08740153
# [omitted]
# iteration 25 ; mu =  0.4605501
# iteration 26 ; mu =  0.4605501
c('mu' = mu_new, 'xi' = quantile(y, probs = p, names = FALSE))
>         mu        xi
> 0.4605501 0.4605501
```

estimate $\hat{\xi}_{0.75}$ in a relatively small number of iterations. A small constant is added to the denominator of the weights to avoid division by zero.

6

Topics in inference

6.1 Introduction

The purpose of this chapter is to explore the use of parametric distributions in quantile-based inference. We consider three distinct sets of problems:

1. we begin with the idea of performing inference on quantiles of the distribution of a random variable Y when the data are assumed to originate from a given model, say F, which depends on a low-dimensional parameter θ. The distribution F is fitted to the data by maximum likelihood estimation of θ. Successively, quantiles are obtained by inverting the fitted distribution. We call this approach 'maximum likelihood for quantiles' (MLQ);

2. we then consider parametric distributions, say G, whose parameters are linked to the quantiles of F, regardless of the form of F. Examples of G include the asymmetric Laplace and the extreme value distributions. We call them *auxiliary* to emphasise that they are used as devices for estimating the quantiles of F;

3. we conclude with a method that makes use of (empirical) quantiles to estimate the parameter θ of a known distribution F, from where the sample is assumed to be drawn.

Throughout this chapter, we assume Y is continuous.

6.2 Maximum likelihood for quantiles

Likelihood-based estimation of quantiles is a natural approach when a parametric distribution is available. This idea has appeared occasionally in the preceding chapters. In this section, we provide a detailed comparison of parametric and nonparametric quantile estimators with regard to their asymptotic properties. We assume that $Y_1, \ldots, Y_n$ is an IID sample of size n from a continuous parametric distribution F with scalar parameter $\theta \in \Theta$, and that the usual regularity conditions for maximum likelihood estimation hold (e.g., Cox and Hinkley, 1997). Here and throughout, $F^{-1}(p;\theta)$ denotes the generalised inverse

$$F^{-1}(p;\theta) = \inf\{y : F(y;\theta) \geq p\}.$$

Under standard smoothness conditions, including differentiability of $F(y;\theta)$ and $f(\xi_p;\theta) > 0$, the derivative of $F^{-1}(p;\theta)$ with respect to θ follows by implicit differentiation.

Let $\hat{\theta}_n$ be the MLE of θ. For large n, we have (Cox and Hinkley, 1997)

$$\hat{\theta}_n \mathrel{\dot\sim} \mathcal{N}\left(\theta, n^{-1}I^{-1}(\theta)\right),$$

DOI: 10.1201/9781003326410-6

where $I(\theta)$ denotes the Fisher information for θ, that is, the negative expected Hessian of the log-likelihood. If we plug $\hat{\theta}_n$ into F^{-1}, then $\hat{F}_n^{-1}(p;\hat{\theta}_n)$ is an estimator of $F^{-1}(p;\theta) = \xi_p$ as a result of the MLE's invariance property. We call this method of estimating quantiles 'maximum likelihood for quantiles' (MLQ) . The approximate sampling distribution of $\hat{F}_n^{-1}$ is obtained by the delta method, that is,

$$\hat{F}_n^{-1}\left(p;\hat{\theta}_n\right) \dot{\sim} \mathcal{N}\left(\xi_p, n^{-1}v^2(\theta)I^{-1}(\theta)\right), \tag{6.1}$$

where

$$v(\theta) = \frac{\partial F^{-1}(p;\theta)}{\partial \theta} = -\frac{\partial F(\xi_p;\theta)}{\partial \theta} \cdot \frac{1}{f(\xi_p;\theta)}$$

by implicit differentiation of $F\left(F^{-1}(p;\theta);\theta\right) = p$.

Sample quantiles based on order statistics are asymptotically normal too (Section 4.3), thus for sufficiently large n

$$\hat{\xi}_{p,n} \dot{\sim} \mathcal{N}\left(\xi_p, \frac{p(1-p)}{n\{f(\xi_p;\theta)\}^2}\right). \tag{6.2}$$

Therefore, both estimators are consistent and asymptotically normal. The asymptotic relative efficiency (ARE) of the distribution-free estimator relative to the likelihood-based estimator, defined as the ratio of asymptotic variances (or, equivalently, the ratio of sample sizes required to attain the same asymptotic variance), is

$$\text{ARE}(\hat{\xi}_{p,n}, \hat{F}_n^{-1}) = \left\{\frac{\partial F(\xi_p;\theta)}{\partial \theta}\right\}^{-2} \frac{p(1-p)}{I^{-1}(\theta)}. \tag{6.3}$$

An ARE equal to 1 indicates that the two estimators have the same asymptotic efficiency, whereas values greater than 1 indicate that the distribution-free estimator requires a larger sample size to attain the same asymptotic variance as the likelihood-based estimator. Since $\hat{\theta}_n$ is asymptotically efficient under the assumed parametric model, the plug-in estimator $\hat{F}_n^{-1}(p;\hat{\theta}_n)$ attains the minimum asymptotic variance among regular estimators based on that model. Hence

$$\text{ARE}(\hat{\xi}_{p,n}, \hat{F}_n^{-1}) \geq 1.$$

This can be verified in specific models, as illustrated below.

Let us consider the following one-parameter models: normal $\mathcal{N}(\theta, 1)$, exponential $\mathcal{E}(\theta)$, logistic Logistic$(\theta, 1)$, and Laplace $\mathcal{L}(\theta, 1)$. For several distributions (e.g., exponential, logistic, Laplace), (6.3) can be expressed in closed form using elementary functions from the expressions of the CDFs and QFs given in Appendix A.3. For instance, in the case of an exponential distribution, we have $F(\xi_p;\theta) = 1 - \exp(-\theta\xi_p)$ and $\xi_p = -\log(1-p)/\theta$. As a function of p, the first factor of (6.3) is given by

$$\frac{\partial F(\xi_p;\theta)}{\partial \theta} = \xi_p \exp(-\theta\xi_p) = -(1-p)\log(1-p)/\theta.$$

Since $I(\theta) = \theta^{-2}$, we obtain $\text{ARE}(\hat{\xi}_{p,n}, \hat{F}_n^{-1}) = \dfrac{p}{1-p}\dfrac{1}{\log^2(1-p)}$.

MLQ clearly does better in all cases considered (Figure 6.1). The lowest ARE, as defined in (6.3), is attained at $p = 0.5$ for the normal and logistic distributions, with values $\pi/2 \approx 1.57$ and $12/\pi^2 \approx 1.21$, respectively, and is approximately 1.54 near $p = 0.8$ for the exponential distribution. However, we see that, for the Laplace distribution, the lowest ARE reaches the unity at the median.[1]

[1]This is simply explained by the fact that the MLE of the location parameter in a Laplace distribution coincides with the sample median. See also Section 6.3.

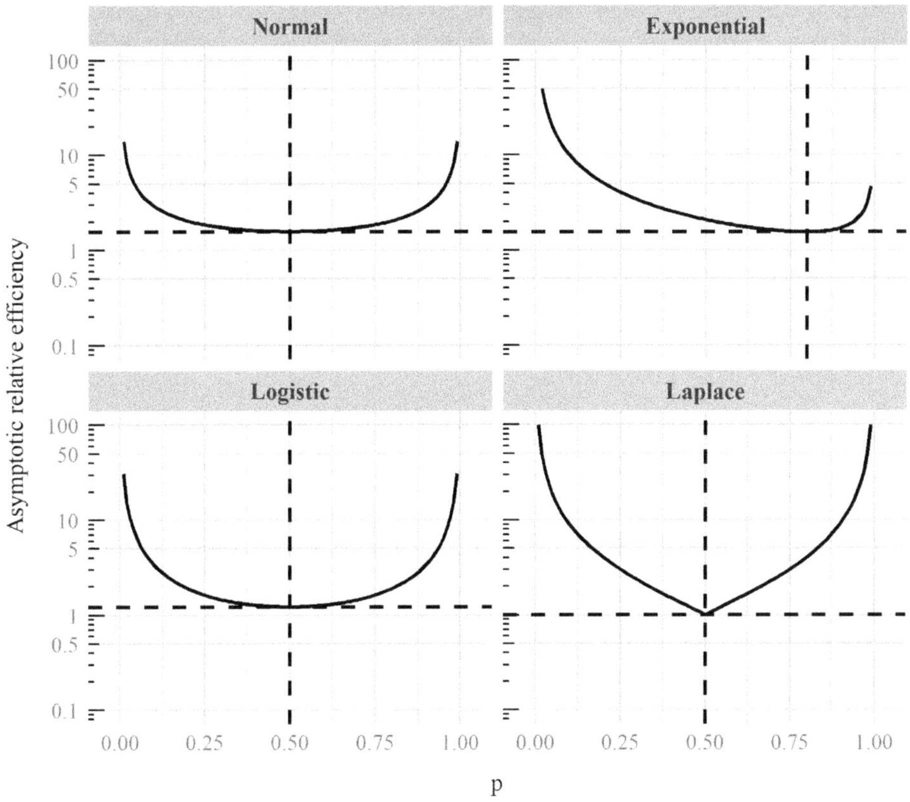

Figure 6.1
Efficiency of the distribution-free quantile estimator relative to the corresponding likelihood-based estimator for four probability models, shown on the $\log_{10}$ scale as a function of p. The coordinates of the minima are marked by dashed lines.

In these examples, we are dealing with IID sampling from continuous distributions with supports that do not depend on θ. In this case, the Pitman–Koopman–Darmois theorem (Pitman, 1936; Koopman, 1936; Darmois, 1935) affirms that there is a sufficient statistic whose dimension does not depend on the parameter if and only if F belongs to the exponential family. Consequently, the likelihood-based quantile estimator $\hat{F}_n^{-1}\left(p; \hat{\theta}_n\right)$, being a function of the MLE, ultimately depends on a low-dimensional summary of the data in these cases.

However, we do not have the same data reduction for the logistic and Laplace distributions, which do not belong to the exponential family. In such cases, there is generally no sufficient statistic of fixed dimension, and the order statistics are sufficient (and minimal). Indeed, outside the exponential family, the minimal sufficient statistic typically has dimension n; under IID sampling from a continuous distribution, it is (up to one-to-one transformations) given by the order statistics, which are essentially the best we can hope for in terms of data reduction (Casella and Berger, 2002). It is also worth noting that sample quantiles, including the sample median, are not sufficient statistics in general, even though in specific models (such as the Laplace) they may coincide with maximum likelihood estimators.

Interestingly, the order statistics contain all the information in the sample (in a nonparametric sense) even when F is not specified, as in nonparametric estimation. This highlights a fundamental difference between parametric and distribution-free approaches, and suggests that comparisons of efficiency cannot rely on classical arguments based on sufficiency alone.

One may wonder whether there is a middle ground for a quantile estimator with an asymptotic variance between that of $\hat{F}_n^{-1}(p;\hat{\theta}_n)$ and $\hat{\xi}_{p,n}$. The answer appears to be negative: *no translation equivariant and asymptotically uniformly median unbiased estimator is asymptotically more concentrated about the distribution quantile than the sample quantile* (Pfanzagl, 1976).

6.3 Auxiliary distributions

6.3.1 Asymmetric Laplace distribution

The asymmetric Laplace (AL) distribution plays a prominent role in quantile estimation. Its use in conditional modelling will be discussed in Chapter 13, with particular reference to clustered data. Here we focus on inferential aspects of unconditional MLE under the AL likelihood when it is misspecified.

Let Z be a continuous random variable with AL density $\mathcal{AL}(\mu,\sigma,\tau)$ (see definition in Appendix A.3), where $\mu \in \mathbb{R}$ (location), $\sigma > 0$ (scale), and $0 < \tau < 1$ (asymmetry). In Figure 6.2, the AL is plotted for $\mu = 0$, $\sigma = 1$, and three different values of τ. The distribution is positively skewed, symmetric, or negatively skewed for $\tau < 0.5$, $\tau = 0.5$, or $\tau > 0.5$, respectively. The quantile function of the AL,

$$Q_Z(p;\mu,\sigma,\tau) = \begin{cases} \mu + \dfrac{\sigma}{1-\tau}\log\left(\dfrac{p}{\tau}\right) & \text{for } 0 < p < \tau \\ \mu - \dfrac{\sigma}{\tau}\log\left(\dfrac{1-p}{1-\tau}\right) & \text{for } \tau \leq p < 1 \end{cases} \tag{6.4}$$

reveals the key property of this distribution and its connection with quantile estimation. Indeed, it is immediate to verify that for $p = \tau$, $Q_Z(\tau;\mu,\sigma,\tau) = \mu$, that is, the location parameter is the τth quantile of Z (as well as its mode).

Now suppose we observe a sample $\boldsymbol{y} = (y_1,\ldots,y_n)$ from $Y \sim F$ with population quantiles $\xi_p = F^{-1}(p)$ and want to fit the AL distribution, say G, with parameter $\theta = (\mu,\sigma,\tau)^\top$.

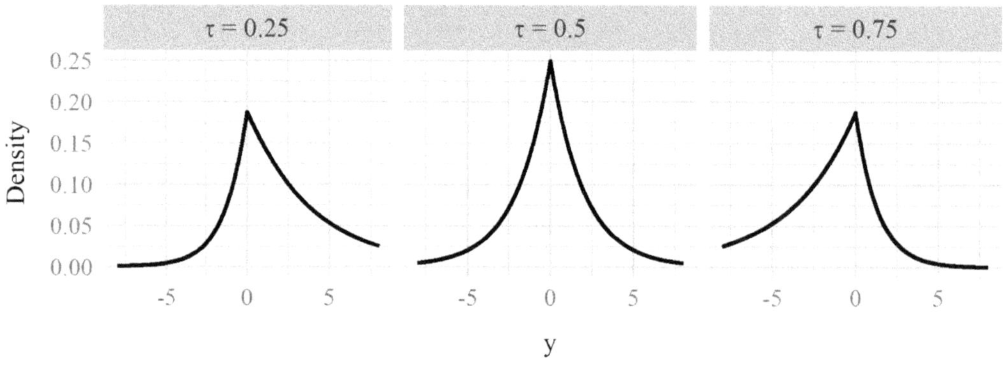

Figure 6.2
Density of the asymmetric Laplace $\mathcal{AL}(0,1,\tau)$ for different values of τ.

A maximum likelihood estimate can be obtained iteratively[2] by using the following estimating equations (Yu and Zhang, 2005):

$$\begin{aligned}
\hat{\mu}_n(\hat{\tau}) &= \underset{\mu}{\operatorname{argmin}} \sum_{i=1}^{n} h_{\hat{\tau}}\left(y_i - \mu\right), \\
\hat{\sigma}_n(\hat{\mu}, \hat{\tau}) &= \frac{1}{n} \sum_{i=1}^{n} h_{\hat{\tau}}\left(y_i - \hat{\mu}\right), \\
\hat{\tau}_n(\hat{\mu}) &= \frac{a + \sqrt{a^2 - (\bar{y} - \hat{\mu})a}}{\bar{y} - \hat{\mu}},
\end{aligned} \tag{6.5}$$

where $h_\tau(u) = \tau \max(u, 0) + (1 - \tau) \max(-u, 0)$, $a = \frac{1}{n} \sum_{i=1}^{n} (y_i - \hat{\mu}) I(y_i \leq \hat{\mu})$, and $\bar{y} = 1/n \sum_{i=1}^{n} y_i$. The function h_τ coincides with the quantile loss function ψ_p in (5.2), with $\tau = p$. Hence, the minimisation problem from the first estimating equation in (6.5) is analogous to that in (5.5), except that $\hat{\tau}$ is not fixed as is p. An illustration is given in R Code 6.1. First, a sample of size $n = 500$ is taken from a standard normal distribution (which, in this example, plays the role of F). Subsequently, the AL distribution is fitted to the sample by means of the function `mleAL` in the `lqmm` package. After 22 iterations, the algorithm gives $\hat{\theta} = (-0.058, 0.384, 0.470)^\top$. The estimated $\hat{\tau}$ indicates that the empirical distribution is fairly symmetric (as we would expect since we sampled observations from a normal distribution).

Note that $\hat{G}^{-1}(\hat{\tau}; \hat{\mu}, \hat{\sigma}, \hat{\tau}) = \hat{\mu}$ by definition of (6.4).[3] It is therefore unsurprising that the $\hat{\tau}$th sample quantile $\hat{\xi}_{\hat{\tau}}$ is numerically close to $\hat{\mu}$. This reflects the fact that, for fixed τ, the AL location parameter coincides with the τth quantile. When τ is estimated jointly with (μ, σ), however, the target quantile level is itself random and converges to $\tau^* = \lim_{n\to\infty} \hat{\tau}_n$.

Clearly, it would not be wise to use the fitted distribution $\hat{G}$ to do inference about F. Say we wanted to estimate the population quantile $\xi_{0.75}$. If we knew that F was normal, then we could proceed as discussed in Section 6.2 to obtain $\hat{F}^{-1}(0.75) = 0.691$, where $\hat{F}$ is the CDF of a normal distribution with mean and variance equal to the sample estimates. The estimated quantile $\hat{F}^{-1}(0.75)$ is obviously close to the sample quantile $\hat{\xi}_{0.75} = 0.685$. If, instead, we were to assume that the sample came from an AL distribution, then our estimate $\hat{G}^{-1}(0.75) = 0.556$ would be rather off the mark, simply because F is not AL.

Now we add a twist. Instead of θ, we consider the parameter $\theta_{\tau_0} = (\mu_{\tau_0}, \sigma_{\tau_0})^\top$ for *fixed* $\tau = \tau_0$ and, instead of solving (6.5), we solve

$$\begin{aligned}
\hat{\mu}_n(\tau_0) &= \underset{\mu}{\operatorname{argmin}} \sum_{i=1}^{n} h_{\tau_0}\left(y_i - \mu\right), \\
\hat{\sigma}_n(\hat{\mu}, \tau_0) &= \frac{1}{n} \sum_{i=1}^{n} h_{\tau_0}\left(y_i - \hat{\mu}\right).
\end{aligned} \tag{6.6}$$

In the estimating equations above, τ is no longer a parameter to be estimated since it is constrained to a fixed value.

R Code 6.2 shows the estimates of μ and σ for the AL distribution $\mathcal{AL}(\mu, \sigma, 0.75)$ obtained with the `mle2` function in the R package `bbmle` (Bolker, 2023). The estimator $\hat{\mu}$

[2]Maximum likelihood estimation of the three-parameter asymmetric Laplace distribution does not need to be iterative. See, for example, Wright (2024) for a finite-step procedure that is faster than the iterative algorithm as well as more reliable when the solution lies on the boundary of the parameter space.

[3]Regardless of whether F is AL, the estimator $\hat{\mu} = \hat{G}^{-1}(\hat{\tau}; \hat{\mu}, \hat{\sigma}, \hat{\tau})$ can be interpreted as an estimate of the modal quantile introduced in Section 2.2.3. This quantity is referred to as the *most likely quantile* by Bera et al. (2015).

R Code 6.1: Maximum likelihood estimation of the asymmetric Laplace parameters.

```
library(lqmm)
set.seed(123)
y <- rnorm(500)
fit <- mleAL(y)
fit
> $m
> [1] -0.05834601
>
> $sigma
> [1] 0.383867
>
> $tau
> [1] 0.4698466
>
> $r
> [1] 22
# quantile 0.4698466 of fitted AL
qal(fit$tau, mu = fit$m, sigma = fit$sigma, tau = fit$tau)
> [1] -0.05834601
# sample quantile 0.4698466
quantile(y, probs = fit$tau, names = FALSE)
> [1] -0.05843686
# quantile 0.75 of fitted normal
qnorm(0.75, mean = mean(y), sd = sd(y))
> [1] 0.6907134
# sample quantile 0.75
quantile(y, probs = 0.75, names = FALSE)
> [1] 0.6852113
# quantile 0.75 of fitted AL
qal(0.75, mu = fit$m, sigma = fit$sigma, tau = fit$tau)
> [1] 0.5558011
```

is effectively 'forced' to coincide with $\hat{\xi}_{0.75}$, since the minimisation problem arising from the first estimating equation in (6.6) coincides with that in (5.5) for $p = \tau_0 = 0.75$. For comparison, the code shows also the results from `lqm` using the algorithm discussed in Section 5.3, which, as expected, gives similar results.[4]

It may be natural to wonder whether we can push this connection a little further and use the AL likelihood for inference beyond point estimation. The large-n standard error of $\hat{\theta}_{\tau_0}$ can be readily obtained (under correct specification of the AL model) from the asymptotic distribution of $\sqrt{n}(\hat{\theta}_{\tau_0} - \theta_{\tau_0})$, which is normal with mean $(0,0)^\top$ and variance-covariance

[4]Unlike `lqm`, which first estimates the quantile parameter by minimising the quantile loss and then obtains the scale residually, `mle2` performs a joint numerical optimisation of the AL negative log-likelihood over μ and $\phi = \log(\sigma)$.

R Code 6.2: Maximum likelihood estimation of the asymmetric Laplace parameters with fixed τ, using the R package `bbmle` on the data generated from R Code 6.1.

```
library(bbmle)

# maximise AL likelihood
loglik_al_constr <- function(mu, phi){
  sigma <- exp(phi)
  -sum(dal(x = y, mu = mu, sigma = sigma, tau = 0.75,
  log = TRUE))
}

fit <- mle2(loglik_al_constr, start = list(mu = 0, phi = 0))
thetahat <- as.numeric(coef(fit))
c(mu = thetahat[1], sigma = exp(thetahat[2]))
>        mu     sigma
> 0.6865766 0.3143648
# similar to lqm
lqm(y ~ 1, tau = 0.75)
> Call: lqm(formula = y ~ 1, tau = 0.75)
>
> Quantile 0.75
> Fixed effects:
> (Intercept)
>   0.6866911
>
> Degrees of freedom: 500 total; 499 residual
> Residual scale parameter: 0.3143631 (standard deviation 1.325471)
> Log-likelihood (Laplace): -758.3849
```

(Kotz et al., 2001, p. 166)[5]

$$\mathbf{\Sigma} = \begin{bmatrix} \dfrac{\sigma^2_{\tau_0}}{\tau_0(1-\tau_0)} & 0 \\ 0 & \sigma^2_{\tau_0} \end{bmatrix}. \tag{6.7}$$

One can immediately verify that the first element of the diagonal of $\mathbf{\Sigma}$, which corresponds to the asymptotic variance of $\hat{\mu}_n(\tau_0)$, is equal to the asymptotic variance in (4.9) by defining

[5] Kotz et al. (2001) give a different parameterisation of the asymmetric Laplace distribution. To facilitate a comparison with (6.4), here we report the quantile function shown in their equation (3.1.32)

$$Q_Z(p;\theta,\phi,\kappa) = \begin{cases} \theta + \dfrac{\phi\kappa}{\sqrt{2}} \log\left(\dfrac{1+\kappa^2}{\kappa^2} p\right) & \text{for } 0 < p \le \dfrac{\kappa^2}{1+\kappa^2} \\ \theta - \dfrac{\phi}{\sqrt{2}\kappa} \log\left((1+\kappa^2)(1-p)\right) & \text{for } \dfrac{\kappa^2}{1+\kappa^2} < p < 1. \end{cases}$$

By setting $\theta = \mu$, $\phi = \sigma\sqrt{2}(1+\kappa^2)/\kappa$, and $\kappa = \sqrt{\tau/(1-\tau)}$ we are back to the parameterisation in (6.4). Kotz et al. (2001, p. 166) discuss the asymptotics of the estimator of (θ, ϕ) when the value of κ is known (which corresponds to our case of known τ). Some asymptotic results for the maximum likelihood estimators of μ, σ, and τ can be found also in Yu and Zhang (2005), although these results seem to contain some errors (see also Gijbels et al., 2019).

f as the density of $\mathcal{AL}(\mu_{\tau_0}, \sigma_{\tau_0}, \tau_0)$ and setting $p = \tau_0$, namely

$$\frac{\tau_0(1-\tau_0)}{\{f(\xi_{\tau_0})\}^2} = \frac{\tau_0(1-\tau_0)}{\{\tau_0(1-\tau_0)/\sigma_{\tau_0}\}^2}.$$

It follows that, in general, the standard errors obtained from (6.7) will be incorrect if the (unknown) true distribution F is not AL. R Code 6.3 shows the results of a small simulation with 500 replications from a standard normal. The constrained AL estimator $\hat{\mu}_n(\tau_0)$ has very small bias as expected, but the standard deviation of the estimates (0.0414) is grossly underestimated by the average estimated standard error (0.0232) as calculated from (6.7). As a consequence, the resulting confidence intervals are very liberal, with an observed coverage equal to 74% versus a 95% nominal level. In contrast, when we use a nonparametric estimate to calculate the asymptotic variance in (4.9), the observed coverage turns out to be 96%, close to the nominal level.

In summary (see Table 6.1), if the data generating distribution F is assumed to be (asymmetric) Laplace, as that might be the case in some applications in, for instance, ocean engineering, finance, and astronomy (Kotz et al., 2001; Geraci and Cortina-Borja, 2018), then it is natural to use the MLE properties of the AL likelihood to the full extent and make inference on the population quantiles ξ_p with $\hat{G}^{-1}(p; \hat{\mu}, \hat{\sigma}, \hat{\tau})$. In contrast, using the AL as a device to estimate ξ_p should be confined to point estimation when no particular assumptions are made about F. The constrained estimator $\hat{G}^{-1}(\tau_0; \hat{\mu}, \hat{\sigma}, \tau_0)$ is consistent for ξ_p, with $p = \tau_0$, even if the AL likelihood is misspecified, because its estimating equation coincides with the empirical quantile estimating equation. In the latter case, however, it would not be wise to measure the uncertainty of the estimates via the large-sample properties of the AL likelihood. Certainly, in case of misspecification, it would be rather pointless to use an estimator such as $\hat{G}^{-1}(p; \hat{\mu}, \hat{\sigma}, \tau_0)$ to estimate ξ_p, for $p \neq \tau_0$.

Then, what makes the constrained AL likelihood worth the trouble? Can we not simply use sample quantiles and the related inferential procedures in the first place? Perhaps the most evident (and only) advantage is in conditional modelling, when the goal is to model μ_{τ_0} as a function of covariates.

6.3.2 Other asymmetric distributions

The AL is not the only auxiliary distribution that can be used for quantile estimation (although it is the only distribution whose negative log-likelihood is equivalent to the loss function in (5.5)). There are several distributions whose location parameter coincides with a specific population quantile. A broad family of such distributions for continuous random variables is given by the *two-piece distributions* (Nassiri and Loris, 2013; Wallis, 2014; Gijbels et al., 2019; Alvarez, 2020),

$$g(y; \mu, \lambda_1, \lambda_2) = \frac{2\lambda_1\lambda_2}{\lambda_1+\lambda_2}\begin{cases} h\{\lambda_1(\mu - y)\} & \text{if } y < \mu, \\ h\{\lambda_2(y-\mu)\} & \text{if } y \geq \mu, \end{cases}$$

where h is a density that is unimodal and symmetric about 0, that is,

$$\int_{-\infty}^{0} h(u)\,\mathrm{d}u = \frac{1}{2}.$$

The parameters λ_1 and λ_2 control the asymmetry of the distribution. It is straightforward to verify that

$$G(\mu; \mu, \lambda_1, \lambda_2) = \int_{-\infty}^{\mu} g(u; \mu, \lambda_1, \lambda_2)\,\mathrm{d}u = \frac{\lambda_2}{\lambda_1+\lambda_2}.$$

R Code 6.3: Simulation study on constrained maximum likelihood estimation of the asymmetric Laplace parameters.

```
loglik_al_constr <- function(mu, phi){
sigma <- exp(phi)
-sum(dal(x = y, mu = mu, sigma = sigma, tau = tau0,
log = TRUE))}

R <- 500
n <- 1000
mu <- 0
sigma <- 1
tau0 <- 0.75

set.seed(123)
nn <- list(NULL, c("Estimate", "SE"), c("Constr. AL",
"Sample quantile"))
res <- array(NA, dim = c(R, 2, 2), dimnames = nn)
CI <- array(NA, dim = c(R, 2, 2), dimnames = nn)

for(i in 1:R){
    y <- rnorm(n)
    fit <- mle2(loglik_al_constr, start = list(mu = 0, phi = 0))
    res[i,1,1] <- m <- coef(fit)[1]
    res[i,1,2] <- xip <- quantile(y, probs = tau0)
    s <- exp(coef(fit)[2])
    d <- density(y)
    res[i,2,1] <- se1 <- s/sqrt(n*tau0*(1-tau0))
    res[i,2,2] <- se2 <- sqrt(tau0*(1 - tau0))/
    (sqrt(n)*approxfun(d$x, d$y)(xip))
    CI[i,,1] <- m + c(-1,1)*qnorm(.975, 0, se1)
    CI[i,,2] <- xip + c(-1,1)*qnorm(.975, 0, se2)
}

# true population quantile
qnorm(tau0)
> [1] 0.6744898
# average estimate and SE
apply(res, 3, colMeans)
>           Constr. AL Sample quantile
> Estimate 0.67446062      0.67367360
> SE       0.02321417      0.04367893
# standard deviation of the estimates
apply(res[,1,], 2, sd)
>       Constr. AL Sample quantile
>       0.04137496      0.04150335
# 95% confidence intervals
apply(CI - qnorm(tau0), 3, function(x) mean(
apply(x, 1, prod) < 0))
>       Constr. AL Sample quantile
>            0.738           0.958
```

Table 6.1
Estimators of quantiles ξ_p of a population F. The asymmetric Laplace (AL) distribution is denoted by G.

Estimator	Consistency	Asymptotic variance
Full AL likelihood $\hat{G}^{-1}(p;\hat{\mu},\hat{\sigma},\hat{\tau})$	If F is AL, consistent for any $p \in (0,1)$. Otherwise, consistent only for $p = \tau^*$, where $\tau^* = \lim_{n\to\infty}\hat{\tau}_n$	Correct if F is AL
Constrained AL likelihood $\hat{G}^{-1}(p;\hat{\mu},\hat{\sigma},\tau_0)$	If $F = G(\cdot;\mu,\sigma,\tau_0)$, consistent for any $p \in (0,1)$. Otherwise, consistent only for $p = \tau_0$	Correct only if $F = G(\cdot;\mu,\sigma,\tau_0)$
Sample quantile $\hat{\xi}_p$	Consistent for any $p \in (0,1)$, for any F	Correct

Therefore, by setting $\lambda_1 = 1 - \tau$ and $\lambda_2 = \tau$, the location parameter μ coincides with the τth population quantile. If, in addition, h is the standard symmetric Laplace density, $h(z) = \frac{1}{2}\exp(-|z|)$, then g reduces to the AL distribution.

Under some regularity conditions, the maximum likelihood estimator $\hat{\mu}$ is consistent for the population quantile ξ_p with $p = \tau_0$, since by construction

$$G^{-1}(\tau_0;\mu,1-\tau_0,\tau_0) = \mu.$$

However, the same considerations regarding asymptotic variance as those in Table 6.1 apply. The remaining question is whether choosing a density h other than the Laplace offers any computational advantage, for which there is currently no clear evidence.

6.3.3 Extreme value distributions

In this section, we consider estimators of quantiles ξ_p when p gets arbitrarily close to 0 or 1 (*extremal* quantiles). We have seen that, to estimate a population quantile accurately, we are faced with the practical problem of having enough information in the vicinity of the quantile itself. The standard theory for quantiles works well as long as we stay reasonably away from the tails of the distribution where, usually, information is more scarce (i.e., the density is lower). Therefore, it may come as no surprise that inference on extremal quantiles requires ad hoc considerations above and beyond those seen for 'central' quantiles. To this end, we tap into the literature of order statistics and extreme value theory (EVT).

To begin with, the definition of what value of p constitutes an 'extremal' quantile needs some elaboration. After all, $p = 0.999$ may appear extremal when $n = 1000$, as this entails, to put it mildly, a rather exiguous number of observations above $\hat{\xi}_{0.999}$. However, the same cannot be said if $p = 0.999$ and, say, $n = 3 \times 10^6$. How far can we go on the tails with the standard inferential theory (Section 4.3) for sample quantiles? It is reasonable to expect that the answer involves both p and n. Indeed, it involves the quantity

$$u_p = \begin{cases} nq & \text{when } q \text{ is close to } 0 \\ np & \text{when } p \text{ is close to } 0, \end{cases}$$

where $q = 1-p$, which is defined as the *order* of the pth quantile in a sample of n independent observations. The left plot of Figure 6.3 shows the smoothed density for $n = 1000$ birth weights from the NCHS Births dataset introduced in Chapter 1. The order of the quantile

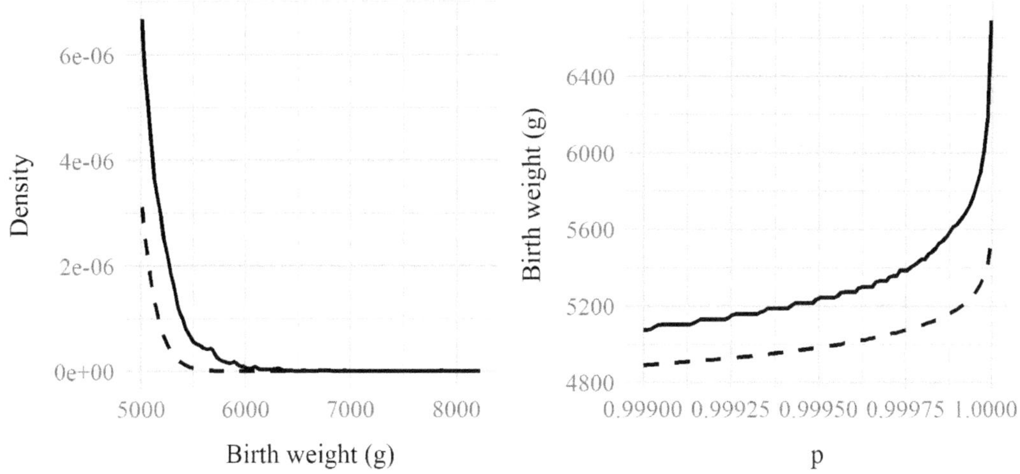

Figure 6.3
Sample-based (solid lines) and normal (dashed lines) estimates of the tail density (left) and quantile (right) functions for birth weights in the United States population, 2001.

0.999 is approximately $u_{0.999} = 3000$ (i.e., about 3000 observations exceed $\hat{\xi}_{0.999} = 5075$ g), while $u_{0.999995} = 15$. The order of the quantile provides a more informative measure of the extremeness of p. Moreover, regardless of how far we move into the right tail of the birth weight distribution, the sample exhibits more mass than that implied by a (parent) normal distribution. Accordingly, the MLQ estimator, as defined in Section 6.2, does not appear to be well suited for estimating such extreme quantiles (right plot in Figure 6.3).

Let p_n be a sequence. In the literature on order statistics, it is customary to distinguish between:

(a) the central (or quantile) case, where $p_n \to p$ with $0 < p < 1$;

(b) the extreme case, where $p_n \to 0$ or $p_n \to 1$ and $np_n \to k < \infty$;

(c) the intermediate case, where $p_n \to 0$ or $p_n \to 1$ and $np_n \to \infty$.

Under case (a), order statistics follow the asymptotic normal laws that have been discussed in the previous chapters. The extreme case (b) implies that, as the sample size grows larger, the amount of information above (when $p_n \to 1$) or below (when $p_n \to 0$) the quantile p_n remains fixed. In such a case, we must resort to the EVT. Under case (c), order statistics are still asymptotically normal under mild regularity conditions imposed on the parent distribution F (see Box 6.1), though they can also be regarded as extremes and studied as such (Reiss, 1989).

Let us consider the maximum of a sample $Y_1, \ldots, Y_n$ (but analogous considerations can be made for the minimum) from a continuous[6] distribution F. If the sample is IID, then the distribution of the nth order statistic $Y_{(n)}$ is given in (4.4). For any fixed value of y and

[6]The case of discrete F is discussed in David and Nagaraja (2003, p. 304). In summary, several common discrete parents do not belong to any domain of maximal attraction. However, limited results for variables that take on sufficiently large positive integer values can be obtained by means of interpolation (Anderson, 1970, 1980).

$n \to \infty$, $F_{(n)}(y)$ is either 1 (if $F(y) = 1$) or 0 (if $F(y) < 1$). That is, $F_{(n)}(y)$ is a degenerate distribution. The problem is two-fold (Beirlant et al., 2004): first find all non-degenerate distributions G that can result from the following limit:

Box 6.1: Intermediate order statistics

The large-n distribution of intermediate order statistics is approximately normal. This is stated in the following theorem which gives regularity conditions on the tail behaviour of F on a neighbourhood of the right endpoint $\omega(F)$.

Theorem 3. *Consider the IID sample* $Y_1, \ldots, Y_n \overset{IID}{\sim} F_Y$. *Assume that* $p_n \to 1$ *and* $np_n \to \infty$ *as* $n \to \infty$. *Moreover, assume that* $\omega(F) < \infty$ *and that* F *has a derivative* f *on the interval* $(\omega(F) - \varepsilon, \omega(F))$ *for some* $\varepsilon > 0$, *where* f *is uniformly continuous and bounded away from zero. Let* $k_n = \lfloor np_n \rfloor$. *Then the order statistic* $Y_{(k_n)}$ *is asymptotically normal, that is,*

$$\Pr\left\{c_n\left(Y_{(k_n)} - F^{-1}\left(\frac{k_n}{n+1}\right)\right) \leq y\right\} \to \Phi(y) \qquad \text{as } n \to \infty,$$

for every y, *where*

$$c_n = \frac{\sqrt{(n+1)^3} f\left(F^{-1}\left(\frac{k_n}{n+1}\right)\right)}{\sqrt{k_n(n - k_n + 1)}}.$$

For a proof of Theorem 3, see Reiss (1989).

Remark 6. *Theorem 3 can be adapted analogously to the case* $p_n \to 0$ *by imposing regularity conditions on a neighbourhood of the left endpoint* $\alpha(F)$.

Remark 7. *Because the regularity conditions above concern the tail behaviour of* F *near* $\omega(F)$ *(or* $\alpha(F)$*), there is a natural connection between intermediate and extreme order statistics. In this respect, EVT provides conditions for asymptotic normality that are often more appropriate for studying the intermediate case than those in Theorem 3. See, for example, Theorem 5.1.7 in Reiss (1989, p. 164).*

$$\lim_{n\to\infty} F^n(a_n + b_n y) = G(y) \tag{6.8}$$

for every continuity point y of G, where a_n and b_n are normalising sequences that may depend on n but not on y (Kotz and Nadarajah, 2000, p. 5). If (6.8) is satisfied, we say that F is in the *domain of maximal attraction* of the limiting distribution G and write $F \in \mathcal{D}(G)$. Secondly, characterise F for which there exist a_n and b_n such that the limit (6.8) holds for any such specific non-degenerate limiting distribution G (Beirlant et al., 2004, p. 46). If $F(y)$ is such that a non-degenerate G exists, then it must be one of three types: Gumbel (G_1), Fréchet (G_2), or Weibull (G_3). Their CDFs are given by

$$G_1(y) = \exp\{-\exp(-y)\}, \quad \text{for } y \in \mathbb{R}, \tag{6.9}$$

$$G_2(y;\alpha) = \begin{cases} 0, & \text{for } y \leq 0, \\ \exp\{-y^{-\alpha}\}, & \text{for } y > 0, \end{cases} \tag{6.10}$$

$$G_3(y;\alpha) = \begin{cases} \exp\{-(-y)^{\alpha}\}, & \text{for } y \leq 0, \\ 1, & \text{for } y > 0, \end{cases} \tag{6.11}$$

where $\alpha > 0$ is a shape parameter. These distributions are also known as type I, II, and III extreme value (EV) distributions, respectively.[7] Moreover, G_1, G_2, and G_3 can be subsumed in the *family* of generalised extreme value (GEV) distributions (see Appendix A.3), with shape parameter γ, whose CDF, in standard form, is given by

$$G_{ev}(y;0,1,\gamma) = \begin{cases} \exp\left\{-(1+\gamma y)^{-\frac{1}{\gamma}}\right\}, & \text{for } 1+\gamma y > 0, \text{ when } \gamma \neq 0, \\ \exp\{-\exp(-y)\}, & \text{for } y \in \mathbb{R},\ \gamma = 0. \end{cases} \tag{6.12}$$

A natural question arises: under what conditions F belongs to the domain of attraction of any of the extreme value distributions G_k, $k = 1, 2, 3$? There are different, often equivalent, ways to answer this question. In some cases, the conditions are difficult to prove, while in others, the task is simplified as, for example, with the conditions given by von Mises (1936). A rigorous and complete treatment of this topic was originally given by Gnedenko (1943) and de Haan (1970), and is typically presented in textbooks on statistical methods for extreme values and order statistics (e.g., Beirlant et al., 2004; Coles, 2001; David and Nagaraja, 2003; de Haan and Ferreira, 2006; Reiss, 1989). Following de Haan and Ferreira (2006), we introduce the extreme value condition

$$\lim_{t \uparrow \omega(F)} \frac{1 - F(t + yh(t))}{1 - F(t)} = (1+\gamma y)^{-\frac{1}{\gamma}}, \quad \text{for all } y \text{ with } 1+\gamma y > 0, \tag{6.13}$$

where $\omega(F)$ is the right endpoint of F. Theorem 1.2.5 in de Haan and Ferreira (2006) states that the distribution function F is in the domain of attraction of the GEV distribution, $F \in \mathcal{D}(G_{ev})$, if and only if (6.13) is satisfied for some positive function $h(t)$. Expression (6.13) suggests that, if Y is a random variable with distribution function $F \in \mathcal{D}(G_{ev})$ for some real γ,

$$\lim_{t \uparrow \omega(F)} \Pr\left(\frac{Y-t}{h(t)} > y \mid Y > t\right) = (1+\gamma y)^{-\frac{1}{\gamma}}, \tag{6.14}$$

for $y \in (0, \max\{0, -\gamma\}^{-1})$. That is, larger observations ($Y > t$) have limiting distribution

$$G_p(y;\gamma) = \begin{cases} 1 - (1+\gamma y)^{-\frac{1}{\gamma}} & \text{for } \gamma \neq 0, \\ 1 - e^{-y} & \text{for } \gamma = 0, \end{cases} \tag{6.15}$$

$0 < y < \infty$ if $\gamma \geq 0$ and $0 < y < -1/\gamma$ if $\gamma < 0$, which is a (standard) generalised Pareto (GP) distribution.

It will be clear by now that the GP distribution can be used as an 'auxiliary' distribution (in the sense introduced in Section 6.1) to make inference about γ and, ultimately, about extreme quantiles of F (whatever F, as long as $F \in \mathcal{D}(G_{ev})$). For this reason, the shape parameter $\gamma \in \mathbb{R}$, which governs the tail behaviour of F, is called *extreme value index* (EVI). For $\gamma > 0$, the distribution is heavy-tailed, being heavier with larger values of γ, and its right endpoint is infinite; for $\gamma = 0$, the distribution is light-tailed and its right endpoint

[7] In the study of sample minima, the extreme value distributions of interest are derived from those used for maxima as follows:

$$\bar{G}_1(y) = 1 - G_1(-y), \quad \text{for } y \in \mathbb{R},$$

$$\bar{G}_2(y;\alpha) = \begin{cases} 1 - G_2(-y;\alpha), & \text{for } y < 0, \\ 1, & \text{for } y \geq 0, \end{cases}$$

$$\bar{G}_3(y;\alpha) = \begin{cases} 0, & \text{for } y < 0, \\ 1 - G_3(-y;\alpha), & \text{for } y \geq 0. \end{cases}$$

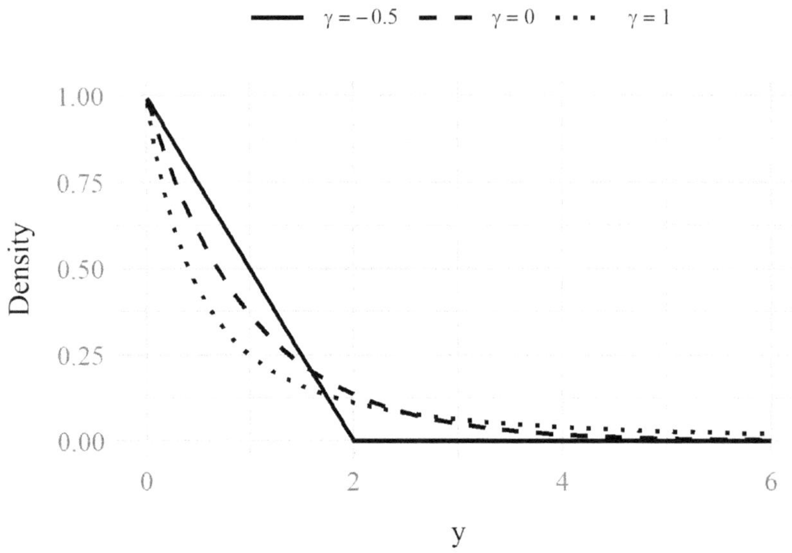

Figure 6.4
Density of the generalised Pareto distribution for different values of γ.

is infinite; finally, for $\gamma < 0$, the distribution is short-tailed and its right endpoint is $-1/\gamma$. The density of the GP distribution is plotted in Figure 6.4 for $\gamma = -0.5$, $\gamma = 0$, and $\gamma = 1$. For example, if $\gamma = 0$, then the domain of attraction of F is type I (Gumbel), which applies when the parent distribution is, for example, normal, exponential, and gamma.

There are several estimators of γ (e.g., Hill, Pickands, and moment estimators). Here we consider the so-called 'maximum likelihood' estimator.[8] Let us define the excess $Z_i = Y_{(n-i+1):n} - Y_{(n-r):n}$, with $Z_0 = Y_{(n-r):n}$, where $r = r_n \to \infty$ and $r/n \to 0$ as $n \to \infty$. Thus $Y_{(n-r):n}$ moves progressively into the far right tail and converges to the right endpoint $\omega(F)$, finite or infinite, in the appropriate sense. In practice, this corresponds to selecting a high (possibly random) threshold and modelling the exceedances above it. By assumption, the Y_is are independent. The excesses Z_i are neither independent nor exactly GP-distributed. The peaks-over-threshold approximation replaces their joint distribution by that of approximately independent GP-distributed exceedances above a high threshold. Then, by virtue of (6.14), the GP estimator, conditionally on $Z_0 = t$, is given by

$$(\hat{\sigma}_n, \hat{\gamma}_n) = \underset{(\sigma,\gamma)\in D}{\operatorname{argmax}} \prod_{i=1}^{r} g_p\left(z_i; 0, \sigma, \gamma\right), \tag{6.16}$$

where $g_p(z; \mu, \sigma, \gamma)$ denotes the GP density with location μ, scale σ, and shape γ. The threshold t is fixed[9] and marks the point above which the tail of F is approximated by

[8]Since the parent distribution F does not necessarily follow a GP distribution, we will more neutrally call it *generalised Pareto (GP) estimator* to avoid any confusion. Here, we mention in passing the interesting result by Falk and Marohn (1993) who showed that the rate of convergence in the von Mises conditions for $F \in \mathcal{D}(G_k)$, $k = 1, 2, 3$, determines the distance of F from the limiting GP distribution, while the limit is ultimately attained if and only if F is GP.

[9]In some applications, there may be the issue of how to determine t (or, equivalently, r). If t is not far enough in the tail, then there will be bias in the estimates of the GP parameter. Conversely, if t is unnecessarily far out, then the variance of the estimator will increase. In other applications, t may be actually predetermined according to a principle or practical consideration. However, the problem of deciding whether the threshold is optimal for fitting a GP distribution still remains. See the detailed discussion in Coles (2001) for threshold selection strategies.

a GP law. Accordingly, t defines the location (threshold) parameter of the limiting GP distribution. The parameters σ and γ are estimated by maximising the resulting conditional (on t) likelihood over the space $D = (0, \infty) \times (-1/2, \infty)$. The reason why the estimation of γ is constrained below by $-1/2$ is due to the fact that the maximum likelihood estimator behaves irregularly if $\gamma \leq -1/2$ (de Haan and Ferreira, 2006). For a given threshold t, the number of exceedances $S_t = \sum_{i=1}^{n} I(Y_i > t)$ has a binomial distribution with parameters n and $q_0 = 1 - p_0$, where $p_0 = F(t)$.

Let $\hat{F}$ be an estimate[10] of the (unknown) distribution F and define $\hat{q}_0 = 1 - \hat{p}_0 = 1 - \hat{F}(t)$. The generalised Pareto estimate of the (extreme) pth quantile of F, with $p > \hat{p}_0$, can be obtained as

$$\hat{\xi}_p^{\mathrm{GP}} = \begin{cases} t + \frac{\hat{\sigma}}{\hat{\gamma}} \left[\left(\frac{\hat{q}_0}{1-p} \right)^{\hat{\gamma}} - 1 \right] & \text{for } |\hat{\gamma}| \geq \epsilon \\ t + \hat{\sigma} \log \left(\frac{\hat{q}_0}{1-p} \right) & \text{for } |\hat{\gamma}| < \epsilon, \end{cases} \tag{6.17}$$

where $\epsilon > 0$ is a small tolerance parameter determined by machine's precision such that $|\hat{\gamma}| < \epsilon$ is interpreted as 0.

Similarly to what we did in (6.1), we can derive the asymptotic distribution of the GP quantile estimator corresponding to (6.17) by using the delta method. It can be shown that such estimator is asymptotically normal, but in general will carry some bias due to the suboptimality of the threshold t. To obtain the standard errors, we need the variance-covariance of $(\hat{q}_0, \hat{\sigma}_n, \hat{\gamma}_n)$ (although, in practice, one can ignore the uncertainty in q_0 (Coles, 2001)). The expression for calculating the asymptotic variance of the GP quantile estimator can be found in Coles (2001, p. 82).

Let us now consider an application using the NCHS Births dataset discussed at the beginning of this section. As a preliminary step, we fixed the threshold t at 5000 g, which corresponds to the quantile at level $p_0 = 0.9985626$ of the empirical distribution. The estimates of the parameters of the GP distribution were obtained by means of the maximum likelihood estimator in (6.16) as implemented in the R package `evd` (Stephenson, 2002). The resulting density (left plot) and quantile (right plot) functions of the GP distribution are illustrated in Figure 6.5, along with the number of exceedances above t and the tail probabilities $1 - \hat{F}_n(t)$, with $t \in [5000, 8000]$, in the secondary axes of the plots. There is an indication that the tail is heavier than normal since $\hat{\gamma} = 0.0763$ (95% confidence interval: 0.0467 to 0.1059). It is rather impressive how close the empirical and GP quantile estimates are, at least up to about $p = 0.9999849$ (Table 6.2). On the other hand, the average length of the 95% confidence intervals was 37.2 g for empirical estimates and 29.1 g for GP estimates.

R Code 6.4 reports a small simulation study comparing the mean squared error (MSE) of the empirical and GP quantile estimators. Samples of size $n = 100{,}000$ are repeatedly (100 times) sampled from a Student's t-distribution, which belongs to the Fréchet domain of attraction. The threshold for the GP model is fixed at the quantile $p_0 = 0.985$ of the parent distribution. A sequence of 100 quantiles is estimated for $p \in [0.99000, 0.99999]$, corresponding to expected upper-tail counts from 1000 ($p = 0.99000$) to 1 ($p = 0.99999$). To ensure meaningful estimates, the code checks whether $\hat{p}_0 < 0.99000$ at each replication. The relative efficiency $\mathrm{RE}(\hat{\xi}_p, \hat{\xi}_p^{\mathrm{GP}}) = \mathrm{MSE}(\hat{\xi}_p)/\,\mathrm{MSE}(\hat{\xi}_p^{\mathrm{GP}})$, plotted in Figure 6.6 as a function of p, is above one throughout, indicating that the estimated MSE of the empirical estimator is larger than that of the GP estimator. The gains are particularly marked at the most extreme values of p.

[10]A natural estimator is the empirical CDF which leads to the estimate $\hat{q}_0 = 1 - r/n$. This is also the maximum likelihood estimate of $\mathrm{E}(S_t)$, the expected value of exceedances above t.

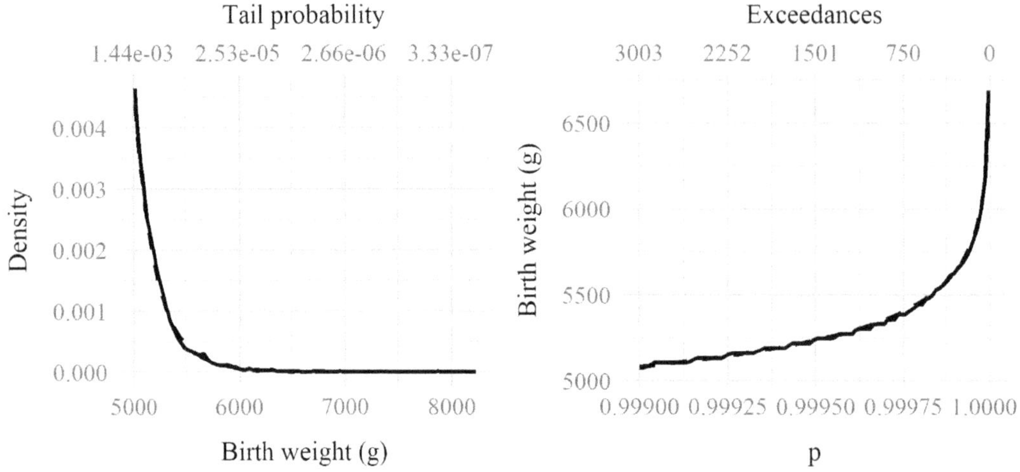

Figure 6.5
Estimated tail density (left) and quantile (right) functions for birth weights in the United States population, 2001. Solid lines denote the empirical estimate, whereas dashed lines denote the fitted generalised Pareto distribution. The corresponding curves are nearly visually indistinguishable.

6.4 Method of quantiles

The method of quantiles or quantile matching (Aitchison and Brown, 1957; Gilchrist, 2000) is akin to the method of moments. The goal is to fit a parametric distribution by matching theoretical with empirical quantiles. Like with moment-based estimation, we need at least as many equations as the dimension of the parameter to estimate. Let $Y_1, \ldots, Y_n$ be an IID sample of size n which we assume drawn from a continuous parametric distribution F_Y with r-dimensional parameter $\boldsymbol{\theta}$. The quantile function $Q_Y(p; \boldsymbol{\theta}) = F_Y^{-1}(p; \boldsymbol{\theta})$ provides the theoretical quantiles for the model of interest and depends on the parameter $\boldsymbol{\theta}$.

Table 6.2
Empirical and generalised Pareto (GP) point and interval (95%) estimates of the birth weight distribution for infants born at term, United States, 2001.

Quantile	Order	Empirical			GP		
p	u_p	Est.	Lower	Upper	Est.	Lower	Upper
0.9990000	3003	5075	5068	5082	5080	5072	5087
0.9992412	2278	5140	5131	5149	5142	5133	5150
0.9994925	1524	5236	5225	5247	5234	5224	5245
0.9997437	769	5387	5370	5404	5398	5384	5413
0.9999849	45	6180	6048	6312	6179	6114	6244
0.9999950	15	6689	6426	6951	6530	6422	6638

R Code 6.4: Simulation study on empirical and generalised Pareto estimation of population quantiles.

```
library(evd)
R <- 100
n <- 1e5
k <- 100
p <- seq(0.99, 0.99999, length = k)
p0 <- 0.985; thresh <- qt(p0, df = 3)
range(n*(1-p))
> [1]    1 1000
xi <- array(NA, dim = c(k,R,2), dimnames =
list(quantile = 1:k,replication = 1:R, estimator =
c("empirical", "GP")))

set.seed(123)
for(i in 1:R){
    y <- rt(n, df = 3)
    Fn <- ecdf(y)
    p0i <- Fn(thresh)
    if(p[1] <= p0i) stop("Decrease threshold")
    fit <- fpot(y, threshold = thresh, model = "gpd")
    sigma <- fit$estimate['scale']
    gamma <- fit$estimate['shape']
    xi[,i,1] <- quantile(y, probs = p)
    xi[,i,2] <- thresh + sigma/gamma*(((1-p0i)/(1-p))^gamma - 1)
}

# calculate MSE
z <- apply(xi, 3, function(x, a) apply(sweep(x, 1, a,
FUN = "-")^2, 1, mean), a = qt(p, df = 3))

head(z)
>          estimator
> quantile    empirical          GP
>        1 0.002535213 0.002332343
>        2 0.002626575 0.002370886
>        3 0.002770171 0.002410578
>        4 0.002840983 0.002451453
>        5 0.002866653 0.002493547
>        6 0.002978955 0.002536898
```

Quantile-matching estimation starts from a system of $k \geq r$ equations

$$
\begin{aligned}
Q_Y(p_1;\boldsymbol{\theta}) &= \hat{\xi}_1 \\
Q_Y(p_2;\boldsymbol{\theta}) &= \hat{\xi}_2 \\
&\cdots \\
Q_Y(p_k;\boldsymbol{\theta}) &= \hat{\xi}_k,
\end{aligned}
\tag{6.18}
$$

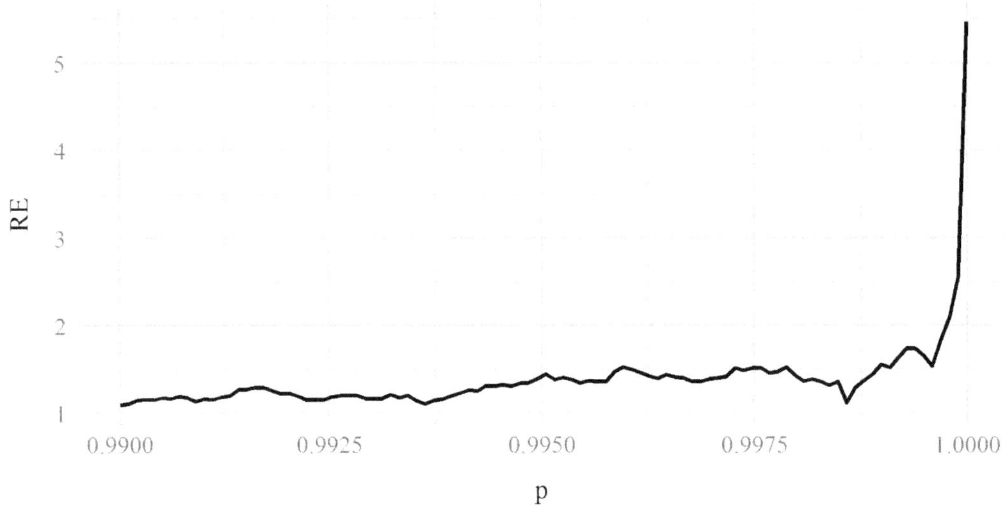

Figure 6.6
Efficiency of the empirical estimator relative to the generalised Pareto estimator using the data generated with R Code 6.4.

where $p_1 < \ldots < p_k$ are probabilities and $\hat{\xi}_j$, $j = 1, \ldots, k$, are (nonparametric) sample quantiles based on Y_i, $i = 1, \ldots, n$. If the system does not have a closed-form solution for $k = r$ or if the system is overdetermined ($k > r$), then we can consider a least-squares-type[11] estimator such as

$$\hat{\boldsymbol{\theta}} = \underset{\boldsymbol{\theta}}{\operatorname{argmin}} \left(\boldsymbol{Q}(\boldsymbol{\theta}) - \hat{\boldsymbol{\xi}}\right)^{\top} \left(\boldsymbol{Q}(\boldsymbol{\theta}) - \hat{\boldsymbol{\xi}}\right), \tag{6.19}$$

where $\boldsymbol{Q}(\boldsymbol{\theta}) = (Q_Y(p_1; \boldsymbol{\theta}), \ldots, Q_Y(p_k; \boldsymbol{\theta}))^{\top}$ and $\hat{\boldsymbol{\xi}}$ is the vector of corresponding sample quantiles. The choice of which quantiles to match is essentially arbitrary. However, we would normally consider quantiles that are most relevant to the nature of the parameter (Tukey, 1965; Park, 1996), that is, central quantiles to estimate a location parameter, tail quantiles to estimate a scale parameter, and so on. In general, we can consider linear combinations of quantiles that compose indices for location, scale and shape (Chapter 3).

One could argue that the estimator in (6.19) is a little naïve as it gives equal weight to all quantiles, whether it be the median or the 90th percentile. Koenker (2005, p. 131) discusses the following weighted estimator:

$$\hat{\boldsymbol{\theta}}_w = \underset{\boldsymbol{\theta}}{\operatorname{argmin}} \left(\boldsymbol{Q}(\boldsymbol{\theta}) - \hat{\boldsymbol{\xi}}\right)^{\top} \boldsymbol{V}^{-1} \left(\boldsymbol{Q}(\boldsymbol{\theta}) - \hat{\boldsymbol{\xi}}\right) \tag{6.20}$$

for some appropriate matrix $\boldsymbol{V}$. Unfortunately, possible gains in efficiency that accompany $\hat{\boldsymbol{\theta}}_w$ may be offset by computational problems (Koenker, 2005).

Let's consider a practical example. Our goal is to fit a lognormal distribution to tetrachlorobenzene (TcCB) concentrations (ppb) in soil samples at a reference area and a cleanup area. We select observations for the reference area only ($n = 47$). The dataset was published by the US Environmental Protection Agency and is available from the `EnvStats` package

[11]The Euclidean norm in (6.19) can be replaced by other norms. See, for example, Chapter 9 in Gilchrist (2000).

R Code 6.5: Fitting a 3-parameter lognormal distribution for the TcCB dataset by quantile matching and maximum likelihood estimation.

```
library(fitdistrplus)
library(EnvStats)

# select data
data("EPA.94b.tccb.df", package = "EnvStats")
dd <- subset(EPA.94b.tccb.df, EPA.94b.tccb.df$Area == "Reference")
y <- dd$TcCB
length(y)
> [1] 47
# fit 3-parameter lognormal with quantile matching (QM)
pj <- c(0.05,.5,.95)
ss <- list(meanlog = mean(log(y)), sdlog = sd(log(y)), threshold = 0)
theta_qm <- qmedist(data = y, distr = "lnorm3",
probs = pj, start = ss)$estimate

# fit 3-parameter lognormal with maximum likelihood (ML)
theta_ml <- mledist(data = y, distr = "lnorm3",
start = ss)$estimate

# estimate quantiles over grid of p
p <- 1:49/50
Q_qm <- do.call(qlnorm3, c(list(p = p), as.list(theta_qm)))
Q_ml <- do.call(qlnorm3, c(list(p = p), as.list(theta_ml)))

# compare estimated quantiles with sample quantiles
xihat <- quantile(y, p)
sum(abs(xihat - Q_qm))
> [1] 1.077343
sum(abs(xihat - Q_ml))
> [1] 1.112663
```

(Millard, 2013). To model the TcCB concentrations, we assume the 3-parameter lognormal distribution which is defined to be the distribution of a random variable $Y = \alpha + \exp(Z)$ where $Z \sim \mathcal{N}(\mu, \sigma^2)$. Fitting this distribution via maximum likelihood is known to be potentially troublesome (Hill, 1963; Johnson and Kotz, 1970) since the likelihood can be driven to infinity by letting $\alpha \to Y_{(1)}$. This suggests that quantiles can be used as stable parameters (Griffiths, 1980).

The parameter of interest is $\boldsymbol{\theta} = (\mu, \sigma, \alpha)^\top$. Quantile-matching (QM) and maximum likelihood (ML) estimation are carried out by using the R functions `qmedist`[12] and `mledist`, respectively, both from the `fitdistrplus` package (Delignette-Muller and Dutang, 2015), as shown in R Code 6.5. For QM estimation, we use $p_1 = 0.05$, $p_2 = 0.5$, and $p_3 = 0.95$. The same starting values are fed into `qmedist` and `mledist` to initiate the algorithms.

[12]The function `qmedist` for quantile matching requires exactly as many quantiles as the number of parameters (i.e., $k = r$). When $k > r$, the R Code to actually implement the estimator $\hat{\boldsymbol{\theta}}$ in (6.19) or $\hat{\boldsymbol{\theta}}_w$ in (6.20) is rather simple.

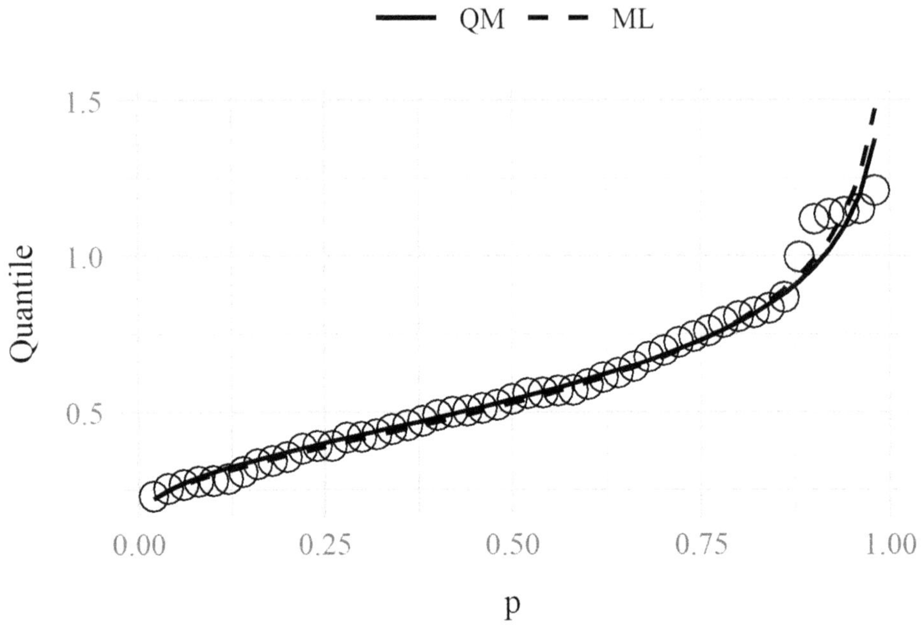

Figure 6.7
Estimated quantiles of tetrachlorobenzene concentrations from a 3-parameter lognormal distribution fitted with quantile matching (QM) and maximum likelihood (ML) using R Code 6.5. Hollow circles represent (nonparametric) sample quantiles.

The estimated quantiles are plotted against p, for $p = i/50$, $i = 1, \ldots, 49$, in Figure 6.7. The method of quantiles performs remarkably well, considering that only 3 values have been used for fitting (also, see Section 7.1.6 and R Code 7.10 for an assessment based on the Q–Q plot). In fact, the total absolute deviation from the sample quantiles is slightly smaller for QM than for ML (1.08 vs 1.11). It is worth stressing that the method of quantiles may turn out to be very handy in meta-analytic research when data are summarised by sample quantiles and, as is often the case, raw data are not available.

7

Standard errors, confidence intervals, and tests of hypotheses

7.1 Standard errors and confidence intervals

In this section, we concern ourselves with computing standard errors of quantile estimators and confidence intervals for population quantiles from observations of a sample $Y_1, \ldots, Y_n \overset{\text{IID}}{\sim} F$, where F is (for the most part) absolutely continuous. The asymptotic results shown in (6.1) and (6.2) tell us that, whether we consider a parametric or nonparametric quantile estimator, we need an estimate of the density f in order to make inference on population quantiles ξ_p, $p \in (0,1)$. However, there are approaches, such as exact inference and the bootstrap, that forgo the use of the density. In Section 7.1.5, we discuss the particular case of a discrete F.

7.1.1 Likelihood-based

In the parametric case, inference can be rather straightforward since the functional form of F is known. This is demonstrated in two examples as shown in R Code 7.1 using the exponential distribution and R Code 7.2 using the gamma distribution. We should stress that the way we optimised the log-likelihoods in R Code 7.1 and R Code 7.2 is not foolproof since the parameters' spaces are $\mathbb{R}_{>0}$ and $\mathbb{R}^2_{>0}$, respectively. In our examples, this does not have any adverse consequence since the values of the parameters are sufficiently far from the boundary. However, in general one should either constrain the optimisation or reparameterise the likelihood. In the latter case, if, say, g is a transformation such that $g(\theta) = \delta \in \mathbb{R}$, then the delta method in (6.1) must be applied to the mapping $\delta \mapsto (g^{-1} \circ F^{-1})(p;\delta)$, not $\theta \mapsto F^{-1}(p;\theta)$.

Suppose $Y \sim \mathcal{E}(\theta)$, that is, exponential with rate θ. Our parametric quantile estimator has a closed form given by $\hat{F}_n^{-1}(p;\hat{\theta}_n) = -\log(1-p)/\hat{\theta}_n$. To calculate the asymptotic standard error based on (6.1), we need

$$\frac{\partial F^{-1}(p;\theta)}{\partial \theta} = \frac{\log(1-p)}{\theta^2}.$$

Knowing that the Fisher information for $\mathcal{E}(\theta)$ is $I(\theta) = 1/\theta^2$, we find that the large-n standard error is

$$\frac{|\log(1-p)|}{\theta\sqrt{n}}. \tag{7.1}$$

A simulated example with $\theta = 2$, $n = 100$, and $p = 0.9$ is shown in R Code 7.1. Although both the MLE of θ, $\hat{\theta}_n = \frac{n}{\sum_i^n Y_i}$, and the asymptotic standard error (7.1) have simple analytical expressions, we employ numerical calculations so as to make the code more flexible for the reader who wants to deal with distributional functions that do not have a tractable form (e.g., this is the case of the gamma distribution considered in R Code 7.2). In

DOI: 10.1201/9781003326410-7

R Code 7.1: Maximum likelihood inference for the 90th percentile of the exponential distribution with rate parameter 2.

```
library(numDeriv)
library(bbmle)
# define exponential log-likelihood and quantiles as functions of theta
loglik_exp <- function(theta){
    -sum(dexp(y, rate = theta, log = TRUE))
}
fn <- function(theta, p){
    qexp(p = p, rate = theta)
}
# generate data
set.seed(123)
n <- 100
p <- 0.9
theta <- 2
y <- rexp(n, rate = theta)
# MLE of theta
fit <- mle2(loglik_exp, start = list(theta = 1/mean(y)))
# estimate quantile, SE and confidence interval
xi <- qexp(p = p, rate = coef(fit))
vtheta <- grad(fn, x = coef(fit), p = p)
xi_se <- sqrt(vtheta^2 * vcov(fit))
CI <- xi + c(-1,1)*qnorm(1-0.05/2, 0, xi_se)
c(xi = xi, se = xi_se, low = CI[1], up = CI[2])
>         xi        se       low        up
> 1.2039282 0.1203928 0.9679626 1.4398937
# compare numerical and analytical calculation of standard error
abs(log(1-p))/(coef(fit)*sqrt(n))
>     theta
> 0.1203928
##############################################################
# Reparameterise to ensure parameter estimates are positive
##############################################################
loglik_exp <- function(lograte){
    -sum(dexp(y, rate = exp(lograte), log = TRUE))
}
fn <- function(lograte, p){
    qexp(p = p, rate = exp(lograte))
}
fit <- mle2(loglik_exp, start = list(lograte = log(1/mean(y))))
xi <- qexp(p = p, rate = exp(coef(fit)))
vlograte <- grad(fn, x = coef(fit), p = p)
xi_se <- sqrt(vlograte^2 * vcov(fit))
CI <- xi + c(-1,1)*qnorm(1-0.05/2, 0, xi_se)
c(xi = xi, se = xi_se, low = CI[1], up = CI[2])
>        xi        se       low        up
> 1.2039282 0.1203928 0.9679626 1.4398937
```

R Code 7.2: Maximum likelihood inference for the 90th percentile of the gamma distribution with shape parameter 2 and rate parameter 2.

```
# define gamma log-likelihood and quantiles as functions of theta
loglik_gamma <- function(shape, rate){
    -sum(dgamma(y, shape = shape, rate = rate, log = TRUE))
}
fn <- function(theta, p){
    qgamma(p = p, shape = theta[1], rate = theta[2])
}
# generate data
set.seed(123)
n <- 100
p <- 0.9
y <- rgamma(n, shape = 2, rate = 2)
ss <- list(shape = mean(y)^2/var(y), rate = mean(y)/var(y))
# MLE of theta
fit <- mle2(loglik_gamma, start = ss)
# estimate quantile, SE and confidence interval
xi <- qgamma(p = p, shape = coef(fit)[1], rate = coef(fit)[2])
vtheta <- matrix(grad(fn, x = coef(fit), p = p))
xi_se <- sqrt(as.numeric(crossprod(vtheta,
vcov(fit)) %*% vtheta))
CI <- xi + c(-1,1)*qnorm(1-0.05/2, 0, xi_se)
c(xi = xi, se = xi_se, low = CI[1], up = CI[2])
>        xi          se        low         up
> 1.6383906 0.1214526 1.4003480 1.8764333
############################################################
# Reparameterise to ensure parameter estimates are positive
############################################################
loglik_gamma <- function(logshape, lograte){
    -sum(dgamma(y, shape = exp(logshape),
    rate = exp(lograte), log = TRUE))
}
fn <- function(theta, p){
    qgamma(p = p, shape = exp(theta[1]),
    rate = exp(theta[2]))
}
ss <- list(logshape = log(mean(y)^2/var(y)), lograte = log(mean(y)/var(y)))
fit <- mle2(loglik_gamma, start = ss)
xi <- qgamma(p = p, shape = exp(coef(fit)[1]),
rate = exp(coef(fit)[2]))
vlogtheta <- matrix(grad(fn, x = coef(fit), p = p))
xi_se <- sqrt(as.numeric(crossprod(vlogtheta,
vcov(fit)) %*% vlogtheta))
CI <- xi + c(-1,1)*qnorm(1-0.05/2, 0, xi_se)
c(xi = xi, se = xi_se, low = CI[1], up = CI[2])
>        xi          se        low         up
> 1.6383919 0.1214524 1.4003495 1.8764343
```

particular, we use the function `mle2` for estimating θ and the function `grad` in the package `numDeriv` (Gilbert and Varadhan, 2019) to approximate numerically the gradient of F^{-1}. The R Code 7.1 shows also that, compared to the analytical computation based on (7.1), the numerical approximation leads to an accurate result.

Now suppose $Y \sim \mathcal{G}(a,b)$, that is, gamma with shape a and rate b. The associated parametric quantile estimator does not have, in general, a closed form. However, we can reliably use numerical calculations as in the previous example. Note that, in this case, the parameter $\boldsymbol{\theta} = (a,b)^\top$ is a vector, therefore the asymptotic variance-covariance matrix of $\hat{F}_n^{-1}(p;\hat{\boldsymbol{\theta}}_n)$ takes the form $v(\boldsymbol{\theta})^\top I^{-1}(\boldsymbol{\theta})v(\boldsymbol{\theta})$, where $v(\boldsymbol{\theta}) = \nabla_{\boldsymbol{\theta}} F^{-1}(p;\boldsymbol{\theta})$. A simulated example with $\boldsymbol{\theta} = (2,2)^\top$, $n = 100$, and $p = 0.9$ is shown in R Code 7.2.

7.1.2 Asymptotic

As we move into the territory of sample quantile estimators, we need to arm ourselves with appropriate tools to build confidence intervals based on (4.9). The latter states that, for a sufficiently large sample size, the distribution of the sample quantile estimator $\hat{\xi}_{p,n}$ is approximately $\mathcal{N}\left(\xi_p, \dfrac{p(1-p)}{n\{f(\xi_p)\}^2}\right)$. Remember, the goal is to estimate $Q = F^{-1}$, where F has density f (Figure 2.7 and Table 2.1). And since no particular shape of the distribution F is assumed, the infinite-dimensional nuisance parameter $f(\xi_p)$ must be estimated to obtain the standard error and build a confidence interval.[1]

There is a rich literature on nonparametric density estimation and smoothing. One may start from the simple Rosenblatt (density) estimator

$$\hat{f}_n(y) = \frac{F_n(y+h) - F_n(y-h)}{2h}, \tag{7.2}$$

that is, the difference quotient of the ECDF. This can be re-written in the form (Tsybakov, 2009)

$$\hat{f}_n(y) = \frac{1}{2nh}\sum_{i=1}^{n} I(y-h < Y_i \leq y+h) = \frac{1}{nh}\sum_{i=1}^{n} K\left(\frac{Y_i - y}{h}\right) \tag{7.3}$$

where $K(u) = \frac{1}{2}I(-1 < u \leq 1)$ is called the *kernel* and $h > 0$ a *bandwidth.*

The Parzen–Rosenblatt (density) estimator generalises (7.2) by letting the kernel K be any function such that $K : \mathbb{R} \to \mathbb{R}$ and $\int K(u)\,du = 1$. Theoretical arguments can be found to justify the choice of one kernel shape over another (Cline, 1988), although in general K is almost always taken to be a symmetric, unimodal density (Wand and Jones, 1995). A much celebrated one is the Epanechnikov kernel $K(u) = 3/4(1-u^2)$, $|u| < 1$ for its optimality in terms of the (asymptotic) mean integrated squared error among nonnegative kernels (Epanechnikov, 1969). However, the choice of K can be based on partial knowledge of F, which may lead to the optimality of other kernels, for example, Gaussian $K(u) = \phi(u)$. In practice, the choice of the kernel has little effect on the performance of the

[1] Rather than estimating $1/f(\xi_p)$ indirectly, that is, by estimating the density f at the quantile of interest and then taking its reciprocal, one may instead obtain a direct kernel estimate of the sparsity $q(p)$. This is essentially the strategy used in Section 8.3.2 for the conditional case. An argument in favour of direct estimation is that its leading bias has a simpler form, involving q'' only, whereas the reciprocal-density approach yields a more complicated bias expression; in addition, direct estimation is often preferable in terms of bias and variability (Jones, 1992). The asymptotically optimal bandwidth depends on the quantile optimality ratio q/q''. Although this quantity is not distribution-free, in practice a rough approximation may suffice, and bandwidths derived from representative families can perform well for distributions of similar shape (see also the finite-sample comparisons in Prendergast and Staudte, 2016).

R Code 7.3: Kernel density estimation for a sample drawn from a standard normal.

```
set.seed(123)
n <- 100
y <- rnorm(n)

# optimal smoothing with SJ
fhat1 <- density(y, bw = "SJ", kernel = "gaussian")
x1 <- fhat1$x
y1 <- fhat1$y

# undersmoothing (one fifth of the SJ bandwidth)
fhat2 <- density(y, bw = fhat1$bw/5, kernel = "gaussian")
x2 <- fhat2$x
y2 <- fhat2$y

# true density
x3 <- seq(-3, 3, by = 0.01)
y3 <- dnorm(x3)

# plot results
g <- ggplot() + geom_line(aes(x = x1, y = y1), linetype = "dashed") +
geom_line(aes(x = x2, y = y2), linetype = "dotted") +
geom_line(aes(x = x3, y = y3), linetype = "solid") + xlab("y") +
ylab("Density")
```

smoother (see, e.g., Table 2.1 in Wand and Jones, 1995, for the relative efficiency of popular kernels).

A much more critical role in determining the optimality of the smoother is played by the bandwidth h as it directly controls the variance-bias trade-off of the smoother. Silverman's rule of thumb is a robust bandwidth selector for unimodal and fairly symmetric distributions (Silverman, 1986). Other bandwidth selectors that may be more versatile when the underlying distribution is not unimodal and symmetric include Sheather & Jones rule-of-thumb (SJ) (Sheather and Jones, 1991) and cross-validation-based methods.

An example is provided in R Code 7.3. The sample y is drawn from the standard normal, that is, $Y \sim \mathcal{N}(0, 1)$. The `R` function `density` allows one to choose the bandwidth, either directly or via a bandwidth selector (default is `nrd0` for Silverman's rule-of-thumb), and the kernel (default is `gaussian`). The results of R Code 7.3 are plotted in Figure 7.1. The solid line depicts the true (unknown) density while the dashed line marks the density estimated by a Gaussian kernel with SJ bandwidth. The latter does a reasonable job at mimicking the underlying true density. By way of comparison, the smoothed density with same kernel but bandwidth equal to one fifth of the optimal SJ bandwidth is also given (dotted line). The excessively wiggly behaviour is indicative of *undersmoothing*, which is of course the intended outcome of the example.

We provide another example in R Code 7.4 where the sample y is drawn from a beta distribution with shape parameters equal to 0.5, that is, $Y \sim \mathcal{B}(0.5, 0.5)$ (also known as arcsine distribution). In this example, there is the added complication of dealing with a distribution on bounded support. To improve the behaviour of the smoother, we first apply

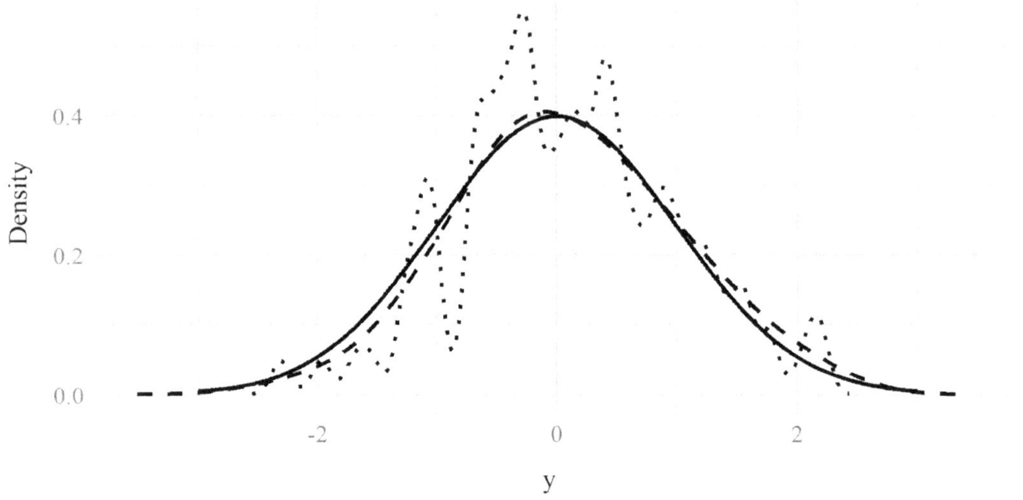

Figure 7.1
Kernel density estimates for a sample from a standard normal distribution generated in R Code 7.3. The solid line is the true density, the dashed line uses the Sheather–Jones bandwidth, and the dotted line uses one fifth of that bandwidth.

the *logit* transformation $z \equiv \text{logit}(y) = \log\{y/(1-y)\}$ so that z is defined over $\mathbb{R}$. We then smooth the distribution of z and transform back to the scale of y with the usual change-of-variable relation $f_Z(z) = f_Y(g^{-1}(z))\left|\frac{\mathrm{d}}{\mathrm{d}z}g^{-1}(z)\right|$, which gives

$$f_Y(\text{logit}^{-1}(z)) = f_Z(z)\left|\frac{\mathrm{d}}{\mathrm{d}z}\{1+\exp(-z)\}^{-1}\right|^{-1} = \frac{f_Z(\text{logit}(y))}{y(1-y)}.$$

For good measure, we truncate smoothing between $\text{logit}(0.01) = -4.60$ and $\text{logit}(0.99) = 4.60$ since the behaviour of the logit is unstable near the extremes.[2] This is achieved by using the arguments `from` and `to` to define the extremes of the grid over which the density is to be estimated. The results of R Code 7.4 are plotted in Figure 7.2, which illustrates the good behaviour of the optimal smoothing as determined by the SJ selector (dashed line) and the impact of *oversmoothing* (dotted line).

Density estimation is only an intermediate step. The goal is to obtain an estimate of the standard error of the sample quantile. Say we want to build a 95% confidence interval for the median of $Y \sim \mathcal{N}(0,1)$. This is shown in R Code 7.5.

We conclude this section with a simulation study to compare the empirical coverage of 95% confidence intervals for the quantiles of three distributions: $\mathcal{N}(0,1)$, χ^2_3 and $\mathcal{B}(0.5, 0.5)$; at two quantile levels $p \in \{0.1, 0.5\}$ and three sample sizes $n \in \{100, 1000, 5000\}$; using three bandwidth selectors: Silverman's (`nrd0`), SJ (`SJ`), unbiased cross-validation (`ucv`); and seven different kernels (including Gaussian and Epanechnikov). For every combination

[2]This approach is very much different from performing estimation directly on y truncated at, say, 0.01 and 0.99, that is, `density(y, from = 0.01, to = 0.99)`, which is not appropriate for bypassing boundary issues (Silverman, 1986; Jones, 1993).

R Code 7.4: Kernel density estimation for a sample drawn from a beta distribution.

```
library(Qtools)
set.seed(123)
n <- 100
y <- rbeta(n, shape1 = 0.5, shape2 = 0.5)
z <- logit(y)
r <- c(0.01, 1-0.01)

# optimal smoothing with SJ
fhat1 <- density(z, bw = "SJ", kernel = "gaussian",
from = logit(r[1]), to = logit(r[2]))
x1 <- invlogit(fhat1$x)
y1 <- fhat1$y/(x1*(1-x1))

# oversmoothing (20 times the SJ bandwidth)
fhat2 <- density(z, bw = fhat1$bw*20, kernel = "gaussian",
from = logit(r[1]), to = logit(r[2]))
x2 <- invlogit(fhat2$x)
y2 <- fhat2$y/(x2*(1-x2))

# true density
x3 <- seq(r[1], r[2], by = 0.01)
y3 <- dbeta(x3, shape1 = 0.5, shape2 = 0.5)

# plot results
g <- ggplot() + geom_line(aes(x = x1, y = y1), linetype = "dashed") +
geom_line(aes(x = x2, y = y2), linetype = "dotted") +
geom_line(aes(x = x3, y = y3), linetype = "solid") + xlab("y") +
ylab("Density")
```

of these five settings, each replicated 10,000 times, the empirical coverage is calculated as the proportion of confidence intervals that contain the true unknown quantile. The results corroborate the general notion that the choice of the kernel function has a limited impact on the performance of the smoother and, consequently in this example, on the coverage of the confidence interval for the quantile (Figure 7.3). The bandwidth selector, in contrast, may have a nonnegligible effect on the coverage as particularly evidenced by the estimates for the 10th percentile under a chi-squared distribution. Particularly striking is the impact of the sample size at the lower quantile, even when the parent distribution is normal, let alone beta. In the latter case, the empirical coverage for $n = 100$ did not go beyond 89.3%.

7.1.3 Exact

We have seen that estimation of sample quantiles (Section 4.3) is intimately related to order statistics. Order statistics can be used also for building (exact) confidence intervals of the type $\Pr\left\{Y_{(r)} \leq \xi_p \leq Y_{(s)}\right\} = \pi$ for some coverage probability $\pi \in (0, 1)$. These can be particularly valuable when asymptotic approximations are not satisfactory as exemplified in Figure 7.3 when using a limited sample.

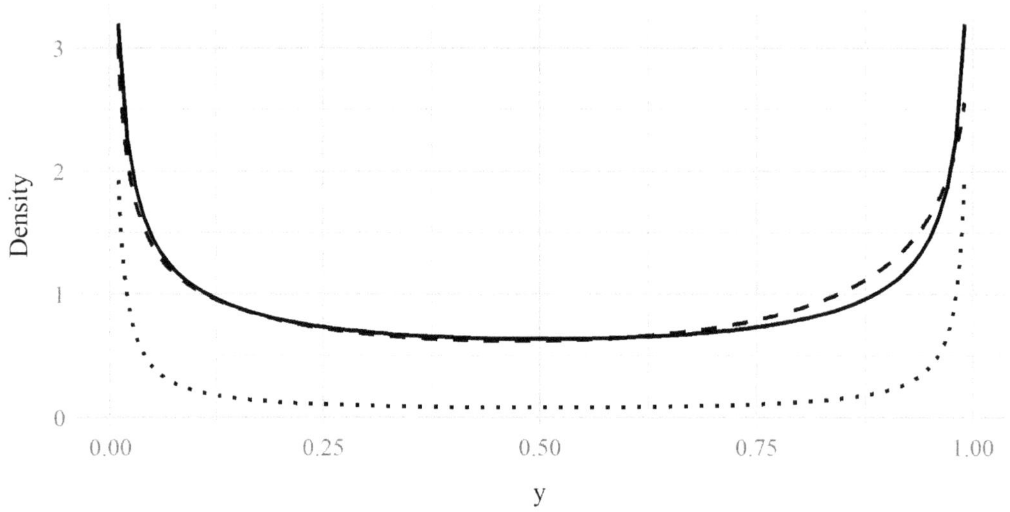

Figure 7.2
Kernel density estimates for a sample from a beta distribution generated in R Code 7.4. The solid line is the true density, the dashed line uses the Sheather–Jones bandwidth, and the dotted line uses a bandwidth twenty times larger.

R Code 7.5: Nonparametric confidence interval for sample quantiles.

```
set.seed(123)
n <- 100
y <- rnorm(n)
p <- 0.5
alpha <- 0.05
# density estimation
fhat <- density(y, bw = "SJ", kernel = "gaussian")
# sample quantile
xi <- quantile(y, probs = p)
# density at quantile
f0 <- approx(fhat$x, fhat$y, xout = xi)$y
#standard error
se <- sqrt(p*(1-p)/n)/f0
# confidence interval
ci <- xi + outer(qnorm(1-alpha/2, 0, se), c(-1,1), FUN = "*")
# print results
c(xi = xi, low = ci[1], up = ci[2])
>      xi.50%         low          up
>  0.06175631 -0.18322060  0.30673322
```

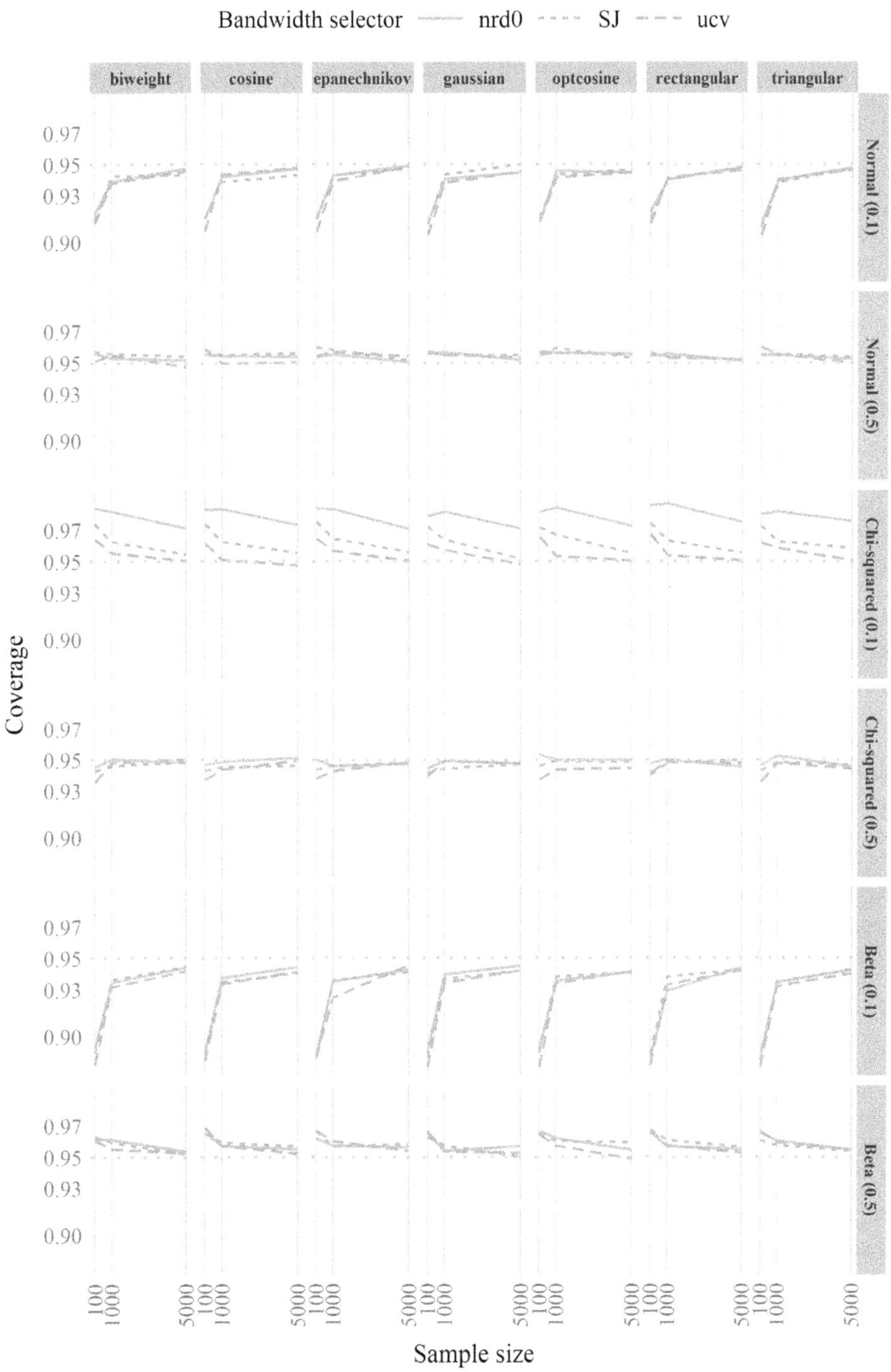

Figure 7.3
Empirical coverage of asymptotic 95% confidence intervals estimated for two quantiles ($p \in \{0.1, 0.5\}$) using different combinations of parent distribution (normal, chi-squared, beta), sample size ($n \in \{100, 1000, 5000\}$), kernel function and bandwidth selector. The nominal level of the intervals is marked by dotted lines.

R Code 7.6: Exact confidence interval at the 95% target level for the 10th percentile of the beta distribution.

```
set.seed(123)
n <- 100
y <- rbeta(n = n, shape1 = 0.5, shape2 = 0.5)
Qtools::qexact(x = y, probs = 0.1, level = 0.95)
>        quantile       lower      upper conf.level
> 10% 0.03365515 0.004452782 0.09207947   0.952273
```

We start with a classical result, due to Thompson (1936), which states that

$$\begin{aligned}\Pr\left\{Y_{(r)} \leq \xi_p \leq Y_{(s)}\right\} &= \Pr\left\{Y_{(r)} \leq \xi_p\right\} - \Pr\left\{Y_{(s)} < \xi_p\right\} \\ &= B(r, n-r+1, p) - B(s, n-s+1, p) \\ &= \sum_{i=r}^{s-1} \binom{n}{i} p^i (1-p)^{n-i} \\ &\equiv \pi(r, s, n, p), \end{aligned} \tag{7.4}$$

where the beta functions after the second equality follow from (4.2). Successively, we determine all the combinations of r and s, $r < s$, for given n and p and calculate $\pi(r, s, n, p)$. Finally, we select r and s to minimise $\pi(r, s, n, p) - (1-\alpha)$ subject to $\pi(r, s, n, p) \geq (1-\alpha)$, for a pre-specified confidence level $1-\alpha$. If the procedure yields more than one such confidence intervals, then one could reasonably choose the interval with smallest width. Since the actual confidence level of the selected interval may or may not be exactly $1-\alpha$, we will refer to the latter as the target level.

The advantage of the exact approach is that the quantity $\pi(r, s, n, p)$ does not depend on F, therefore the resulting confidence interval is distribution-free.[3] On the other hand, computation becomes burdensome with large n since the number of combinations is of the order $\mathcal{O}(n^2)$. Variants of the exact approach include normal approximation of the binomial probabilities and interpolation of adjacent order statistics (Nyblom, 1992). Details can be found in David and Nagaraja (2003, Chapter 7).

In R Code 7.6, we use the function `qexact` from the `Qtools` package to obtain the exact confidence interval for a sample of size $n = 100$ drawn from the beta distribution. The output gives the estimate of the 10th percentile obtained with the $\lceil np \rceil = 10$th order statistic (corresponding to `type = 1` in the function `quantile`), that is, $\hat{\xi}_{0.1} = 0.034$. By pre-specifying the confidence level at 95%, we find that the interval $(0.004, 0.092)$ has the closest level at 95.2%. The function `qexact` is not recommended for a large sample size (for which an exact inferential approach is not intended anyway) because it calculates the probability $\pi(r, s, n, p)$ for $n(n-1)/2$ combinations of r and s (via the function `combinations` in the `gtools` package, Warnes et al. 2023). For example, this amounts to listing $n(n-1)/2 = 4950$ combinations when $n = 100$.

In the simulation study presented in the previous section, we remarked the poorer performance of the asymptotic confidence interval at $p = 0.1$ and $n = 100$ for beta-distributed data. Just for comparison, R Code 7.7 shows an analogous simulation to calculate the empirical coverage of the exact confidence interval (at an actual nominal level equal to 95.2%).

[3]Independence from F is no longer the case if Y is discrete. Still, in the discrete case $\pi(r, s, n, p)$ can be taken as a lower bound for $\Pr\left\{Y_{(r)} \leq \xi_p \leq Y_{(s)}\right\}$ or as an upper bound for $\Pr\left\{Y_{(r)} < \xi_p < Y_{(s)}\right\}$ (David and Nagaraja, 2003, p. 160).

R Code 7.7: Empirical coverage of the exact confidence interval at the 95% target level for the 10th percentile of the beta distribution.

```
set.seed(123)
p <- 0.1
Y <- replicate(10000, rbeta(n = 100, shape1 = 0.5, shape2 = 0.5))
Q <- qbeta(p, shape1 = 0.5, shape2 = 0.5)
out <- t(apply(Y, 2, function(y, probs, level)
Qtools::qexact(y, probs, level), probs = p, level = 0.95))
# CIs that include true quantile Q have discordant signs of
# bounds centered around Q
I <- ((out[,2] - Q)*(out[,3] - Q)) < 0
# empirical coverage
mean(I)
> [1] 0.9526
```

Box 7.1: Consistency of the bootstrap and confidence intervals for quantiles

Let F be the study population and $T(F)$ the unknown parameter of interest. Consider a sample $Y_1, \ldots, Y_n \overset{\text{IID}}{\sim} F$ and let $\hat{F}_n$ be the corresponding ECDF. Draw a (bootstrap) sample $Z_1, \ldots, Z_n$ of size n independently from $\hat{F}_n$ and let $\tilde{F}_n$ be the ECDF corresponding to the bootstrap sample. We define the cumulative distributions $\tilde{G}_n(y) = \Pr_n\left\{\sqrt{n}\left(T\left(\tilde{F}_n\right) - T\left(\hat{F}_n\right)\right) \leq y\right\}$ and $\hat{G}_n(y) = \Pr\left\{\sqrt{n}\left(T\left(\hat{F}_n\right) - T(F)\right) \leq y\right\}$, where the first probability is indexed by n as it depends on the outcome $Y_1, \ldots, Y_n$.

Theorem 4.

$$\sup_{y \in \mathbb{R}} \left|\tilde{G}_n(y) - \hat{G}_n(y)\right| \to 0 \tag{7.5}$$

in probability, as $n \to \infty$.

See Bickel and Freedman (1981) for a proof. Now, if $\hat{G}_n$ converges to a continuous and strictly increasing distribution G, it follows from (7.5) that (Falk and Kaufmann, 1991)

$$\tilde{G}_n^{-1}(\alpha) \xrightarrow{p} G^{-1}(\alpha),$$

for $\alpha \in (0, 1)$. If we let $T(F)$ be the quantile ξ_p of a given order $p \in (0, 1)$, then we have the (ordinary) bootstrap interval

$$I_{1,n} = \left\{\hat{\xi}_p - \tilde{G}_n^{-1}(1 - \alpha/2)n^{-1/2}, \hat{\xi}_p - \tilde{G}_n^{-1}(\alpha/2)n^{-1/2}\right\} \tag{7.6}$$

and

$$\Pr\left\{\xi_p \in I_{1,n}\right\} \to 1 - \alpha.$$

Interestingly, Falk and Kaufmann (1991) showed that the *backward* confidence interval

$$I_{2,n} = \left\{\hat{\xi}_p + \tilde{G}_n^{-1}(\alpha/2)n^{-1/2}, \hat{\xi}_p + \tilde{G}_n^{-1}(1 - \alpha/2)n^{-1/2}\right\}, \tag{7.7}$$

which coincides with the interval obtained with the percentile method (Efron, 1982), has a higher coverage probability than $I_{1,n}$.

The better empirical coverage equal to 95.3% is a sufficient motivation for adopting the exact interval in this specific scenario. Additionally, order statistics are immune to boundary issues and, thus, no transformation is needed.

7.1.4 Bootstrap

The bootstrap (Efron, 1979) is an invaluable statistical technique used for estimating the sampling distribution of a statistic of interest and, hence, as a by-product, standard errors and confidence intervals. Not only does the bootstrap come in many variants, but it also belongs to the larger family of resampling methods, any of which may have variants of its own.[4]

The principle that drives bootstrap estimation is rather simple (yet ingenious). Say we calculated the quantile $\hat{\xi}_p$ based on $y = (y_1, \ldots, y_n)$, an IID sample of size n from a continuous distribution F_Y (also suppose that $y_i \neq y_j$, for $i \neq j$, without loss of generality). We now need to quantify the uncertainty of our estimate which depends on the sampling distribution, say G_n, of the quantile estimator $\hat{\xi}_{n,p}$. Information about this uncertainty can be taken from the sample itself. More precisely, the empirical distribution

$$\hat{F}_Y(y) = \frac{1}{n}\sum_{i=1}^{n} I\left(y_i \leq y\right) \tag{7.8}$$

can be seen as a small-scale 'copy' of F_Y. Since $\hat{F}_Y$ is discrete, one could list all the possible $B = n^n$ samples (with replacement) $y^{*(b)}$, $b = 1, \ldots, B$, calculate the quantile $\hat{\xi}_p^{*(b)}$ for each sample, and determine

$$G_n^*(x) = \frac{1}{B}\sum_{b=1}^{B} I\left\{\sqrt{n}(\hat{\xi}_p^{*(b)} - \hat{\xi}_p) \leq x\right\}. \tag{7.9}$$

The idea is that G_n^* should mimic G_n. Indeed, provided some suitable regularity conditions hold (Bickel and Freedman, 1981), the bootstrap consistently estimates the sampling distribution of many statistics.[5] Then $[\hat{\xi}_p - G_n^{*-1}(1-\alpha/2)n^{-1/2}, \hat{\xi}_p - G_n^{*-1}(\alpha/2)n^{-1/2}]$ is a confidence interval for ξ_p of (asymptotic) level $1-\alpha$. The theoretical equivalent of this interval is denoted by $I_{1,n}$ in Equation (7.6) (Box 7.1). Unfortunately, the coverage error of $I_{1,n}$ is of order $O(n^{-1/2})$, hence lower than the $O(n^{-1})$ found for intervals in more classical problems. As explained by its proponents (Falk and Kaufmann, 1991), the *backward* confidence interval $I_{2,n}$ in Equation (7.7) theoretically outperforms $I_{1,n}$, at least when $T(F)$ is not a smooth functional of the mean of F (Hall, 1988; Falk and Kaufmann, 1991).[6] Another bootstrap interval, often used in practical applications, assumes normality for the sampling distribution and estimates the associated standard error as the standard deviation of the bootstrap samples,[7] that is

$$I_{3,n} = \left\{\hat{\xi}_p - z_{1-\alpha/2}\sqrt{\operatorname{var}(\hat{\xi}_p^*)}, \hat{\xi}_p + z_{1-\alpha/2}\sqrt{\operatorname{var}(\hat{\xi}_p^*)}\right\}, \tag{7.10}$$

where $z_{1-\alpha/2} = \Phi^{-1}(1-\alpha/2)$ is the familiar standard normal critical value.

[4]For example, another well-known resampling method is the jackknife. When applied to estimating standard errors of sample quantiles, the delete-1 jackknife is known to fail since it yields a variance that does not converge to the correct asymptotic variance (Efron, 1982; Koenker, 2005). In contrast, the delete-d jackknife ($1 < d < n$) enjoys a better fate (Shi, 1991).

[5]Intuitively, as $n \to \infty$, sampling from $\hat{F}_Y$ becomes equivalent to sampling from F_Y.

[6]A subtle point raised by De Angelis et al. (1993) concerns the *inherently* bad behaviour of confidence intervals based on order statistics. Those authors show that, the coverage error of either $I_{1,n}$ or $I_{2,n}$ can be reduced, although not beyond $O(n^{-1})$, by smoothing the bootstrap.

[7]In the R package `boot` (Davison and Hinkley, 1997; Canty and Ripley, 2025), this interval, which is obtained with `boot.ci(..., type = "norm")`, is adjusted for the bootstrap estimate of the bias $\mathrm{E}(\hat{\xi}_p^*) - \hat{\xi}_p$ (Davison and Hinkley, 1997, p. 199).

R Code 7.8: Bootstrap confidence interval for the median.

```
# sample from a population
y <- c(1.76, 5.98, 11.90, 12.66, 15.43)
n <- length(y)
print(xihat <- median(y))
> [1] 11.9
# take B = 900 bootstrap samples with replacement from y
set.seed(321)
Y_star <- replicate(900, sample(y, n, replace = TRUE))

# calculate the median of each bootstrap sample
xi_star <- apply(Y_star, 2, median)

# first two samples
Y_star[,1:2]
>        [,1]  [,2]
> [1,]  5.98  1.76
> [2,] 15.43 12.66
> [3,] 15.43 11.90
> [4,]  1.76  1.76
> [5,]  5.98 12.66
# first two medians
xi_star[1:2]
> [1]  5.98 11.90
# bootstrap estimate of the sampling distribution of the sample median
z <- sqrt(n)*(xi_star - xihat)
Gstar <- ecdf(z)

# 95% confidence interval for the population median (ordinary)
c(xihat - quantile(z, probs = 0.975, names = FALSE)/sqrt(n),
xihat - quantile(z, probs = 0.025, names = FALSE)/sqrt(n))
> [1]  8.37 22.04
# 95% confidence interval for the population median (percentile method)
c(xihat + quantile(z, probs = 0.025, names = FALSE)/sqrt(n),
xihat + quantile(z, probs = 0.975, names = FALSE)/sqrt(n))
> [1]  1.76 15.43
# 95% confidence interval for the population median (normal approximation)
c(xihat - qnorm(0.975)*sd(xi_star), xihat + qnorm(0.975)*sd(xi_star))
> [1]  4.768389 19.031611
```

In reality, G_n^* is never obtained by enumeration since n^n can be very large. Instead, B bootstrap samples are generated via Monte Carlo sampling for a suitable value of B. There is no universal rule to determine B as this will in general depend on the context. If we use the so-called 3σ-rule, $\Pr\{|\delta| < 3/(2\sqrt{B})\} \geq 0.99$, where δ is the Monte Carlo approximation, then for $B = 900$ the error will most likely be less than 0.05 (Reiss, 1989).

The bootstrap procedure to build a 95% confidence interval for the median is illustrated in R Code 7.8. The original sample consists of $n = 5$ observations $y =$

Table 7.1
Empirical coverage and mean width (mean standard errors) of 95% confidence intervals for the median of the exponential distribution for asymptotic, bootstrap normal, and bootstrap percentile methods. The last column shows the width of the interval based on the true sampling distribution.

n	Coverage			Width (mean standard error)			
	Asymp.	Normal	Percentile	Asymp.	Normal	Percentile	Sampling
5	0.758	0.925	0.939	1.345 (0.34)	2.062 (0.53)	2.070	1.761
11	0.869	0.928	0.932	1.085 (0.28)	1.317 (0.34)	1.219	1.187
51	0.927	0.938	0.947	0.533 (0.14)	0.574 (0.15)	0.554	0.550
101	0.932	0.941	0.949	0.382 (0.10)	0.402 (0.10)	0.393	0.390

$(1.76, 5.98, 11.90, 12.66, 15.43)$. The observations in y are sampled with replacement $B = 900$ times, and the median is calculated for each bootstrap sample. The first bootstrap sample gives a median estimate equal to 5.98, the second 11.90, and so on. At this point, one can obtain an estimate of the bootstrap distribution $\hat{G}_n^*$ (as, e.g., by means of the `ecdf` function) and any related quantity, including confidence intervals.

We conducted a simulation study to compare the bootstrap intervals as defined in (7.7) (bootstrap percentile) and (7.10) (bootstrap normal), and the interval based on the asymptotic approximation as illustrated in R Code 7.5. We calculated the empirical coverage at the 95% nominal level as well as the mean width of the intervals for four sample sizes $n = 2k+1$ with $k \in \{2, 5, 25, 50\}$ (which entails the $(k+1)$th order statistic as median). Samples were drawn independently from a standard exponential $\mathcal{E}(1)$, while the bootstrap intervals were each based on $B = 900$ bootstrap replications. The bootstrap percentile intervals were computed using the `boot.ci` function with `type = "perc"`[8] from the `boot` package (Davison and Hinkley, 1997; Canty and Ripley, 2025). The results, based on 10,000 simulation replications, are shown in Table 7.1. For comparison, the last column of the table reports the width $G_n^{-1}(0.975) - G_n^{-1}(0.025)$, where G_n was calculated using Equation (4.1), that is,

$$G_{(k+1)}(x) = \sum_{r=k+1}^{n} \binom{n}{r} \{1 - \exp(-x)\}^r \{\exp(-x)\}^{n-r},$$

based on the *true* (unknown) parent distribution F. At smaller sample sizes, the asymptotic confidence intervals have too narrow widths and, thus, poor coverage. This is not so much a consequence of the presumption of the sampling distribution's normality – shared by both asymptotic and bootstrap-normal intervals – as it is a reflection of underestimated standard errors. In contrast, the bootstrap-percentile method is free from both the particular shape of the sampling distribution and the particular variance estimation method.

7.1.5 Discrete distributions

When the parent distribution F is discrete, the application of inferential methods calls for additional precautions. We have already seen that, in this case, the consistency of sample quantiles is not guaranteed in general (see Section 4.4), hence the asymptotic results of Section 7.1.2 do not apply. It should also come as no surprise that Efron's (1979) bootstrap does not consistently mimic the sampling distribution of sample quantiles. In this context, a few solutions have been suggested. Jentsch and Leucht (2016) considered an m-out-of-n bootstrap for ordinary quantiles as well as mid-quantiles. More recently, Chernozhukov

[8]For two-sided $(1-\alpha)$ intervals, the `"perc"` method in `boot.ci` obtains the confidence bounds by linear interpolation on the normal quantile scale whenever $\alpha(B+1)/2$ is not an integer (see Davison and Hinkley, 1997, p. 195).

Table 7.2
Ordinary quantiles and mid-quantiles with 95% confidence intervals of number of fatal vehicle crashes as estimated in R Code 7.9.

p	Ordinary quantile			Mid-quantile		
	Est.	Lower	Upper	Est.	Lower	Upper
0.25	2	2	2	1.7	1.6	1.8
0.50	4	3	4	3.7	3.5	3.9
0.75	8	7	9	8.0	7.5	8.4
0.90	17	15	19	16.6	15.3	17.8
0.95	28	24	32	27.8	23.7	31.9
0.99	84	60	117	84.2	73.9	94.5

et al. (2020) have proposed an algorithm to build *uniform* confidence bands[9] for quantile functions and quantile treatment effects of possibly discrete random variables. In a nutshell, a confidence band for the entire empirical CDF $\hat{F}(y)$, $y \in [a, b] \subset \mathbb{R}$ is built from B bootstrap samples using a formula that follows the general idea of Equation (7.10). However, the critical value is taken from the bootstrapped Kolmogorov-Smirnov maximal statistic (rather than from the standard normal), while the standard error is estimated robustly using the interquartile range of the bootstrap estimates (rather than the bootstrap standard deviation). A confidence band for the function $\hat{Q}(p)$, $p \in [0, 1]$ is obtained by inversion.

As an alternative to resampling, one can rely on the asymptotic results for mid-quantiles, especially if the sample size is large and computation time is an issue. Of course, the choice should also take into consideration that mid-quantiles are not ordinary quantiles in the strict sense. Yet, estimates of ordinary quantiles can be recovered from mid-quantiles as discussed in Section 4.4.

We conclude this section with an application of the methods discussed above using the `fars` dataset. In this dataset, restricted to the conterminous United States (US), each observation represents the total number of fatal motor vehicle crashes recorded in a US county during 2020, separately for weekdays and weekend days. In total, there are 5662 counts, two for each of the 2831 counties. Since counts for weekdays and weekend days refer to unequal temporal windows, we restrict the present analysis to weekdays only.

The counts distribution, ranging from 0 to 473, exhibits a long right tail with many zero or small counts, which justifies an analysis based on quantiles (see also Section 10.3). In R Code 7.9, a 95% confidence band for the (ordinary) quantile function (with $B = 1000$ bootstrap replications) is obtained using the `R` function `discreteQ` from the namesake package (Melly et al., 2025), while (asymptotic) 95% pointwise confidence intervals for mid-quantiles (see (4.13)) are obtained with the function `Qtools::confint`. As evidenced by the computational time required by the former procedure, there is a small price to pay when using the bootstrap for inference.[10] Overall, the results, summarised in Table 7.2, show that the two methods lead to similar conclusions. The last line of R Code 7.9 compares the estimates of $\hat{Q}(p)$ obtained from `discreteQ` with those returned by `Qtools::midq2q`.

[9]Such a confidence band seeks to achieve the *simultaneous* nominal coverage for the entire function. In contrast, a sequence of pointwise confidence intervals at a given level is not guaranteed to have the same nominal coverage for all the points simultaneously. It follows that the former tends to be wider than the latter.

[10]The function `discreteQ`, which calls the routine `discreteQ:::dq_univariate`, actually implements a Bayesian bootstrap (Rubin, 1981) with exponential weights, which is often more computationally efficient than crude resampling.

R Code 7.9: Confidence intervals for quantiles of the number of fatal vehicle crashes in counties of the conterminous United States, 2020.

```
# pak::pak("bmelly/discreteQ")
library(Qtools)
library(discreteQ)

data("fars", package = "Qtools")
y <- fars$count[fars$weekend == 0]
# sample size
length(y)
> [1] 2831
# summary statistics
summary(y)
>    Min. 1st Qu.  Median    Mean 3rd Qu.    Max.
>    0.00    2.00    4.00    8.29    8.00  473.00
# quantile levels
p <- c(0.25, .5, .75, .9, .95, .99)
# estimate quantiles and 95% CI based on 1000 bootstrap replications
set.seed(321)
system.time(Qhat <- discreteQ(y, bsrep = 1000, alpha = 0.05))
>    user  system elapsed
>    6.06    0.25    6.37
# estimate mid-quantiles and 95% CIs
system.time({midQ <- midquantile(y, probs = p);
  midQci <- confint(midQ, level = 0.95)})
>    user  system elapsed
>    0.00    0.00    0.02
# recover quantiles from mid-quantiles
as.numeric(midq2q(midQ))
> [1]  2  4  8 17 28 84
# and compare with Qhat
as.numeric(Qhat$Q(p))
> [1]  2  4  8 17 28 84
```

7.1.6 Q–Q plots

A useful application of confidence intervals for quantiles concerns Q–Q plots. In Equation (2.3), we defined the comparison quantile function $\kappa(y; F, G) = G^{-1}(F(y))$, for two distributions F and G. Say $Y_1, \ldots, Y_n \sim F$ and we wish to compare F to a theoretical model G. We do not observe F directly, so we use $\hat{F}$ instead after observing $y_1, \ldots, y_n$, whose unique values are $z_1, \ldots, z_k$, say. The plot with points $(\kappa(z_i; \hat{F}, G), z_i)$ is a Q–Q plot. The closer such points lie to a straight line, the stronger the evidence is that the sample comes from G. However, since F is estimated, we might want to add a measure of the uncertainty in our conclusions (Doksum and Sievers, 1976).

We consider the example that is given in R Code 6.5 where a lognormal distribution is fitted to TcCB concentrations via quantile matching. In R Code 7.10, we assess whether the observations are consistent with this theoretical distribution by calculating the coordinates of the Q–Q plot and the 95% confidence interval. The reference line is

R Code 7.10: Q–Q plot and 95% pointwise confidence intervals using the Tetrachlorobenzene dataset.

```
library(reshape2)
library(fitdistrplus)
library(EnvStats)
data(EPA.94b.tccb.df)
dd <- subset(EPA.94b.tccb.df, EPA.94b.tccb.df$Area == "Reference")
y <- dd$TcCB
n <- length(y)

# fit lognormal with quantile matching
ss <- list(meanlog = mean(log(y)), sdlog = sd(log(y)),
threshold = min(y)-0.1)
theta_qm <- qmedist(data = y, distr = "lnorm3",
probs = c(0.05,.5,.95), start = ss)$estimate

# Q-Q plot ordinates
z <- sort(y)

# probabilities (shifted) corresponding to z
p <- (sort(rank(y, ties.method = "average")) - 0.5)/n

# Q-Q plot abscissas
xit <- do.call(qlnorm3, c(list(p = p), as.list(theta_qm)))

# reference line
que <- quantile(y, c(0.25, 0.75), names = FALSE)
qut <- do.call(qlnorm3, c(list(p = c(0.25, 0.75)),
as.list(theta_qm)))
slope <- diff(que)/diff(qut)
int <- que[1] - slope * qut[1]

# 95% confidence interval (asymptotic)
f <- do.call(dlnorm3, c(list(x = xit), as.list(theta_qm)))
se <- slope*sqrt(p*(1-p)/n)/f
zhat <- int + slope * xit
ci <- melt(zhat + outer(qnorm(0.975, 0, se), c(-1,1), FUN = "*"))
ci$xit <- rep(xit, 2)

# plot results
g <- ggplot() + geom_point(aes(x = xit, y = z)) +
geom_line(data = ci, aes(x = xit, y = value, group = Var2),
linetype = "dashed") + geom_abline(intercept = int, slope = slope) +
xlab("Theoretical") + ylab("Empirical") + xlim(0, 2.45) + ylim(0, 2.45)
```

anchored to the first and third quartiles, which determine its intercept and slope. Note also that the slope of such a line is used to rescale the density of the theoretical distribution. The latter is calculated to obtain an (asymptotic) confidence interval (ignor-

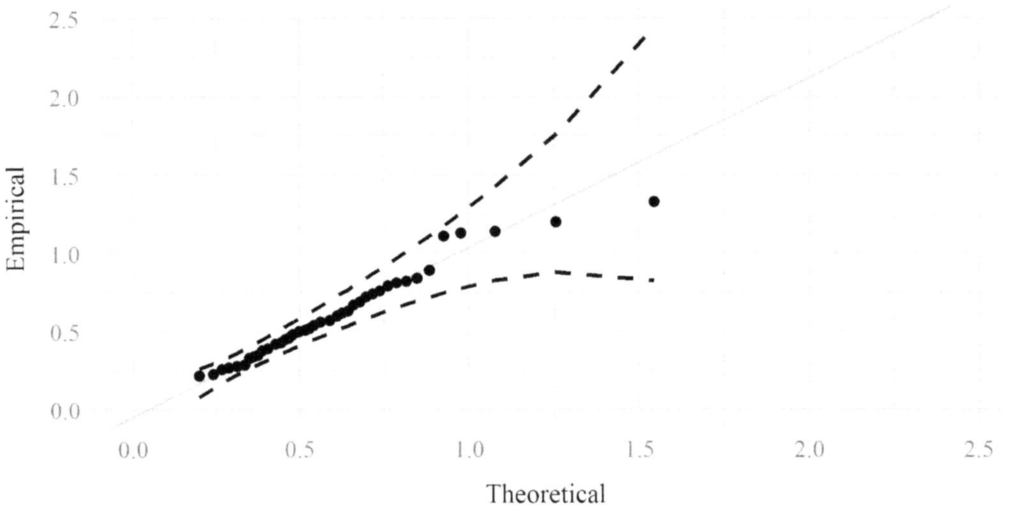

Figure 7.4
Q–Q plot with 95% pointwise confidence intervals (dashed lines) for the 3-parameter lognormal distribution fitted to tetrachlorobenzene concentrations as shown in R Code 7.10.

ing the additional uncertainty due to estimation of the distributional parameters and of the reference line). The resulting plot (Figure 7.4) shows that the sample reasonably follows a lognormal distribution with those estimated parameters (which are treated as *given*).

7.2 Statistical tests

7.2.1 One sample and one quantile

Say we have an IID sample $Y_1, \ldots, Y_n$ from a population F, with (unknown) density $f = F'$. The simplest statistical test for quantiles is one where we are interested in the null hypothesis $H_0 : \xi_p = \xi_0$, where ξ_p, $p \in (0, 1)$, is the pth quantile of F. The test of this hypothesis can be accomplished by means of (6.2), which leads to the Wald-type test statistic

$$T_w = \frac{\hat{\xi}_p - \xi_0}{\hat{\sigma}}, \tag{7.11}$$

where $\hat{\sigma}^2 = \dfrac{p(1-p)}{n\{\hat{f}(\hat{\xi}_p)\}^2}$. Under H_0, the (asymptotic) distribution against which we assess T_w is the standard normal. Given $T_w = t$, the p-value for the two-sided alternative $H_1 : \xi_p \neq \xi_0$ can be found as $2(1 - \Phi(|t|))$.

As in the case of the asymptotic confidence intervals discussed in Section 7.1.2, the performance of T_w depends on a sufficiently large sample size for the normal approximation to be satisfactory as well as for a reliable estimation of f. To remedy these limitations, an exact test for $H_0 : \xi_p = \xi_0$ can be performed by calling on the empirical cumulative distribution function. Equation (4.5) gives us some clues since the quantity of interest is

R Code 7.11: One-sample test of the median in the `PlantGrowth` dataset.

```
data(PlantGrowth)
y <- PlantGrowth$weight[PlantGrowth$group == "ctrl"]
n <- length(y)
p <- 0.5
print(xihat <- quantile(y, probs = p, names = FALSE))
> [1] 5.155
xi0 <- 5

# Wald test
fhat <- density(y, bw = "SJ", kernel = "gaussian")
f <- approx(fhat$x, fhat$y, xout = xihat)$y
sigma <- sqrt(p*(1-p))/(sqrt(n)*f)
Tw <- (xihat - xi0)/sigma
2*(1 - pnorm(abs(Tw)))
> [1] 0.5521873
# exact binomial
Tb <- sum(y <= xi0)
2*min(pbinom(Tb, n, p), 1 - pbinom(Tb, n, p) + dbinom(Tb, n, p))
> [1] 0.7539063
```

the number (proportion) of observations that fall below the threshold ξ_0, that is,

$$T_b = \sum_{i=1}^{n} I\left(Y_i \leq \xi_0\right), \tag{7.12}$$

which, under H_0, follows a binomial distribution with parameters n and p. Given $T_b = t$, the p-value for the two-sided alternative $H_1 : \xi_p \neq \xi_0$ can be found as $2\min\{G(t; n, p), 1 - G(t; n, p) - g(t; n, p)\}$ (assuming that $\Pr(Y_i = \xi_0) = 0$), where G and g denote the CDF and PMF, respectively, of a binomial distribution with parameters n and p.

An example of the two tests described above is given in R Code 7.11. The `PlantGrowth` dataset contains measurements of plant yield (dried weight in grams) obtained under a control and two different treatment conditions (Dobson, 1983). In our example, we define y as the weight for the control group. The null hypothesis is $H_0 : \xi_{0.5} = 5$. The sample median gives 5.15. Both tests yield large p-values, thus supporting the hypothesis that the population median is not different from 5.

7.2.2 Two or more samples and one quantile

In research studies, it is often of interest to test for differences between samples. The much-celebrated two-sample t-test is a popular tool for testing differences between means in groups comparisons. There might be reasons that make a sample quantile preferable to the sample mean in such comparisons.

Suppose $Y_{1,1}, Y_{1,2}, \ldots, Y_{1,n_1}$ and $Y_{2,1}, Y_{2,2}, \ldots, Y_{2,n_2}$ are two independent samples of size n_1 and n_2, respectively, from two populations F_1 and F_2, respectively. The null hypothesis we want to test is $H_0 : \xi_{1,p} = \xi_{2,p}$, where $\xi_{j,p} = F_j^{-1}(p)$, $j = 1, 2$, or, equivalently, $H_0 : \xi_{2,p} - \xi_{1,p} = 0$, assuming that the two samples come from populations with possibly different densities f_js (hence, NID). The difference $\xi_{2,p} - \xi_{1,p}$ corresponds to the quantile treatment effect (QTE) defined in Section 3.2.2. We may consider the test statistic

$$T_w = \frac{\hat{\xi}_{2,p} - \hat{\xi}_{1,p}}{\hat{\sigma}_d}, \tag{7.13}$$

where $\hat{\sigma}_d^2 = \hat{\sigma}_1^2 + \hat{\sigma}_2^2$ and $\hat{\sigma}_j^2 = \dfrac{p(1-p)}{n_j\{\hat{f}_j(\hat{\xi}_{j,p})\}^2}$, $j = 1, 2$. Under H_0, the distribution against which we assess T_w is the standard normal. Given $T_w = t$, the p-value for the two-sided alternative $H_1 : \xi_{1,p} \neq \xi_{2,p}$ can be found as $2(1 - \Phi(|t|))$. More conservatively, if n_1 and n_2 are modest, the CDF of the standard normal, Φ, may be heuristically replaced by that of a Student's $t_{n_1+n_2-2}$ to account for the larger variability associated with small sample sizes.

The test statistic in (7.13) suffers from the same limitations as those remarked for (7.11), with an even more onerous requirement as both n_1 and n_2 must be sufficiently large, if not to yield a reasonable estimate of the f_js. A simple yet effective alternative comes from the bootstrap (e.g., see Wilcox et al., 2014), briefly described as follows. The two groups are bootstrapped B times, where each bootstrap sample is the same size as that of the original sample. The statistic $D^* = \hat{\xi}_{2,p}^{*(b)} - \hat{\xi}_{1,p}^{*(b)}$, whose sampling distribution is denoted by, say, G^*, is computed for each pair of bootstrap quantile estimates $\hat{\xi}_{1,p}^{*(b)}$ and $\hat{\xi}_{2,p}^{*(b)}$, $b = 1, \ldots, B$. The ECDF of D^* is evaluated at the null, that is $\hat{G}^*(0)$ (the estimation of G^* is akin to the procedure followed in Section 7.1.4). Finally, the (generalised) bootstrap p-value (Liu and Singh, 1997) for the two-sided alternative $H_1 : \xi_{1,p} \neq \xi_{2,p}$ is $2\min\{\hat{G}^*(0), 1 - \hat{G}^*(0)\}$.

We give a practical example in R Code 7.12 using the `PlantGrowth` dataset, where we define y_1 and y_2 as the plant yield for the control group and for one of the treatment conditions, respectively. The observed difference between medians is -0.60 grams (i.e., the median plant yield in the treatment group is slightly lower than that in the control group). Nine hundred bootstrap samples are drawn independently from each experimental group. The sample sizes of the bootstrap samples are the same as those of the original samples. For each replication, we calculate the difference between medians (900 in total) and then obtain $\hat{G}^*$. The resulting p-value is 0.20, which, consistently with the t-test for the mean, supports the null hypothesis of identical medians. Note that the t-test in R Code 7.12 is carried out under the assumption of unequal variances to put it on a similar footing to the test for the medians in (7.13) (which assumes a more general NID setting).

When there are more than two groups, say $k+1$, $k > 1$, then the typical null hypothesis in many analyses (as, for example, in the analysis of variance) is that the $k+1$ samples come from the same population. Here, we are more specific in the sense that the hypothesis we consider is for a specific quantile p. We define group 1 as the 'control' group (this is an arbitrary choice that has no bearing on the test) and contrast the other k groups to the control. Let $\delta_j = \xi_{j+1,p} - \xi_{1,p}$, $j = 1, \ldots, k$, be the QTE that contrasts the pth quantile of population $j+1$ against the pth quantile of the control population. Then, the null hypothesis is $H_0 : \delta_1 = \delta_2 = \ldots = \delta_k = 0$, against the alternative that $\delta_j \neq 0$ for at least one of the contrasts. We maintain the NID assumption as in the two-sample case discussed above.

Let $\hat{\boldsymbol{\delta}}$ be the sample equivalent of the vector $\boldsymbol{\delta}$ with elements $\hat{\delta}_j = \hat{\xi}_{j+1,p} - \hat{\xi}_{1,p}$, $j = 1, \ldots, k$. Analogously to (7.13), we consider

$$T_f = \frac{1}{k}\hat{\boldsymbol{\delta}}^\top \hat{\boldsymbol{S}}^{-1}\hat{\boldsymbol{\delta}}, \tag{7.14}$$

R Code 7.12: Two-sample test of the median in the `PlantGrowth` dataset.

```
data(PlantGrowth)
y1 <- PlantGrowth$weight[PlantGrowth$group == "ctrl"]
y2 <- PlantGrowth$weight[PlantGrowth$group == "trt1"]
n1 <- length(y1)
n2 <- length(y2)
print(d <- median(y2) - median(y1))
> [1] -0.605
# take B = 900 bootstrap samples with replacement from y1 and y2
set.seed(111)
B <- 900
Y1_star <- replicate(B, sample(y1, n1, replace = TRUE))
Y2_star <- replicate(B, sample(y2, n2, replace = TRUE))

# calculate the median of each bootstrap sample for y1 and y2
xi1_star <- apply(Y1_star, 2, median)
xi2_star <- apply(Y2_star, 2, median)

# take the difference
d_star <- xi2_star - xi1_star

# calculate p-value
Gstar <- (sum(d_star < 0) + 0.5*sum(d_star == 0))/B
print(pv <- 2*min(Gstar, 1 - Gstar))
> [1] 0.2011111
# t-test
t.test(y1, y2, var.equal = FALSE)
>
>  Welch Two Sample t-test
>
> data:  y1 and y2
> t = 1.1913, df = 16.524, p-value = 0.2504
> alternative hypothesis: true difference in means is not equal to 0
> 95 percent confidence interval:
>  -0.2875162  1.0295162
> sample estimates:
> mean of x mean of y
>     5.032     4.661
```

which, under H_0, is assessed against an F-distribution with k and $n - (k+1)$ degrees of freedom, where $n = \sum_{j=1}^{k+1} n_j$ and n_j is the sample size for group j. The calculation of $\hat{\boldsymbol{S}}$, which plays the same role as $\hat{\sigma}_d^2$ in (7.13), is a little more laborious than that of $\hat{\sigma}_d^2$ but, luckily, calculations are amenable to simple coding. First, let us define the following quantities. Let $\hat{f}_{j,p} = \hat{f}_j(\hat{\xi}_{j,p})$ be the group-specific estimated density evaluated at the group-specific pth sample quantile $\hat{\xi}_{j,p}$, $j = 1, \ldots, k+1$ and let $m_j = n_j \hat{f}^2_{j,p}$. We have

$$\hat{\boldsymbol{S}} = \frac{p(1-p)}{m_1} \begin{bmatrix} \frac{m_1+m_2}{m_2} & 1 & \cdots & \cdots & 1 \\ 1 & \frac{m_1+m_3}{m_3} & 1 & \cdots & 1 \\ 1 & \cdots & \ddots & \cdots & 1 \\ 1 & \cdots & \cdots & \ddots & 1 \\ 1 & \cdots & \cdots & 1 & \frac{m_1+m_{k+1}}{m_{k+1}} \end{bmatrix}. \tag{7.15}$$

These calculations rely on asymptotic results for NID settings (Koenker, 2005, p. 74). In the IID case, that is, under the assumption that the densities f_js differ only by a location-shift from a common density, say, f, then the m_js in (7.15) are replaced by the n_js and $\hat{\boldsymbol{S}}$ will be proportional to f^2 (Koenker, 2005, p. 73). The decision to assume IID or NID can be bypassed with bootstrap.

One can easily verify that the test statistic T_w in (7.13) is a special case of T_f in (7.14). With two groups, $\hat{\boldsymbol{S}}$ reduces to $p(1-p)\dfrac{m_1+m_2}{m_1 m_2} = p(1-p)\dfrac{n_1\hat{f}^2_{1,p} + n_2\hat{f}^2_{2,p}}{n_1\hat{f}^2_{1,p} n_2\hat{f}^2_{2,p}} = \hat{\sigma}^2_d$, while $\boldsymbol{\delta} = \hat{\xi}_{2,p} - \hat{\xi}_{1,p}$; hence, $T_f = \dfrac{(\hat{\xi}_{2,p} - \hat{\xi}_{1,p})^2}{\hat{\sigma}^2_d} = T^2_w$, while inference is carried out based on an F-distribution with 1 and $n_1 + n_2 - 2$ degrees of freedom (equivalent to that of a squared random variable distributed as a Student's $t_{n_1+n_2-2}$).

We conclude with the analysis of the `PlantGrowth` dataset in R Code 7.13, where the medians of all three samples are now tested simultaneously. The medians of the first and second treatment groups differ by -0.73 and 0.23 grams, respectively. The p-value associated with the test in (7.14) is 0.02, which, at the 5% significance level, supports the alternative hypothesis that at least one of the δ's is different from 0.[11] Analogously, at the 5% significance level, the analysis of variance rejects the null hypothesis of equality in mean yields.

7.2.3 Two samples and multiple quantiles

Consider again the case with two independent samples $(Y_{1,1}, Y_{1,2}, \ldots, Y_{1,n_1}) \sim F_1$ and $(Y_{2,1}, Y_{2,2}, \ldots, Y_{2,n_2}) \sim F_2$. However, we now want to test the null hypothesis $H_0 : \delta_1 = \delta_2 = \ldots = \delta_m$, where $\delta_r = \xi_{2,p_r} - \xi_{1,p_r}$, $r = 1, \ldots, m$, $m \geq 2$, with $p_r \neq p_s$ for $r \neq s$. That is, we investigate whether the QTE contrasting the two groups is constant across selected values of p (see Section 3.2.2).

The test statistic of interest is akin to the statistic T_f in (7.14), which we now write in a slightly different form. Let $\hat{\boldsymbol{V}}$ be the matrix with generic rsth element

$$v_{rs} = \frac{\min\{p_r, p_s\} - p_r p_s}{n_1 \hat{f}_{1,p_r} \hat{f}_{1,p_s}} \cdot \frac{n_1 \hat{f}_{1,p_r} \hat{f}_{1,p_s} + n_2 \hat{f}_{2,p_r} \hat{f}_{2,p_s}}{n_2 \hat{f}_{2,p_r} \hat{f}_{2,p_s}}, \qquad r, s = 1, \ldots, m.$$

Note that the first element of the diagonal of the matrix in (7.15) is exactly v_{rr} for $p = p_r$.

[11]An analogous Wald-type test can be carried out using `anova.rqlist` from package `quantreg`; see Section 8.3.3 for a more general treatment of hypothesis testing for quantile regression models. Although the present formulation and the regression-based approach are asymptotically equivalent, finite-sample differences should be expected because the calculations are not identical. In particular, `rq()` solves the quantile regression optimisation problem directly and, at $\tau = 0.5$, may return a different admissible median in the presence of non-uniqueness, whereas the empirical quantiles in R Code 7.13 are computed using `quantile(..., type = 1)`. Differences may also arise from the estimation of the covariance matrix (or sparsity function).

R Code 7.13: Test of equality of medians among three samples from the PlantGrowth dataset.

```
data(PlantGrowth)

p <- 0.5
y <- split(PlantGrowth$weight, PlantGrowth$group)
k <- length(y) - 1
print(xi <- sapply(y, quantile, probs = p, names = FALSE, type = 1))
> ctrl trt1 trt2
> 5.14 4.41 5.37
# estimate quantile differences (treatments vs control)
print(delta <- kronecker(xi[-1], xi[1], "-"))
> [1] -0.73  0.23
# estimate group-specific densities
fj <- sapply(y, function(z, p)
{fhat <- density(z, bw = "SJ", kernel = "gaussian")
return(approx(fhat$x, fhat$y, xout = quantile(z, probs = p))$y)}, p = p)

# calculate S
nj <- table(PlantGrowth$group)
n <- sum(nj)
mj <- nj*fj^2
S <- matrix(1, k, k)
diag(S) <- kronecker(mj[1], mj[-1], "+")/mj[-1]
S <- S*p*(1-p)/mj[1]

# test for equality of medians of the three groups
print(Tf <- as.numeric(t(delta)%*%solve(S)%*%delta)/k)
> [1] 4.323944
print(pv <- 1 - pf(Tf, df1 = k, df2 = n - (k + 1)))
> [1] 0.02349387
# F-test (analysis of variance)
summary(aov(weight ~ group, PlantGrowth))
>             Df Sum Sq Mean Sq F value Pr(>F)
> group        2  3.766  1.8832   4.846 0.0159 *
> Residuals   27 10.492  0.3886
> ---
> Signif. codes:  0 '***' 0.001 '**' 0.01 '*' 0.05 '.' 0.1 ' ' 1
```

The test statistic for the hypothesis of interest is then

$$T_f = \frac{(\boldsymbol{C}\hat{\boldsymbol{\delta}})^\top(\boldsymbol{C}\hat{\boldsymbol{V}}\boldsymbol{C}^\top)^{-1}\boldsymbol{C}\hat{\boldsymbol{\delta}}}{m-1}, \tag{7.16}$$

where $\boldsymbol{C}$ is a $(m-1)\times m$ matrix of contrasts. Under H_0, the test statistic T_f is assessed against an F-distribution with $m-1$ and $nm-(m-1)$ degrees of freedom, with $n = n_1+n_2$. The alternative hypothesis is $\delta_r \neq \delta_s$, with $r \neq s$, for at least a pair of QTEs.

Following the example in R Code 7.12, we define y_1 and y_2 as the plant yield for the control group and for one of the treatment conditions, respectively. Suppose we want to test

R Code 7.14: Test of equality of quantile treatment effects for two samples from the PlantGrowth dataset.

```
data(PlantGrowth)

y1 <- PlantGrowth$weight[PlantGrowth$group == "ctrl"]
y2 <- PlantGrowth$weight[PlantGrowth$group == "trt1"]
n1 <- length(y1)
n2 <- length(y2)
n <- n1 + n2

p <- c(0.25, 0.75)
m <- length(p)
xi1 <- quantile(y1, probs = p, names = FALSE)
xi2 <- quantile(y2, probs = p, names = FALSE)

# estimate quantile differences (treatment vs control)
print(delta <- xi2 - xi1)
> [1] -0.3425 -0.4225
# estimate group-specific densities
fj <- sapply(list(y1, y2), function(z, p)
{fhat <- density(z, bw = "SJ", kernel = "gaussian")
return(approx(fhat$x, fhat$y, xout = quantile(z, probs = p))$y)},p = p)
rownames(fj) <- p

# calculate V
pp <- outer(p, p, FUN = pmin) - outer(p, p, FUN = "*")
ff1 <- outer(fj[,1], fj[,1], FUN = "*")
ff2 <- outer(fj[,2], fj[,2], FUN = "*")
print(V <- pp/ff1 * (n1*ff1 + n2*ff2)/(n2*ff2))
>           0.25      0.75
> 0.25 1.4756097 0.4805006
> 0.75 0.4805006 1.4321655
# define C
C <- matrix(c(-1,1), nrow = 1)

# test for equality of QTEs
Cd <- C%*%delta
CVC <- C%*%V%*%t(C)
print(Tf <- as.numeric(t(Cd)%*%solve(CVC)%*%(Cd)))
> [1] 0.00328749
print(pv <- 1 - pf(Tf, df1 = m - 1, df2 = n * m - (m - 1)))
> [1] 0.9545698
```

the quantiles $p_1 = 0.25$ and $p_2 = 0.75$, hence $\boldsymbol{C} = \begin{pmatrix} -1 & 1 \end{pmatrix}$. Testing $H_0 : \hat{\xi}_{2,0.75} - \hat{\xi}_{1,0.75} = \hat{\xi}_{2,0.25} - \hat{\xi}_{1,0.25}$ is equivalent to testing that the IQR in the two populations are equal (e.g., see Koenker, 2005, p. 75), that is, $H_0 : \hat{\xi}_{2,0.75} - \hat{\xi}_{2,0.25} = \hat{\xi}_{1,0.75} - \hat{\xi}_{1,0.25}$. The procedure to build the test and the test results are reported in R Code 7.14. The estimated QTEs

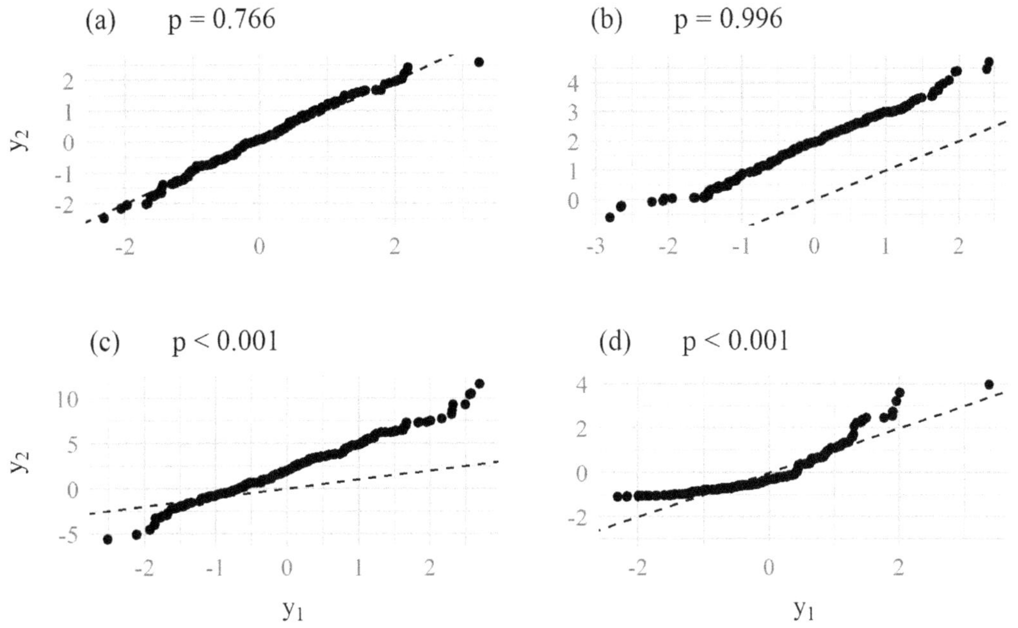

Figure 7.5
Q–Q plots of two simulated samples, each of size 200, from: (a) the same population; (b) populations differing in location only; (c) populations differing in location and scale only; and (d) populations differing in shape only. The top of each panel reports the p-value for the null hypothesis of equality of the quantiles $p_r = r/(n+1)$, $r = 1, \ldots, n$, based on the test in (7.16). The dashed diagonal line has slope one and intercept zero.

are -0.34 grams and -0.42, which are not much dissimilar. The test indeed returns a large p-value.

It is worth remarking that the test in (7.16) can be seen as an inferential summary of a Q–Q plot. To see this, consider two samples of size n_1 and n_2 and, for the sake of simplicity, assume $n_1 = n_2 = n$. Building a Q–Q plot for two samples requires first sorting their values, that is, $y_{1,(1)}, y_{1,(2)}, \ldots, y_{1,(n)}$ and $y_{2,(1)}, y_{2,(2)}, \ldots, y_{2,(n)}$, and then assessing whether the points $(y_{1,(i)}, y_{2,(i)})$, $i = 1, \ldots, n$, scatter along the line $y_2 = by_1 + a$. The values of the slope b and intercept a of the reference line are set according to the particular aspect of the comparison one wants to evaluate. For example, departures from the line with $b = 1$ and $a = 0$ can be potentially ascribed to differences in either location or scale or shape, or any combination thereof (see Figure 7.5). On the other hand, if we set $b = (\hat{\xi}_{2,0.75} - \hat{\xi}_{2,0.25})/(\hat{\xi}_{1,0.75} - \hat{\xi}_{1,0.25})$ and $a = \hat{\xi}_{2,0.25} - b \cdot \hat{\xi}_{1,0.25}$ (e.g., see R Code 7.10), which is usually the default type of line in statistical software (e.g., in `R`), then location and scale differences are accounted for and any residual shift should be attributed to differences in shape. Now, if we calculate a as in the previous example but force b to be unitary, then testing whether the Q–Q plot significantly departs from such a line is equivalent to testing the null hypothesis $H_0 : \delta_r = \bar{\delta}$ for all $r = 1, \ldots, n$, with $\delta_r = \xi_{2,p_r} - \xi_{1,p_r}$ and, say, $p_r = r/(n+1)$. The p-values for the four scenarios in Figure 7.5 are shown at the top of each plot.

7.2.4 Two or more samples and multiple quantiles

In this section, we briefly introduce a generalisation of (7.14) and (7.16), whose discussion is deferred to Chapter 8. Rather than comparing multiple groups for a single quantile or multiple quantiles for two groups, we wish to compare multiple (say, m, $m \geq 2$) quantiles between $k + 1$ groups, $k \geq 1$. The null hypothesis of interest is a joint hypothesis of the k hypotheses

$$\begin{aligned} &H_{0,1} : \delta_{1,1} = \delta_{1,2} = \ldots = \delta_{1,m} \\ &H_{0,2} : \delta_{2,1} = \delta_{2,2} = \ldots = \delta_{2,m} \\ &\ldots \\ &H_{0,k} : \delta_{k,1} = \delta_{k,2} = \ldots = \delta_{k,m}, \end{aligned} \tag{7.17}$$

where $\delta_{j,r} = \xi_{j+1,p_r} - \xi_{1,p_r}$, for $j = 1, \ldots, k$ and $r = 1, \ldots, m$. The test statistic can be better appreciated in the context of conditional quantiles (Section 8.3.3). Analogously to the T_f statistics introduced in the previous sections, the distribution we need for testing the null hypothesis specified above is F with $k(m-1)$ and $nm - k(m-1)$ degrees of freedom.

For the right choice of the p_rs, a test for multiple groups and multiple quantiles can be seen as a Q–Q plot generalised to multiple distributions as well as a distribution-free alternative to the F-test in the analysis of variance (ANOVA) under the assumption of normality and homoskedasticity.

Part III

Conditional quantiles

8

Linear quantile regression

8.1 Modelling

In Chapter 3, we introduced conditional modelling as a problem of comparing two or more groups defined according to a variable X in relation to a response Y. In this chapter, we expand this notion to the case of multiple covariates (whereas the response Y is taken to be univariate throughout the rest of the book). Moreover, Y is assumed to be a continuous random variable. Quantile regression for discrete responses is discussed in Chapter 10.

Let $(Y_i, \boldsymbol{X}_i)$, $i = 1, \ldots, n$, be a random sample of size n. The responses Y_i are assumed to come from a continuous distribution F_Y, while inference is conducted conditionally on the observed values of $\boldsymbol{X}_i = (1, X_{i,1}, X_{i,2}, \ldots, X_{i,k})^\top$, namely $\boldsymbol{x}_i = (1, x_{i,1}, x_{i,2}, \ldots, x_{i,k})^\top$, $i = 1, \ldots, n$, which may contain both discrete and continuous variables, and where the first element is equal to the constant 1 to allow for a regression's intercept. We define the pth[1] linear quantile regression model as (Koenker and Bassett, 1978)

$$Q_{Y_i|\boldsymbol{X}_i=\boldsymbol{x}_i}(p) \equiv Q_{Y_i}(p \mid \boldsymbol{x}_i) = \boldsymbol{x}_i^\top \boldsymbol{\beta}(p) = \beta_0(p) + \beta_1(p)x_{i,1} + \ldots + \beta_k(p)x_{i,k}, \tag{8.1}$$

for fixed $p \in (0,1)$, where $\boldsymbol{x}_i^\top \boldsymbol{\beta}(p)$ is the linear predictor. The $(k+1) \times 1$ vector $\boldsymbol{\beta}(p) = (\beta_0(p), \beta_1(p), \ldots, \beta_k(p))^\top$ collects the regression coefficients associated with $\boldsymbol{x}_i$. Model (8.1) is semiparametric in the sense that it specifies the conditional quantile function (QF) without imposing a parametric form on the conditional distribution of the response variable (see also Powell, 1994).

The apparent simplicity of model (8.1) conceals a considerable flexibility and potential for gauging complex relationships between the covariates and the response. We now provide a few hypothetical examples.

For the first example, consider $Y = \mu_x + \varepsilon$, where $\mu_x = \boldsymbol{x}^\top \boldsymbol{\beta}$, $\boldsymbol{\beta} = (\beta_0, \beta_1, \ldots, \beta_k)^\top$ and $\varepsilon \sim t_2$. The QF of this heavy-tail location-shift (L-shift) model (with infinite variance) is then given by

$$\begin{aligned} Q_Y(p \mid \boldsymbol{x}) =& \boldsymbol{x}^\top \boldsymbol{\beta} + Q_\varepsilon(p) \\ =& \beta_0 + \beta_1 x_1 + \ldots + \beta_k x_k + (2p-1)\sqrt{\frac{1}{2p-2p^2}}, \end{aligned} \tag{8.2}$$

where $Q_\varepsilon(p)$ denotes the QF of a t_ν, which for $\nu = 2$ reduces to the expression above (see definition in Appendix A.3). Model (8.2) has the form (8.1) if we let $\beta_0(p) = \beta_0 + (2p - 1)/\sqrt{2p - 2p^2}$ and $\beta_j(p) = \beta_j$, $j = 1, \ldots, k$. An instance of the L-shift model in (8.2) is

[1]We use p to denote the quantile level. Readers accustomed to the traditional symbol τ should mentally substitute one for the other. Fortunately, statistical theory is invariant to notational preferences.

DOI: 10.1201/9781003326410-8

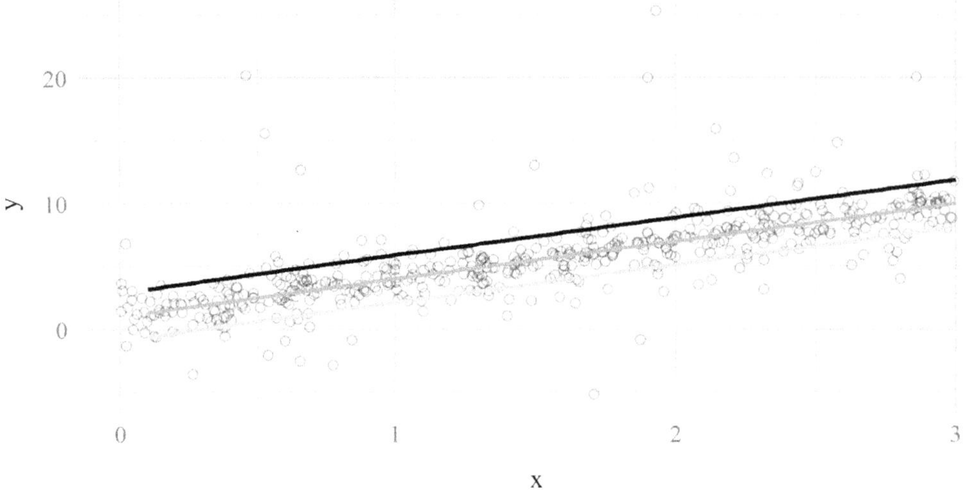

Figure 8.1
Data generated from a Student's t_2 location-shift model along with three conditional quantiles ($p \in \{0.1, 0.5, 0.9\}$) of the underlying true model.

given in Figure 8.1, which plots a sample from $Y = 1 + 3x + \varepsilon$ along with three conditional quantiles of the underlying true model for $p \in \{0.1, 0.5, 0.9\}$. The covariate x affects only the location of the response and therefore the quantile regression lines are parallel vertical translations. This highlights the importance of robustness in the presence of heavy-tailed errors.

Suppose now $Y \sim \mathcal{N}(\mu_x, \sigma_x^2)$, where $\mu_x = \boldsymbol{x}^\top \boldsymbol{\beta}$ and $\sigma_x = \boldsymbol{x}^\top \boldsymbol{\gamma} > 0$ are both functions of $\boldsymbol{x}$ with parameters $\boldsymbol{\beta} = (\beta_0, \beta_1, \ldots, \beta_k)^\top$ and $\boldsymbol{\gamma} = (\gamma_0, \gamma_1, \ldots, \gamma_k)^\top$, respectively. The QF of this normal location–scale-shift model (LS-shift) is then given by

$$\begin{aligned} Q_Y(p \mid \boldsymbol{x}) =& \boldsymbol{x}^\top \boldsymbol{\beta} + \boldsymbol{x}^\top \boldsymbol{\gamma} \Phi^{-1}(p) \\ =& \beta_0 + \beta_1 x_1 + \ldots + \beta_k x_k \\ & + \gamma_0 \Phi^{-1}(p) + \gamma_1 \Phi^{-1}(p) x_1 + \ldots + \gamma_k \Phi^{-1}(p) x_k, \end{aligned} \tag{8.3}$$

where Φ^{-1} is the QF of a standard normal or probit function (see definition in Appendix A.3). Model (8.3) has the form (8.1) if we let $\beta_j(p) = \beta_j + \Phi^{-1}(p)\gamma_j$, $j = 0, 1, \ldots, k$. An instance of the LS-shift model in (8.3) is given in Figure 8.2, which plots a sample from $\mathcal{N}(\mu_x, \sigma_x^2)$, where $\mu_x = 1 + 3x$ and $\sigma_x = 1 + 2x$, along with three conditional quantiles of the underlying true model for $p \in \{0.1, 0.5, 0.9\}$. Since x affects both the location and the scale of the response, the quantile regression lines are not simply vertical translations of one another but they have different slopes that reflect an increasing variability of the response with increasing x. In this particular case, one can appreciate the ability of conditional quantiles to naturally reflect the heteroskedasticity of the model.

In our third and last example, we consider $Y \sim \mathcal{W}(a_x, b_x)$, where $a_x = (\boldsymbol{x}^\top \gamma)^{-1} > 0$ and $b_x = \exp(\boldsymbol{x}^\top \boldsymbol{\beta})$. The QF of this Weibull distribution (see definition in Appendix A.3) is then given by $Q_Y(p \mid \boldsymbol{x}) = \exp(-\boldsymbol{x}^\top \boldsymbol{\beta}) \{c(p)\}^{\boldsymbol{x}^\top \gamma}$, where $c(p) = -\log(1 - p)$. This is a nonlinear quantile model, which, however, can be reduced to linear by exploiting the

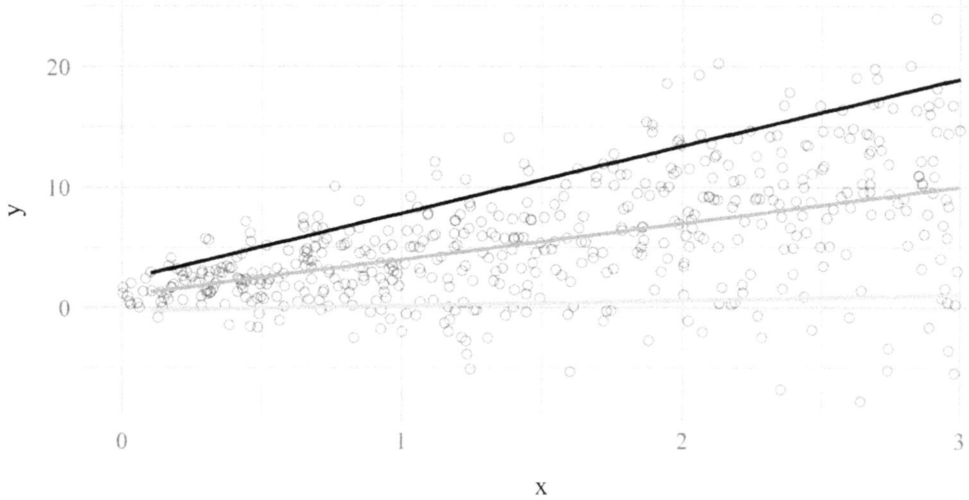

Figure 8.2
Data generated from a normal location–scale-shift model along with three conditional quantiles ($p \in \{0.1, 0.5, 0.9\}$) of the underlying true model.

Q-transformation rule (Section 2.2.2). After log-transforming Y, we have

$$\begin{aligned} Q_{\log(Y)}(p \mid \boldsymbol{x}) &= \log\{Q_Y(p \mid \boldsymbol{x})\} \\ &= -\boldsymbol{x}^\top\boldsymbol{\beta} + \boldsymbol{x}^\top\boldsymbol{\gamma}\log c(p) \\ &= -\beta_0 - \beta_1 x_1 - \ldots - \beta_k x_k \\ &\quad + \gamma_0 \log c(p) + \gamma_1 \log c(p) x_1 + \ldots + \gamma_k \log c(p) x_k. \end{aligned} \tag{8.4}$$

Again, model (8.4) has the form (8.1) if we let $\beta_j(p) = -\beta_j + \gamma_j \log c(p)$, $j = 0, 1, \ldots, k$. An instance of the location–scale–shape-shift (LSS-shift) model in (8.4) is given in Figure 8.3, which plots a sample from $\mathcal{W}(a_x, b_x)$, where $a_x = (0.5 + 0.1x)^{-1}$ and $b_x = 1 + 2x$, along with three conditional quantiles of the underlying true model for $p \in \{0.1, 0.5, 0.9\}$. Now every aspect of the response distribution (location, scale and shape) depends on x. Starting from the lower end of the plot in Figure 8.3 at $x = 0.01$, the median, IQR and Bowley skewness are 0.300, 0.232 and 0.149, respectively. However, at $x = 3$, the median, IQR and Bowley skewness are 0.001, 0.001 and 0.349, respectively. That is, as x increases, the distribution appears more concentrated around lower values and skewed to the right.

In summary, the linear model (8.1) is able to accommodate for a variety of situations. These examples highlight that linear quantile regression is not a restriction on the distribution of Y, but rather a flexible representation of the conditional QF. Even when the data-generating mechanism is nonlinear in Y, it may be linear in $Q_{Y|\mathbf{X}}(p)$. Of course, a functional form may not always be amenable to obvious transformations towards linearity as in the ad-hoc example given above. The topic of nonlinear models is covered in Chapter 9.

8.2 Interpretation

A quantile treatment effect (QTE) that is not constant over different quantile levels can be seen as a form of interaction (Cox, 1984). In a regression framework, this corresponds to

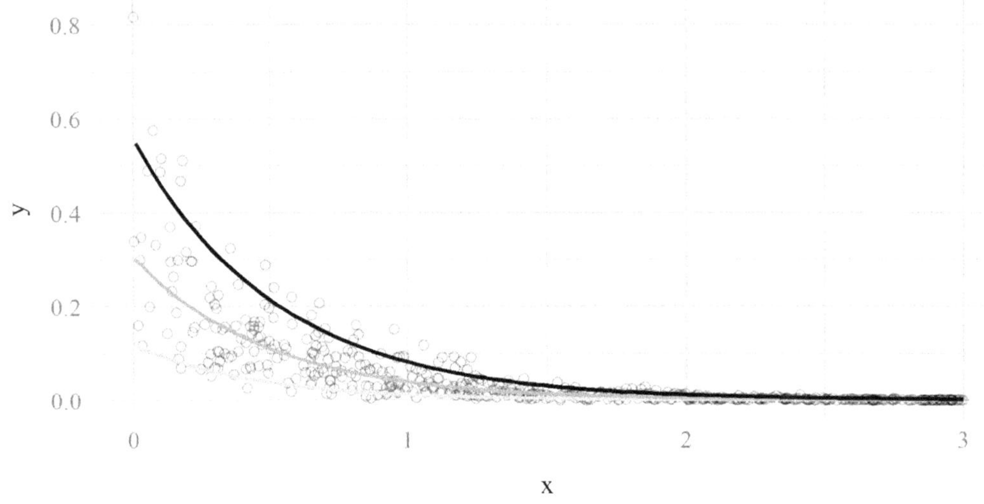

Figure 8.3
Data generated from a Weibull location–scale–shape-shift model along with three conditional quantiles ($p \in \{0.1, 0.5, 0.9\}$) of the underlying true model.

allowing the slope coefficients $\beta_j(p)$ to vary with p, which is equivalent to an interaction between covariates and the latent rank indexed by p.

Consider a confounding-free experiment with two treatment groups in which units are assigned randomly to either group 0 (control) or group 1 (active treatment). Let $Y^{(j)} \sim F_j(y)$ denote the response in group j, $j = 0, 1$. Under a L-shift model, the response obtained on a particular experimental unit assigned to group 1 differs by a constant amount Δ, say, from the response it would have been obtained if that unit were assigned to group 0. That is, for *any response value* y the relationship between F_0 and F_1 is

$$F_1(y) = F_0(y - \Delta). \tag{8.5}$$

If we apply F_0^{-1} to both sides of (8.5), we get

$$\begin{aligned} F_0^{-1}\{F_1(y)\} &= y - \Delta \\ F_0^{-1}\{p\} &= F_1^{-1}(p) - \Delta \\ \Delta &= F_1^{-1}(p) - F_0^{-1}(p), \end{aligned}$$

where the second equation follows from the change of variable $F_1(y) = p$, hence $y = F_1^{-1}(p)$. This result is equivalent to a constant QTE as introduced in Section 3.2.2. If (8.5) does not hold because the distance between $F_0(y)$ and $F_1(y)$ depends on the particular value of y, then it means that there is an interaction between the QTE and the response; hence there is a function $\Delta(p)$ such that $F_1(y) = F_0\{y - \Delta(p)\}$ (provided only that F_1 is continuous, Doksum, 1974).

The QTE can be interpreted in terms of a latent characteristic of subjects under the control distribution (Doksum, 1974). The quantile index p identifies a subject's rank in that distribution and thus indexes this unobserved characteristic. Since the characteristic itself is not directly observable, $\Delta(p)$ measures the effect of treatment for subjects located at the p-th quantile of the control response, that is, at $F_0^{-1}(p)$ (Koenker and Geling, 2001). As an example, consider children's growth. More specifically, let Y be the weight (kg) of boys, whose quantiles are modelled as a function of age (years). Our treatment variable, say x, is therefore 'time'. Figure 8.4 shows a chart developed by the Centers for Disease Control and Prevention (CDC) with estimated percentiles of both weight and height for boys aged 2 years or older in the United States (US) population. It should be borne in mind that this chart (like many others) was estimated from successive cross-sectional surveys (Kuczmarski et al., 2002). This means that children were not followed over time and that each survey was independent from the others.

One can appreciate from the appearance of the weight and height trajectories in Figure 8.4 that these would be best described by some smooth function of age. For now, we will not concern ourselves with what is supposed to be the most appropriate model.[2]

Now suppose that we wish to examine how weight changes between ages 9 and 10 at the 5th and 95th percentiles, therefore the QTE is $\Delta(p) = Q_{Y|X=10}(p) - Q_{Y|X=9}(p)$, $p \in \{0.05, 0.95\}$. This is illustrated in Figure 8.5. The QTE $\Delta(0.05)$ is about 2 kg but $\Delta(0.95)$ is 3 times as large. That is, a 9-year old boy ranking at the 95th percentile will need a bigger 'effort' in terms of weight gain to maintain the same ranking at age 10 as compared to one of his peers ranking at the 5th. Clearly, when seen prospectively at age 9, the response at age 10 is a random variable and there is no guarantee that future measurements will follow those trajectories. The QTE is simply a natural guess[3] one can make about the conditional distribution of future responses (not individual trajectories), reflecting a combination of genetic and environmental factors (the latent characteristic).

8.3 Inference

8.3.1 Estimation and prediction

The goal is to estimate $\boldsymbol{\beta}(p)$ in model (8.1) for a given $p \in [0, 1]$. We want to solve

$$\min_{\boldsymbol{\beta} \in \mathbb{R}^{k+1}} \frac{1}{n} \sum_{i=1}^{n} \psi_p(y_i - \boldsymbol{x}_i^\top \boldsymbol{\beta}), \tag{8.6}$$

where $\psi_p(u) = u\{p - I(u < 0)\}$ is the loss function defined in Equation (5.2). The optimisation problem in (8.6) is the sample analogue of minimising the expected check loss. The solution $\hat{\boldsymbol{\beta}}(p)$ is called pth *regression quantile* (Koenker and Bassett, 1978). Since $\hat{\boldsymbol{\beta}}(p)$ need not be unique for a given p, in general it would be more appropriate to refer to the *regression quantile set*

$$B_p = \left\{ \boldsymbol{\beta}^* \in \mathbb{R}^{k+1} : \frac{1}{n} \sum_{i=1}^{n} \psi_p(y_i - \boldsymbol{x}_i^\top \boldsymbol{\beta}^*) \text{ is minimised} \right\}.$$

Koenker and Bassett (1978, Theorem 3.3) provides the conditions under which the solution is unique.

If we look at the optimisation problem in Equation (8.6) from the perspective of solving it for different values of p, then the solution set $\{\hat{\boldsymbol{\beta}}(p) : p \in [0, 1]\}$ is called *quantile regression*

[2]The CDC growth chart in Figure 8.4 was estimated by means of the LMS procedure (Cole, 1988; Cole and Green, 1992), which pertains to the methods that we labelled as maximum likelihood for quantiles (see Sections 6.2 and 9.4).

[3]Arguably, the *best* guess in terms of the asymmetric loss function (Gneiting, 2011).

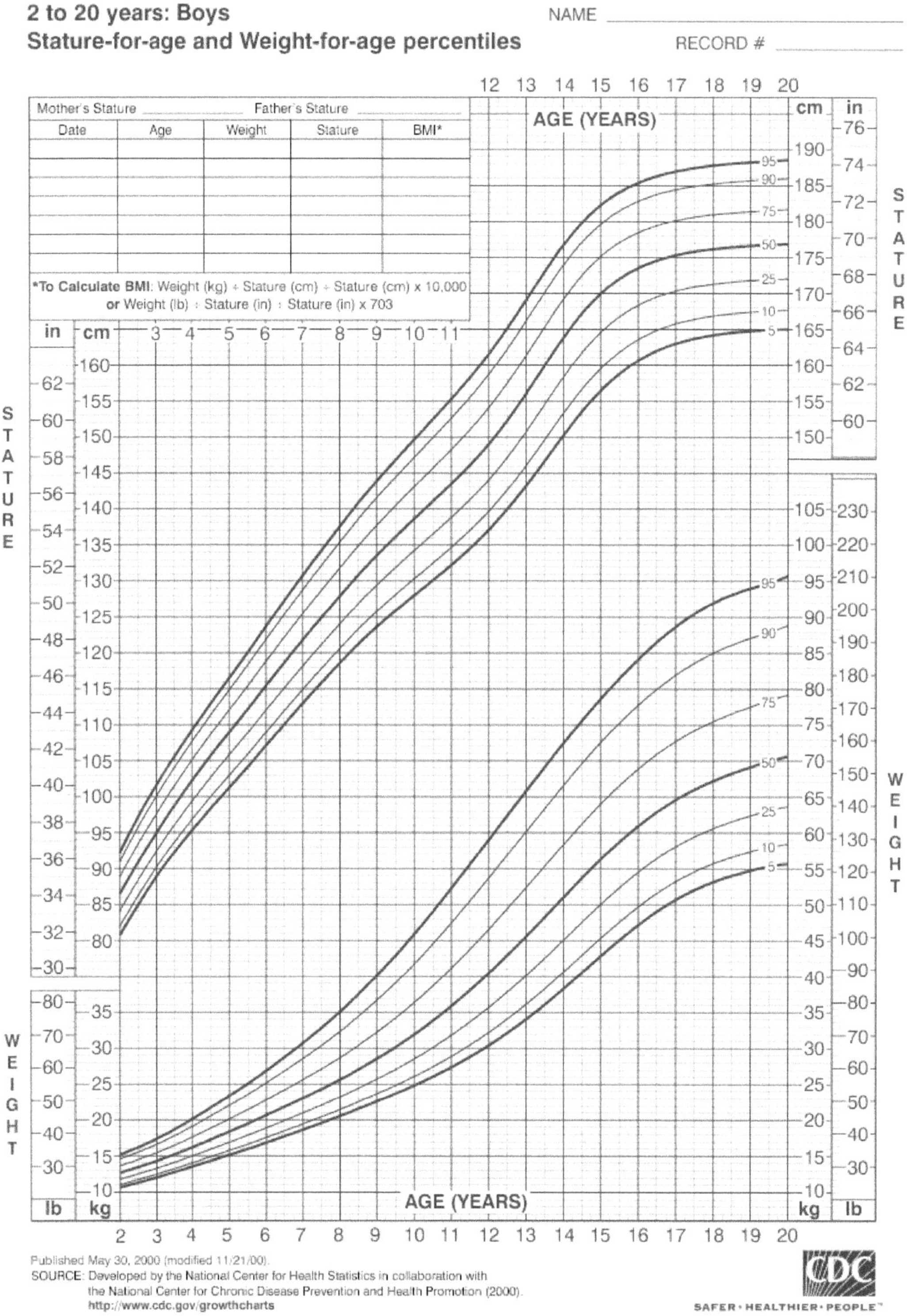

Figure 8.4
Height and weight growth chart for boys aged 2 years and older in the United States population. Source: Developed by the National Center for Health Statistics in collaboration with the National Center for Chronic Disease Prevention and Health Promotion (2000). Public-domain material available free of charge from https://www.cdc.gov/growthcharts.

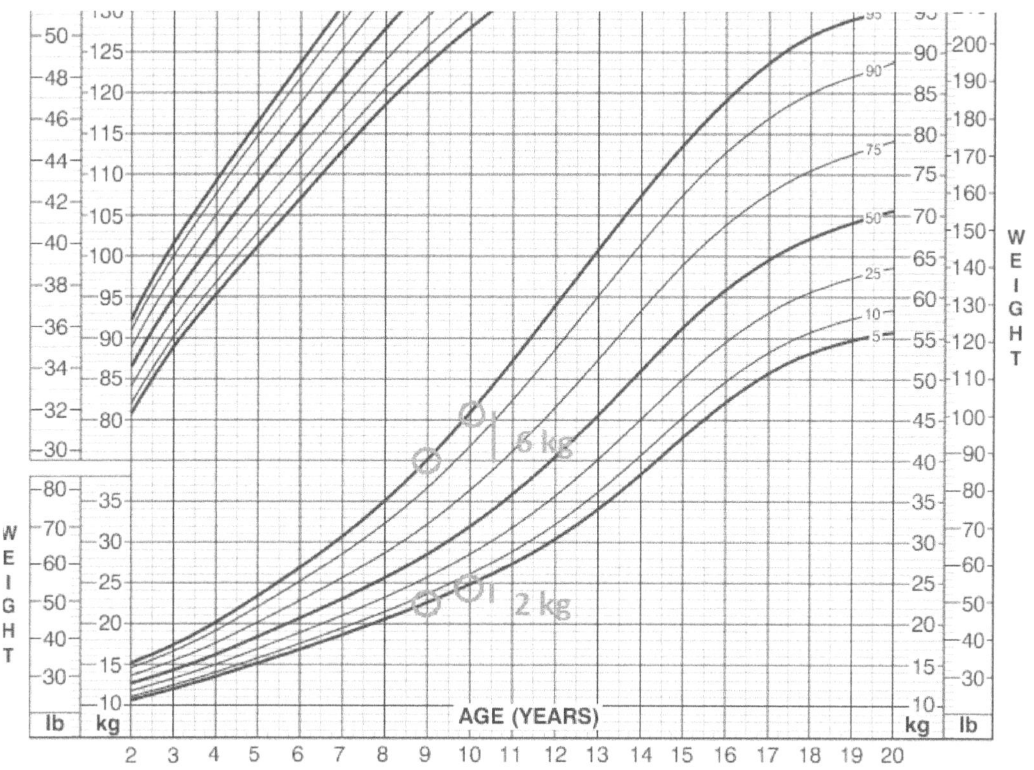

Figure 8.5
Enlarged detail of Figure 8.4, highlighting the quantile treatment effect between ages 9 and 10 years at the 5th and 95th percentiles of weight.

process. In general, there will be m_n breakpoints $0 = p_0 < p_1 < \ldots < p_{m_n} = 1$ such that B_p has a single estimator for fixed p, $p_r < p < p_{r+1}$ (Portnoy, 1991). The number m_n clearly depends on n: intuitively, with few sample observations there is only a limited number of distinct quantiles we can estimate, while if n is large (infinite) the number of breakpoints necessarily increases with continuous $F_{Y|X}$. Indeed m_n grows (roughly linearly) with the sample size[4] and, according to evidence from simulations, this behaviour seems to be fairly independent from the distribution of $\boldsymbol{x}$ and Y (Portnoy, 1991).

Now we turn to how in practice one can obtain a regression quantile by using the algorithms discussed in Chapter 5. We begin with the classical LP approach.

By following steps analogous to those in Section 5.2, the problem in (8.6) can be written as

$$\min_{(\boldsymbol{\beta},\boldsymbol{u},\boldsymbol{v})\in R^{k+1}\times\mathbb{R}^{2n}_{\geq 0}} \left\{p\mathbf{1}_n^\top \boldsymbol{u} + (1-p)\mathbf{1}_n^\top \boldsymbol{v} \mid \boldsymbol{X\beta} + \boldsymbol{u} - \boldsymbol{v} = \boldsymbol{y}\right\}, \tag{8.7}$$

where $\boldsymbol{y} = (y_1, \ldots, y_n)^\top$, $\boldsymbol{X}$ is the $n \times (k+1)$ design matrix with ith row vector $\boldsymbol{x}_i$, and $(\boldsymbol{u}, \boldsymbol{v})$ are slack variables. This problem can be solved with efficient algorithms, including the simplex method of Barrodale and Roberts (Koenker and D'Orey, 1987) for small to moderate problems (a few thousands observations), and the Frisch–Newton interior point method (Portnoy and Koenker, 1997) or the Frisch–Newton algorithm after preprocessing for larger problems. In the past few years, there have notable advances in computational

[4]Portnoy (1991) provides the upper bound $O(n \log n)$ (in probability), for a fixed number of regression coefficients.

R Code 8.1: Linear quantile regression via a simplex algorithm with the R package `lpSolve`.

```
library(lpSolve)
library(quantreg)
# generate data
set.seed(123)
n <- 500
x <- runif(n, 0, 3)
b <- c(1,3)
g <- c(1,2)
y <- cbind(1,x)%*%b + rnorm(n, mean = 0, sd = cbind(1,x)%*%g)
p <- 0.9
# define the LP problem
obj <- c(rep(p, n), rep(1 - p, n), rep(0, 2))
S <- cbind(diag(2*n), 0, 0)
A <- cbind(diag(n), -diag(n), 1, x)
b <- c(rep(0, 2*n), y)
constr <- c(rep(">=", 2*n), rep("=", n))
# solve
fit.lp <- lp("min", obj, rbind(S, A), constr, b)
fit.lp
> Success: the objective function is 329.8739
# compare solution with function rq from quantreg package
print(bhat <- fit.lp$solution[-c(1:(2*n))])
> [1] 2.130219 5.510370
fit.rq <- rq(y ~ x, tau = p, method = "br")
coef(fit.rq)
> (Intercept)          x
>    2.130219   5.510370
# check linear constraints xi + u - v = y
u <- fit.lp$solution[1:n]
v <- fit.lp$solution[(n+1):(2*n)]
all.equal(cbind(1,x)%*%bhat + u - v, y)
> [1] TRUE
# calculate residuals and count 0s
r <- u - v
sum(mapply(identical, x = r, y = 0))
> [1] 2
# calculate loss function 'psi'
sum(r * (p - ifelse(r < 0, 1, 0)))
> [1] 329.8739
fit.rq$rho
> [1] 329.8739
```

methods for conditional quantiles to cope with an ever-increasing demand for the analysis of very large datasets (e.g., see He et al., 2023b). The interested reader can refer to Koenker (2005) and Davino et al. (2013) who give a more technical exposition of linear programming for conditional quantile estimation.

Analogously to what was done in Section 5.2, in R Code 8.1 we first formulate (8.7) in canonical LP form and then solve it using the `lpSolve` package. The data are generated by using the LS-shift model of Section 8.1 for a sample with $n = 500$ observations and x

R Code 8.2: Comparison between the simplex and interior point methods (R package **quantreg**), and the gradient descent algorithm (R package **lqmm**).

```
library(quantreg)
library(lqmm)
set.seed(123)
n <- 200e3
x <- runif(n, 0, 3)
b <- c(1,3)
g <- c(1,2)
y <- cbind(1,x)%*%b + rnorm(n, mean = 0, sd = cbind(1,x)%*%g)
p <- 0.9

system.time(bhat.br <- rq(y ~ x, tau = p, method = "br")$coef)
>    user  system elapsed
>   12.14    0.67   13.06
system.time(bhat.fn <- rq(y ~ x, tau = p, method = "fn")$coef)
>    user  system elapsed
>    0.62    0.04    0.65
system.time(bhat.lqm <- lqm(y ~ x, tau = p)$theta)
>    user  system elapsed
>    0.26    0.01    0.29
# compare estimates
cbind(bhat.br, bhat.fn, bhat.lqm)
>                bhat.br  bhat.fn bhat.lqm
> (Intercept) 2.291532 2.291532 2.291724
> x           5.571668 5.571668 5.571419
```

randomly drawn from a uniform between 0 and 3 (the actual data are plotted in Figure 8.2). We consider the estimation of the quantile $p = 0.9$. In the order provided in our example, the solution is a vector of length $2n + 2 = 1002$ that contains the estimates of the slack variables $\hat{\boldsymbol{u}}$ and $\hat{\boldsymbol{v}}$, followed by $\hat{\boldsymbol{\beta}}(0.9)$. The latter is exactly the estimate we obtain from the function `rq` in the R package `quantreg` (Koenker, 2025) with the option `method = br` which stands for the simplex method of Barrodale and Roberts (currently the default). The residuals are computed as $\hat{\boldsymbol{u}} - \hat{\boldsymbol{v}}$, two of which are exactly 0 since the number of unknowns in $\boldsymbol{\beta}(p)$ is precisely two. As in the unconditional case, we can verify that $\sum \psi_p(\hat{u}_i - \hat{v}_i)$ is equal to the value of the LP objective function at the optimum.

Naturally, as already remarked in Section 5.2, obtaining the solution via a 'crude' LP solver becomes unmanageable as soon as the sample size reaches a few thousands, and hence R Code 8.1 is essentially for didactic purposes. Not only the algorithms implemented in `quantreg` are very efficient, but the command `rq(..., tau)` allows one to estimate multiple quantiles at once by providing a vector of quantile indices via the argument `tau`. It should be borne in mind that since these multiple quantiles are estimated *sequentially* without any constraint on monotonicity, *quantile crossing* may ensue. This issue is addressed in Section 8.3.5. At the extreme end of estimating multiple quantiles, we obtain the quantile process with `rq(..., tau = -1)` (see Section 8.3.4 for an application).

In R Code 8.2, we provide a limited comparison between the simplex (Barrodale–Roberts) and interior point (Frisch–Newton) methods available via the `rq` function (`method = br` and `method = fn`, respectively), and the gradient descent algorithm of Bottai et al. (2015) via the `lqm` function. Again, the data are generated from the normal LS-shift model

R Code 8.3: Comparison between the interior point method (R package **quantreg**) and the convolution-type smoothed quantile regression algorithm (R package **conquer**).

```
library(quantreg)
library(conquer)
set.seed(123)
n <- 500e3
k <- 20
x <- matrix(runif(n*k, 0, 3), n, k)
b <- rep(1,k+1)
g <- rep(1,k+1)
y <- cbind(1,x)%*%b + rnorm(n, mean = 0, sd = cbind(1,x)%*%g)
p <- 0.9

system.time(bhat.fn <- rq(y ~ x, tau = p, method = "fn")$coef)
>     user  system elapsed
>     7.86    0.53    8.95
system.time(bhat.cq <- conquer(x, y, tau = p)$coef)
>     user  system elapsed
>     0.64    0.07    0.84
# true beta
beta <- b + qnorm(p)*g
# mean absolute relative error (average)
mean(abs(bhat.fn - beta)/beta)
> [1] 0.03403264
mean(abs(bhat.cq - beta)/beta)
> [1] 0.03737192
```

as in R Code 8.1 but with a much larger sample size ($n = 200{,}000$). It is apparent that the Barrodale–Roberts algorithm becomes inefficient relative to the other two algorithms. All three methods yield similar estimates.

Finally, in R Code 8.3, we scale up the sample to $n = 500{,}000$ observations and $k = 20$ continuous variables (so that $\boldsymbol{\beta}(p)$ contains 20+1 coefficients). We avoid the simplex method, as it would require considerable computing time (though still less than brewing a moka coffee), and instead consider two algorithms suited to larger problems: the interior point method and the convolution-type smoothed quantile regression algorithm (conquer) of He et al. (2023b), implemented in the `conquer` package (He et al., 2023a). The latter belongs to the class of estimators based on smooth approximations introduced in Section 5.4. The conquer algorithm is suitable not only when n is large, but also when p grows with n (provided p/n remains small). This example shows that both algorithms are fast, although the interior point method begins to lose ground to its competitor. By contrast, they are comparable in terms of mean absolute relative error.

Having obtained an estimate $\hat{\boldsymbol{\beta}}(p)$, it is straightforward to obtain the predictions

$$\hat{Q}_{Y_i}(p \mid \boldsymbol{x}_i) = \boldsymbol{x}_i^\top \hat{\boldsymbol{\beta}}(p), \qquad i = 1, \ldots, n, \tag{8.8}$$

and the residuals

$$\hat{r}_i(p) = y_i - \boldsymbol{x}_i^\top \hat{\boldsymbol{\beta}}(p), \qquad i = 1, \ldots, n. \tag{8.9}$$

Box 8.1: Consistency and asymptotic normality of the linear quantile regression estimator (regularity conditions)

The regularity conditions (as well as the demonstration of asymptotic results) are given by Koenker (2005, p. 120).

1. The conditional distribution functions F_{Y_i} are absolutely continuous, with continuous densities f_{Y_i} bounded away from 0 and ∞ in a neighbourhood of the conditional quantiles $Q_Y(p \mid \boldsymbol{x}_i)$.

2. There exist positive-definite matrices $\boldsymbol{H}(p)$ and $\boldsymbol{J}$ such that:

$$\boldsymbol{H}(p) = \lim_{n\to\infty} \boldsymbol{H}_n(p),$$
$$\boldsymbol{J} = \lim_{n\to\infty} \boldsymbol{J}_n,$$

where

$$\boldsymbol{H}_n(p) = n^{-1}\sum_{i=1}^{n} \boldsymbol{x}_i\boldsymbol{x}_i^\top f_{Y_i}\left(Q_Y(p \mid \boldsymbol{x}_i)\right), \qquad \boldsymbol{J}_n = n^{-1}\sum_{i=1}^{n} \boldsymbol{x}_i\boldsymbol{x}_i^\top,$$

and

$$\max_{i=1,\ldots,n} \|\boldsymbol{x}_i\|/\sqrt{n} \to 0.$$

By construction, $n^{-1}\sum_{i=1}^{n} I(\hat{r}_i(p) < 0) \approx p$ since $\Pr(Y_i \leq Q_{Y_i}(p \mid \boldsymbol{x}_i) \mid \boldsymbol{x}_i) = p$. The quantile regression residuals have a role in several tests and procedures, some of which we will encounter in the remainder of this chapter.

8.3.2 Standard errors and confidence intervals

In Chapter 7, we discussed different methods to estimate standard errors and confidence intervals for unconditional quantiles. In this section, we present analogous results for conditional quantiles. We start with large-sample approximations and then briefly introduce rank and bootstrap confidence intervals.

Asymptotics

We begin with large-n approximations of the sampling distribution of $\hat{\boldsymbol{\beta}}(p)$. To do so, we need to make a distinction between (i) independent and identically distributed observations (IID) and (ii) independent but not identically distributed observations (NID). Let $Y_i \sim F_{Y_i}$, $i = 1, \ldots, n$, where F_{Y_i} is a distribution indexed by a parameter $\theta_i = \theta(\boldsymbol{x}_i)$. We further assume that the Y_is are independent of one another. We introduced a few examples in Section 8.1 starting from the classical L-shift model $Y_i = \boldsymbol{x}_i^\top\boldsymbol{\beta} + \varepsilon_i$. If the error $\varepsilon_i \sim F_\varepsilon$ for all $i = 1, \ldots, n$, then the distribution of Y_i depends on $\boldsymbol{x}_i$ only through its location and $Q_{Y_i}(p \mid \boldsymbol{x}_i) = \boldsymbol{x}_i^\top\boldsymbol{\beta} + Q_\varepsilon(p)$. In this instance, the errors are IID. On the other hand, if $\varepsilon_i \sim F_{\varepsilon_i}$, with $F_{\varepsilon_i} \neq F_{\varepsilon_j}$ for some $i \neq j$, then we say that the errors are NID.

The distinction between IID and NID has consequences on the sampling distribution of $\hat{\boldsymbol{\beta}}(p)$. Let's see the NID case first. Given the linear quantile regression model $Q_{Y_i}(p \mid \boldsymbol{x}_i) =$

$\boldsymbol{x}_i^\top \boldsymbol{\beta}(p)$ and some regularity conditions (see Box 8.1 for technical details)

$$\sqrt{n}\left(\hat{\boldsymbol{\beta}}_n(p) - \boldsymbol{\beta}(p)\right) \xrightarrow{d} \mathcal{N}(\mathbf{0}, \boldsymbol{\Omega}), \tag{8.10}$$

where $\boldsymbol{\Omega} = p(1-p)\boldsymbol{H}^{-1}(p)\boldsymbol{J}\boldsymbol{H}^{-1}(p)$, and $\boldsymbol{H}(p)$ and $\boldsymbol{J}$ are the limiting matrices defined in Box 8.1.

In the IID case, $\boldsymbol{\Omega}$ simplifies since $f_{Y_i}(Q_{Y_i}(p \mid \boldsymbol{x}_i)) = f_\varepsilon(Q_\varepsilon(p))$ for all $i = 1, \ldots, n$ and $\boldsymbol{H}(p) = f_\varepsilon(Q_\varepsilon(p))\,\boldsymbol{J}$, hence $\boldsymbol{\Omega} = \dfrac{p(1-p)}{\{f_\varepsilon(Q_\varepsilon(p))\}^2}\boldsymbol{J}^{-1}$. The unconditional case corresponds to $\boldsymbol{x}_i \equiv 1$ (i.e., an intercept-only model), which leads to $\boldsymbol{J} = 1$ and thus $\boldsymbol{\Omega}$ reduces to the variance in expression (4.9).

To obtain an estimate of the variance-covariance matrix of $\hat{\boldsymbol{\beta}}(p)$, we start from the finite-sample quantity

$$\boldsymbol{V} = \frac{p(1-p)}{n}\boldsymbol{H}_n^{-1}(p)\boldsymbol{J}_n\boldsymbol{H}_n^{-1}(p) \tag{8.11}$$

where $\boldsymbol{H}_n(p)$ and $\boldsymbol{J}_n$ are defined in Box 8.1. As in the case of sample (unconditional) quantiles, estimation of the variance for regression quantiles requires an estimate of the density f. From a semiparametric point of view, we are in a situation analogous to that seen in Section 7.1.2. On that occasion, we saw how to obtain a nonparametric estimate of f from the difference quotient of the ECDF and then evaluate it at the quantile of interest, ξ_p, to finally obtain the sparsity $q(p) = 1/f(\xi_p)$ (see Equation (2.1)). Rather than following the same steps using the (conditional) CDF, it is more natural to use model (8.1) as a starting point since the ultimate goal is to estimate the sparsity anyway. Provided that the density exists and is strictly positive at $Q_{Y_i}(p \mid \boldsymbol{x}_i)$, the aim is to estimate

$$\frac{\partial}{\partial p}Q_{Y_i}(p \mid \boldsymbol{x}_i) = \frac{1}{f_{Y_i}(Q_{Y_i}(p \mid \boldsymbol{x}_i))} = \boldsymbol{x}_i^\top \frac{\partial}{\partial p}\boldsymbol{\beta}(p). \tag{8.12}$$

We use the difference quotient of the conditional QF at p, that is,

$$\begin{aligned}\frac{\mathrm{d}\hat{Q}_{Y_i}(p \mid \boldsymbol{x}_i)}{\mathrm{d}p} &= \frac{\hat{Q}_{Y_i}(p+h \mid \boldsymbol{x}_i) - \hat{Q}_{Y_i}(p-h \mid \boldsymbol{x}_i)}{2h} \\ &= (2h)^{-1}\boldsymbol{x}_i^\top\{\hat{\boldsymbol{\beta}}(p+h) - \hat{\boldsymbol{\beta}}(p-h)\}.\end{aligned} \tag{8.13}$$

From (8.13) it becomes clear why working directly with Q rather than F is advantageous, since the estimates $\hat{\boldsymbol{\beta}}(p \pm h)$ are obtained from the same fitting procedure used to estimate $\hat{\boldsymbol{\beta}}(p)$. We must of course select the bandwidth h. This is discussed in detail by Koenker (2005, p. 139), who considers the bandwidth proposed by Bofinger (1975), which is optimal in the mean-squared-error (MSE) sense,

$$h_n = n^{-1/5}\left[4.5q^2(p)/\{q''(p)\}^2\right]^{1/5}.$$

This expression is not directly useful because it depends on the unknown quantile density function through the ratio $q(p)/q''(p)$. However, as noted by Koenker (2005), this ratio is often not very sensitive to the precise form of F, so one may approximate it using a known, well-behaved reference distribution such as the normal. In the latter case, one obtains

$$h_n = n^{-1/5}\left\{\frac{4.5\phi^4(z_p)}{(2z_p^2+1)^2}\right\}^{1/5},$$

R Code 8.4: Linear quantile regression analysis of waiting time between eruptions and duration of eruption in the **faithful** dataset.

```
library(quantreg)
data(faithful)
cutoff <- 3
x <- as.numeric(faithful$eruptions >= cutoff)
y <- faithful$waiting
p <- c(0.1, 0.5, 0.9)
fit <- rq(y ~ x, tau = p, method = "br")
fit
> Call:
> rq(formula = y ~ x, tau = p, method = "br")
>
> Coefficients:
>              tau= 0.1 tau= 0.5 tau= 0.9
> (Intercept)        47       54       63
> x                  26       26       25
>
> Degrees of freedom: 272 total; 270 residual
```

where $z_u = \Phi^{-1}(u)$. Instead of the MSE as the optimality criterion, Hall and Sheather (1988) considered a bandwidth that is optimal in terms of the coverage accuracy of tests and confidence intervals at level α. The resulting bandwidth (HS) is

$$h_n = n^{-1/3} z_{1-\alpha/2}^{2/3} \left\{1.5q(p)/q''(p)\right\}^{1/3}.$$

Again, if we substitute q with the Gaussian sparsity, we obtain

$$h_n = n^{-1/3} z_{1-\alpha/2}^{2/3} \left\{\frac{1.5\phi^2(z_p)}{2z_p^2 + 1}\right\}^{1/3}.$$

In the event of $p + h > 1$ or $p - h < 0$, one can halve h until $p \pm h$ is within the unit interval.

We revisit R Code 3.3 to illustrate the application of quantile regression in the IID case. In R Code 8.4, we fitted three linear quantile regressions for $p \in \{0.1, 0.5, 0.9\}$. The intercept $\hat{\beta}_0(p)$ is interpreted as the estimated pth quantile of waiting time between eruptions when the duration of the last eruption is less than 3 minutes. The slope $\hat{\beta}_1(p)$ is interpreted as the QTE associated with durations of 3 minutes or longer. One can see that the QTEs are approximately constant across the values of p, consistently with what we observed in R Code 3.3.

For didactic purposes, it might be instructive to follow on from the example in R Code 8.4 with a step-by-step calculation of an estimate of the variance-covariance matrix for $\hat{\beta}(p)$. As illustrated in R Code 8.5, we start with the matrix $\boldsymbol{J}_n$, which, in the case of a binary x, has a very simple form, namely

$$\boldsymbol{J}_n = n^{-1} \sum_{i=1}^{n} \begin{bmatrix} 1 \\ x_i \end{bmatrix} \begin{bmatrix} 1 & x_i \end{bmatrix} = \begin{bmatrix} 1 & u \\ u & u \end{bmatrix},$$

R Code 8.5: Step-by-step computation of the variance-covariance matrix and standard errors for regression quantiles estimated for the **faithful** dataset.

```
# calculate J and xbar for generic x
n <- length(y)
X <- cbind(1, x)
J <- crossprod(X)/n
xbar <- c(1,mean(x))
# alternatively with binary x
# u <- mean(x == 1); J <- matrix(c(1,u,u,u), 2, 2); xbar <- c(1,u)

# Hall-Sheather bandwidth (hs = TRUE)
print(h <- quantreg::bandwidth.rq(p = p, n = n, hs = TRUE, alpha = 0.05))
> [1] 0.05340063 0.14994991 0.05340063
# obtain dQ from two quantile models
bhat_low <- rq(y ~ x, tau = p - h, method = "br")$coef
bhat_up <- rq(y ~ x, tau = p + h, method = "br")$coef
dQ <- t(xbar) %*% (bhat_up - bhat_low)
as.numeric(dQ)
> [1] 3.286765 4.356618 4.356618
# difference quotient
sparsity <- dQ/(2*h)
as.numeric(sparsity)
> [1] 30.77459 14.52691 40.79182
# calculate variance-covariance
V <- outer(solve(J), as.numeric(p*(1-p)*sparsity^2/n), "*")
dimnames(V) <- list(c("intercept", "slope"),
c("intercept", "slope"), p)
V
> , , 0.1
>
>            intercept      slope
> intercept  0.8787296 -0.8787296
> slope     -0.8787296  1.3657968
>
> , , 0.5
>
>            intercept      slope
> intercept  0.5438946 -0.5438946
> slope     -0.5438946  0.8453676
>
> , , 0.9
>
>           intercept     slope
> intercept  1.543892 -1.543892
> slope     -1.543892  2.399649
# standard errors
apply(V, 3, function(x) sqrt(diag(x)))
>                 0.1       0.5      0.9
> intercept 0.9374058 0.7374921 1.242534
> slope     1.1686731 0.9194387 1.549080
```

where $u = n^{-1}\sum_{i=1}^{n} I(x_i = 1)$. Next, we estimate the sparsity, which, in the IID case, is $1/f_\varepsilon(Q_\varepsilon(p))$. One way to obtain it is to use Equation (8.13) evaluated at $\bar{\boldsymbol{x}} \equiv n^{-1}\sum_{i=1}^{n} \boldsymbol{x}_i$ (see Koenker, 2005, p. 141). Again, with binary x, this simplifies to $\bar{\boldsymbol{x}} = (1, u)^\top$. The rationale for using $\bar{\boldsymbol{x}}$ is that

$$\begin{aligned} \mathrm{d}\hat{Q}_{Y_i}(p \mid \bar{\boldsymbol{x}})/2h &= \{\mathrm{d}\hat{\beta}_0(p) + u\mathrm{d}\hat{\beta}_1(p)\}/2h \\ &= [(1-u)\mathrm{d}\hat{\beta}_0(p) + u\{\mathrm{d}\hat{\beta}_0(p) + \mathrm{d}\hat{\beta}_1(p)\}]/2h \\ &= (1-u)\mathrm{d}\hat{Q}_{Y_i}(p \mid \boldsymbol{w}_0)/2h + u\mathrm{d}\hat{Q}_{Y_i}(p \mid \boldsymbol{w}_1)/2h, \end{aligned}$$

where $\mathrm{d}\hat{\beta}_j(p) = \hat{\beta}_j(p+h) - \hat{\beta}_j(p-h)$ and $\boldsymbol{w}_j = (1, j)^\top$, $j = 0, 1$. That is, the sparsity evaluated at $\bar{\boldsymbol{x}}$ is a weighted average of the conditional sparsities evaluated at $x = 0$ and $x = 1$, with weights given by the respective sample proportions. It also turns out that the centroid of the design, $\bar{\boldsymbol{x}}$, is the point at which the QF is monotone in p. This guarantees that $\mathrm{d}\hat{Q}_{Y_i}(p \mid \bar{\boldsymbol{x}}) > 0$ and, hence, that the resulting sparsity estimate is positive.[5] In R Code 8.5, we used the HS bandwidth with $\alpha = 0.05$ as provided by the function `quantreg::bandwidth.rq` by setting the argument `hs = TRUE` (when `hs = FALSE` the function returns the Bofinger's bandwidth). The sparsity calculated at the three quantiles $p \in \{0.1, 0.5, 0.9\}$ is not much different from what we would have obtained by assuming a traditional L-shift model with normal errors as shown in the code fragment below (consistently with the analysis presented in Section 3.2.2).

```
sigma <- summary(lm(y ~ x))$sigma
1/dnorm(qnorm(p, 0, sigma), 0, sigma)
> [1] 33.84586 14.88910 33.84586
```

At the end of R Code 8.5, the standard errors of $\hat{\beta}0(p)$ and $\hat{\beta}1(p)$ are calculated, for each value of p, as the square roots of the diagonal elements of the variance-covariance matrix, and the corresponding 95% confidence intervals are subsequently obtained.

In practice, one would obtain the variance-covariance matrix directly from the `summary` function applied to the fitted quantile model as shown in R Code 8.6 (to extract this from the output, the argument `covariance` must be set to `TRUE`). With the option `"iid"`, the function will return standard errors under the IID assumption. However, differently from the procedure presented in R Code 8.5, `summary(..., se = "iid")` estimates the sparsity by making use of the residuals.[6] The basic idea is to regress the absolute residuals $z_i(p) = |\hat{r}_i(p)|$ against their relative ranks p_i as a smoothing device by means of, for example, least absolute deviations or least squares. Then the slope b of the model

$$z_i(p) = a + b \cdot p_i + \zeta_i, \qquad i = 1, \ldots, n,$$

is taken as the residual sparsity (i.e., as the derivative of the residual QF with respect to p). A caveat concerns the exclusion of the zero residuals that the fitting of quantile regression naturally brings about (see Koenker, 2005, p. 140).

This section concludes with the conditional equivalent of the generalisation in Equation (4.10), namely the case of multiple quantiles. For a sequence of quantile levels p_r, $r = 1, \ldots, m$, the estimators $\hat{\boldsymbol{\beta}}(p_1), \ldots, \hat{\boldsymbol{\beta}}(p_m)$ are consistent and asymptotically jointly normal. The finite-sample variance-covariance matrix, say $\boldsymbol{\Sigma}$, is given by a $(k+1)m \times (k+1)m$

[5] This is not guaranteed at points other than $\bar{\boldsymbol{x}}$. See *quantile crossing* in Section 8.3.5.

[6] Numerical differences between these two procedures should be expected, as evidenced by the discrepancy between the estimated standard errors in R Codes 8.5 and 8.6. However, such discrepancies will fade away as the sample size gets larger.

R Code 8.6: Obtaining the variance-covariance matrix with the **quantreg** package for regression quantiles estimated in R Code 8.4 and calculation of 95% confidence intervals in the **faithful** dataset.

```
[!t]

fit.s <- summary(fit, se = "iid", covariance = TRUE)
# extract variance-covariance matrix for each quantile
lapply(fit.s, function(x) x$cov)
> [[1]]
>                [,1]       [,2]
> [1,]  2.705567 -2.705567
> [2,] -2.705567  4.205224
>
> [[2]]
>                [,1]       [,2]
> [1,]  1.462785 -1.462785
> [2,] -1.462785  2.273586
>
> [[3]]
>                [,1]       [,2]
> [1,]  5.521565 -5.521565
> [2,] -5.521565  8.582090
# calculate 95% CI for p = 0.9
betahat <- coef(fit)[,3]
se <- sqrt(diag(fit.s[[3]]$cov))
betahat
> (Intercept)          x
>          63         25
cbind(betahat - qnorm(0.975)*se, betahat + qnorm(0.975)*se)
>                   [,1]     [,2]
> (Intercept) 58.39447 67.60553
> x           19.25825 30.74175
```

block matrix with rsth block

$$\frac{\min(p_r, p_s) - p_r p_s}{n} \boldsymbol{H}_n(p_r)^{-1} \boldsymbol{J}_n \boldsymbol{H}_n(p_s)^{-1}. \tag{8.14}$$

Note that (8.14) reduces to (8.11) for $p_r = p_s = p$.

Other methods

Alternative inferential methods for conditional quantiles include those based on ranks and resampling. The former are rooted in a large literature spanning early foundational contributions such as those by Wilcoxon (1945) and Mann and Whitney (1947), to later developments linking regression quantiles and ranks (Gutenbrunner and Jurečková, 1992). Resampling methods encompass several bootstrap schemes, each with different strengths and weaknesses, including the classical xy-pair bootstrap and wild bootstrap (Feng et al., 2011), as well as resampling based on pivotal estimating functions (Parzen et al., 1994).

A very comprehensive collection of methods is implemented in the `summary.rq` function in the `quantreg` package.

As discussed previously, order statistics and ranks are intimately connected with quantiles. Such a connection can be found when examining the solution to (8.7) and, more specifically, the corresponding dual program (see Section 5.2). The dual to the LP problem (8.7) is

$$\max_{\boldsymbol{a}\in[0,1]^n} \left\{\boldsymbol{y}^\top \boldsymbol{a} \mid \boldsymbol{X}^\top \boldsymbol{a} = (1-p)\boldsymbol{X}^\top \mathbf{1}_n\right\}. \tag{8.15}$$

The above linear program reduces to the one in (5.12) when the model has only an intercept (hence $\boldsymbol{X} \equiv \mathbf{1}_n$). The rankscores $\hat{a}_i$ provide the basis for building rank-based confidence intervals for $\boldsymbol{\beta}$, the details of which are not discussed here but can be found elsewhere (Koenker and Bassett, 1978, p. 91).

8.3.3 Hypothesis testing

We now pick up where we left off in Section 7.2 with the problem of testing two or more samples and multiple quantiles. On that occasion, we remarked that such a problem is more pertinent to the conditional setting and we now show why.

Suppose we have a continuous response variable Y and a single categorical covariate with $k+1$ levels, $k \geq 1$. Define $\boldsymbol{x} = (1, x_1, \ldots, x_k)^\top$, where x_j is the dummy variable for the $(j+1)$th category, $j = 1, \ldots, k$. The linear quantile regression model

$$Q_Y(p \mid \boldsymbol{x}) = \boldsymbol{x}^\top \boldsymbol{\beta}(p) = \beta_0(p) + \sum_{j=1}^{k} \beta_j(p) x_j \tag{8.16}$$

includes the intercept $\beta_0(p)$, which is the quantile p for the reference category, and k 'slopes' contrasting the pth quantile for the reference category against the pth quantile for the other categories. That is, $\beta_j(p)$, $j = 1, \ldots, k$ is nothing other than the QTE $\delta_j = \xi_{j+1,p} - \xi_{1,p}$, $j = 1, \ldots, k$ in Section 7.2.

Clearly, the advantage of working within a regression framework is that we can generalise (8.16) to covariates of any type. Another advantage is that we can also adjust the analysis for other variables as appropriate. So, suppose that $\boldsymbol{x}$ is now a vector of k generic variables (with a leading constant equal to 1) and let $\boldsymbol{\beta}(p)$ be the corresponding $(k+1)$ regression parameter, which includes the intercept $\beta_0(p)$. The (joint) null hypothesis of interest

$$\begin{aligned}
&H_{0,1} : \beta_1(p_1) = \beta_1(p_2) = \ldots = \beta_1(p_m)\\
&H_{0,2} : \beta_2(p_1) = \beta_2(p_2) = \ldots = \beta_2(p_m)\\
&\ldots\\
&H_{0,k} : \beta_k(p_1) = \beta_k(p_2) = \ldots = \beta_k(p_m)
\end{aligned} \tag{8.17}$$

is that *all* the k slopes are constant across the m quantiles p_r, $r = 1, \ldots, m$. Of course, if testing the joint hypothesis is too broad, one can investigate individual $H_{0,j}$s.

The test statistic is

$$T_f = \frac{1}{k(m-1)} \hat{\boldsymbol{\beta}}(p)^\top \boldsymbol{C}^\top (\boldsymbol{C}\hat{\boldsymbol{\Sigma}}\boldsymbol{C}^\top)^{-1} \boldsymbol{C}\hat{\boldsymbol{\beta}}(p), \tag{8.18}$$

where $\boldsymbol{C}$ is a $(m-1)\times m$ matrix of contrasts and $\hat{\boldsymbol{\Sigma}}$ is an estimate of the matrix $\boldsymbol{\Sigma}$ introduced in the previous section. Under H_0, the statistic T_f is assessed against an F-distribution with $k(m-1)$ and $nm - k(m-1)$ degrees of freedom. Given that this test is the distribution-free alter ego of the classical analysis-of-variance test, we will generically refer to (8.18) as an ANOVA test, correspondingly to the `R` function `anova.rq` in the `quantreg` package.

R Code 8.7: Summary statistics for the Childhood Respiratory Disease Study dataset.

```
summary(fev)
>        age              fev              height            sex
>   Min.    : 3.000   Min.    :0.791   Min.    :46.00   female:318
>   1st Qu.: 8.000   1st Qu.:1.981   1st Qu.:57.00   male   :336
>   Median :10.000   Median :2.547   Median :61.50
>   Mean    : 9.931   Mean    :2.637   Mean    :61.14
>   3rd Qu.:12.000   3rd Qu.:3.119   3rd Qu.:65.50
>   Max.    :19.000   Max.    :5.793   Max.    :74.00
```

We illustrate the application of (8.18) in the following example. Forced expiratory volume (FEV) in one second (l/s) was measured in 654 individuals aged 3 to 19 years, with heights ranging from 46 to 74 inches (117 to 188 cm), who were seen in the Childhood Respiratory Disease Study in East Boston, Massachusetts. Numerical and graphical summaries of the dataset are given in R Code 8.7 and Figure 8.6, respectively. One can appreciate there is a positive association of FEV with the three variables of interest: age (years), height (in) and sex. Also, these associations do not seem to be simply L-shift as evidenced, for example, by the funnel-shaped scatterplots.

The model for the pth conditional quantile of FEV is specified as follows:

$$Q_Y(p \mid \boldsymbol{x}) = \beta_0(p) + \beta_1(p)x_1 + \beta_2(p)x_2 + \beta_3(p)x_3, \tag{8.19}$$

where x_1 is age centred at the sample mean age (10 years), x_2 is height centred at the sample mean height (61 in) and x_3 is the indicator for male sex. Centring facilitates interpretation of the intercept, which corresponds to the conditional quantile for a female with average age and height. For our test of the L-shift effect of age, height and sex, we choose to estimate the quantiles $p \in \{0.1, 0.25, 0.5, 0.75, 0.9\}$, which provide a reasonable range to detect differences in location, scale and shape (see Section 3.1.4). The test is carried out with the function `quantreg::anova.rq`, which allows one to specify whether the test should be run jointly on all slopes or individually. The results in R Code 8.8 suggest rejecting, at the 10%, the

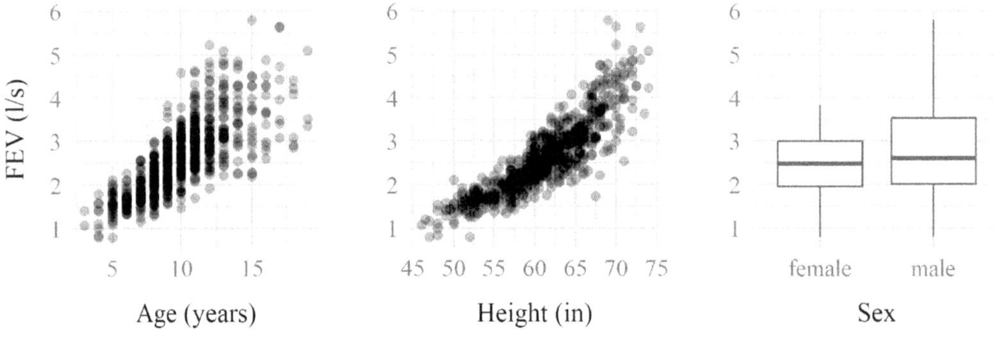

Figure 8.6
Left to right: plot of forced expiratory volume versus age, height and sex, respectively, using the Childhood Respiratory Disease Study dataset.

R Code 8.8: Regression quantiles for forced expiratory volume as a function of age, height and sex, with ANOVA test for equality of slopes, using the Childhood Respiratory Disease Study dataset.

```
fit <- rq(fev ~ I(age - 10) + I(height - 61) + sex,
    tau = c(0.1, 1:3/4, 0.9), data = fev)
fit
> Call:
> rq(formula = fev ~ I(age - 10) + I(height - 61) + sex, tau = c(0.1,
>     1:3/4, 0.9), data = fev)
>
> Coefficients:
>                    tau= 0.10  tau= 0.25  tau= 0.50  tau= 0.75  tau= 0.90
> (Intercept)      2.09514943 2.34065217 2.55727273 2.80644681 3.02705785
> I(age - 10)      0.05227586 0.04930435 0.04890909 0.06334043 0.05117355
> I(height - 61) 0.07852874 0.09191304 0.10672727 0.10778723 0.12409917
> sexmale          0.07239080 0.04243478 0.12827273 0.16680851 0.19367769
>
> Degrees of freedom: 654 total; 650 residual
anova(fit, joint = TRUE)
> Quantile Regression Analysis of Deviance Table
>
> Model: fev ~ I(age - 10) + I(height - 61) + sex
> Joint Test of Equality of Slopes: tau in {  0.1 0.25 0.5 0.75 0.9  }
>
>   Df Resid Df F value    Pr(>F)
> 1 12     3258  8.5213 4.441e-16 ***
> ---
> Signif. codes:  0 '***' 0.001 '**' 0.01 '*' 0.05 '.' 0.1 ' ' 1
anova(fit, joint = FALSE)
> Quantile Regression Analysis of Deviance Table
>
> Model: fev ~ I(age - 10) + I(height - 61) + sex
> Tests of Equality of Distinct Slopes: tau in {  0.1 0.25 0.5 0.75 0.9  }
>
>                  Df Resid Df F value    Pr(>F)
> I(age - 10)      4     3266  0.3702    0.82999
> I(height - 61)  4     3266  8.2627 1.255e-06 ***
> sexmale          4     3266  2.1055    0.07754 .
> ---
> Signif. codes:  0 '***' 0.001 '**' 0.01 '*' 0.05 '.' 0.1 ' ' 1
```

joint hypothesis, as well as the individual null hypotheses for height and sex. On the other hand, the test for age shows a large p-value, indicating that this variable has a L-shift effect on FEV.

Another test that can be used to investigate hypotheses about the type of relationship between the covariates and the quantiles of the response is the Khmaladze test (Koenker and Xiao, 2002). Here we give the gist of it and refer the reader to Koenker and Xiao (2002) and Koenker (2005) for technical details. The Khmaladze test considers the quantile process

$\{\hat{\boldsymbol{\beta}}(p) : p \in \mathcal{P}\}$ where $\mathcal{P} = [p_0, p_1] \subset (0, 1)$. The null hypothesis can be of two types, either that the effect of the covariates is L-shift or that the effect is LS-shift. The alternative hypothesis is that the effect is more complex than that of the null. The interval $[p_0, p_1]$ may be restricted to a specific region of the conditional distribution or cover (almost) entirely its range. In the latter case, since estimation on the tails can be troublesome in the presence of high sparsity, it is recommended not to go too far on the tails, for example, $\mathcal{P} = [0.05, 0.95]$.

The ANOVA and Khmaladze tests are primarily aimed at assessing whether the effects of the covariates vary across quantiles or along the quantile process. In practice, however, one is often faced with a broader model selection problem, namely whether the chosen linear predictor is adequate at a given quantile or whether additional terms should be introduced. In this respect, three complementary approaches deserve a brief mention.

First, the function `anova.rq` in the `quantreg` package can also be used in a more conventional model-comparison sense. Besides the multi-quantile setting discussed above, it allows one to compare nested models fitted at the same quantile, thereby testing whether a smaller model is adequate relative to a larger one in a likelihood-ratio-type framework (Koenker, 2005). Conversely, when the linear predictor is kept fixed, the same function can be used to compare fits across different quantiles. Thus, `anova.rq` is useful both for testing constancy of effects over p and for deciding whether additional covariates improve the fit at a specific quantile.

Second, one may assess adequacy of a fitted pth regression quantile by examining whether the proportion of negative residuals is indeed close to p across levels of the covariates. If the fitted model is correct, then the binary indicator

$$I\{y_i - \hat{Q}_Y(p \mid \boldsymbol{x}_i) < 0\}$$

should have conditional mean approximately equal to p, with no residual association with the explanatory variables. This suggests a simple diagnostic test based on logistic (binomial) regression of the indicator above on candidate covariates omitted from, or added to, the quantile model (Redden et al., 2004). A significant likelihood-ratio test in this auxiliary regression indicates that the fitted quantile fails to maintain the nominal proportion p uniformly over the design space, hence pointing to model misspecification or missing terms. An attractive feature of this approach is its simplicity and modest computational burden.

Third, one may resort to omnibus lack-of-fit tests based on cumulative sums of suitable score-like quantities. He and Zhu (2003) proposed a CUSUM process built from the gradient vector of the quantile objective function, with the aim of testing a parametric quantile regression model against general nonparametric alternatives. The resulting procedure does not require nonparametric smoothing, is suitable for both linear and nonlinear quantile regressions, and is consistent against broad classes of alternatives. In this sense, it plays a role analogous to classical omnibus specification tests in mean regression, but tailored to conditional quantiles. However, this approach can be computationally demanding, as the null distribution must be obtained via simulation; an implementation is available in the `Qtools::LOFTest` function.

These procedures should be viewed as complementary rather than competing. The ANOVA-type test is especially convenient for comparing nested parametric specifications, the binomial-regression approach directly checks whether the fitted quantile achieves the intended conditional coverage, and the CUSUM test provides a broader omnibus assessment of lack of fit.

Criteria such as the Akaike Information Criterion (AIC) and its variants are not directly derived in ordinary quantile regression, since estimation is not, in general, likelihood-based. Nonetheless, information criteria can be extended beyond the classical likelihood framework. In particular, Machado (1993) showed that Schwarz Information Criterion (SIC) can be formulated for general M-estimators (Huber, 1981) by replacing the log-likelihood with the

Table 8.1
Median, mean and cumulative probability at the mean of faecal calprotectin (mg/kg) in controls and disease-free first degree relatives (FDRs) of Crohn's disease patients.

Group	Sample size	Median	Mean	$F(\mu)$
Controls	9	21	20	0.44
FDRs	29	31	44	0.76

sample objective function, yielding consistent selection of the most parsimonious pseudo-true model under suitable conditions (see also White, 1982). In the context of quantile regression, related ideas arise from goodness-of-fit and specification testing procedures based on the regression quantile process, which provide global diagnostics over ranges of quantiles rather than relying on a single scalar criterion (Koenker and Machado, 1999).

8.3.4 Which quantiles?

A common question about the application of quantile regression is the determination of which quantiles are to be estimated. In one situation, the interest may be in a single quantile that is a robust alternative to the mean when in the presence of, say, non-normal features of the data. An obvious choice is the median as it sits in the middle of the distribution. In contrast, in a study specifically concerning obesity or stunting, one would naturally focus on, respectively, a higher or a lower quantile of the weight distribution, and ultimately select a conventional level (e.g., top or bottom 10%).

In the following example, we consider data from a clinical trial aimed at investigating early signs of Crohn's disease (CD), an inflammatory bowel condition, in first degree relatives (FDRs) of CD patients (Sorrentino et al., 2014). Faecal calprotectin (mg/kg – FC) is a surrogate marker of inflammation often used during treatment and as a diagnostic marker. In this particular study, disease-free FDRs were compared to healthy controls based on a variety of markers, including FC. Table 8.1 gives summary statistics for FC in the two groups while Figure 8.7 compares the estimated group-specific density using a Gaussian kernel with SJ bandwidth.[7]

The distribution of FC in the two groups shows heteroskedasticity, bimodality and extreme skewness, with values ranging from 6 to 34 mg/kg in the control group, and from 5 to 191 mg/kg in the FDR group. The ECDF calculated at the mean (denoted by $F(\mu)$ in Table 8.1) gives 0.44 for the controls and 0.76 for the FDRs. It is fair to say that a t-test would be rather inappropriate in this case. Instead, we can test the null hypothesis that the median FC is equal in the two groups by fitting

$$Q_Y(0.5 \mid x) = \beta_0(0.5) + \beta_1(0.5)x,$$

where Y denotes FC and x is a group indicator ($x = 0$, if control; $x = 1$, if FDR). Hence, $\beta_1(0.5)$ is interpreted as the difference in median FC between the two groups. According to its p-value as shown in R Code 8.9, we fail to reject, at that study significance level of 5%, the null hypothesis that the group medians are equal. R Code 8.9 shows also the results of both the Wilcoxon rank-sum test and the Student's t test, which seem to be on the liberal

[7]Since we treated FC as a strictly positive variable, we used an approach similar to that in R Code 7.4 by log-transforming FC prior to estimation, and then by back-transforming the output, that is, $f_Y(\exp(z)) = f_Z(z)/y$, where $z = \log(y)$ and Y denotes FC.

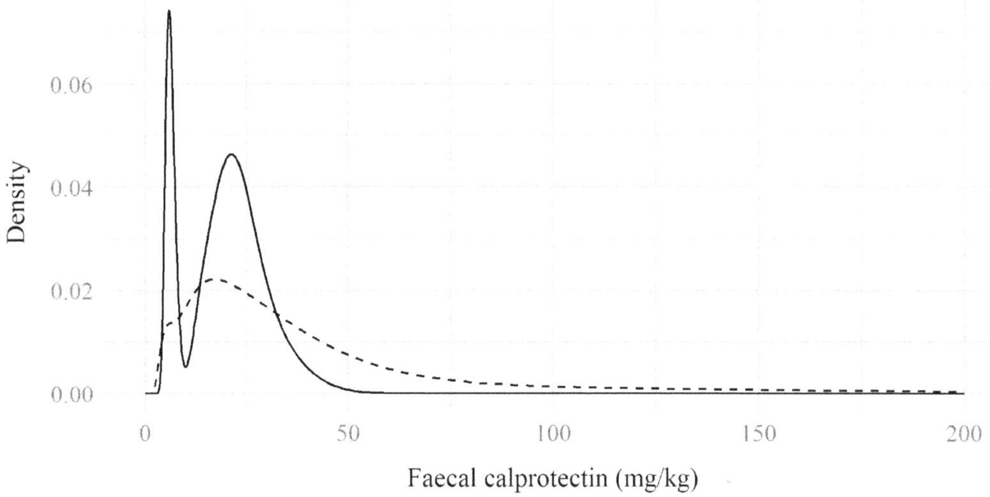

Figure 8.7
Estimated density of faecal calprotectin in healthy controls (solid line) and disease-free first degree relatives of Crohn's disease patients (dashed line).

side. The test of the median effect $\beta_1(0.5)$ accounts for the different distributions in the two groups by assuming that the errors are not identically distributed (`summary(..., se = "nid")`). On the other hand, the t-test with the option `var.equal = FALSE` accounts for heteroskedasticity but not for shape differences. Finally, the Wilcoxon test, like the median test, is nonparametric and makes no assumption on the shape of the distribution, but the null hypothesis is not specific to differences in medians,[8] hence its interpretation is not as unequivocal as for the median test. The contrasting results of this analysis highlights the importance of clearly stating at the outset the hypothesis to be tested and the associated distributional assumptions.

In the second example, we consider small-for-gestational age (SGA), a risk factor for infant mortality and morbidity (Boghossian et al., 2018). An infant is defined to be SGA when their birth weight is less than the 10th percentile of the birth weight distribution for infants born at the same gestational age in the reference population (which is typically determined in relation to sex and birth multiplicity). Therefore the definition of SGA requires conditional quantiles. Let Y represent birth weight (g) and $\boldsymbol{x}$ a vector that includes gestational age (weeks), among other variables. Then SGA is defined as the variable Z such that

$$Z = I(Y < Q_Y(0.1 \mid \boldsymbol{x})) = I(Y - Q_Y(0.1 \mid \boldsymbol{x}) < 0). \tag{8.20}$$

In R Code 8.10, we estimate the 10th birth weight percentile, conditional on gestational age (weeks) and sex, for about 3.3 million singletons using the NCHS Births dataset. Gestational age is categorised into four groups (e.g., see World Health Organization definition): (i) at term (after 37 weeks), (ii) moderate-to-late preterm (32 to 37 weeks), (iii) very preterm

[8]Contrary to a common misconception, the null hypothesis of the Wilcoxon rank-sum test is *not* that the medians of the two populations are equal but, rather, that the probability of one random variable being greater than the other is 0.5, that is, $H_0 : \Pr(Z_0 > Z_1) = 0.5$ where Z_j is the conditional variable $Y \mid x = j$. This hypothesis may or may not be consistent with equality in medians depending on the shape of the conditional distributions (see, e.g., Conroy, 2012).

R Code 8.9: Test of the difference in median faecal calprotectin (mg/kg) between controls and disease-free first-degree relatives (FDRs) of Crohn's disease patients.

```
summary(rq(FC ~ group, data = crohn), se = "nid")
>
> Call: rq(formula = FC ~ group, data = crohn)
>
> tau: [1] 0.5
>
> Coefficients:
>             Value    Std. Error t value  Pr(>|t|)
> (Intercept) 21.00000  5.40686    3.88395  0.00042
> groupFDR    10.40000  7.48275    1.38986  0.17311
wilcox.test(FC ~ group, data = crohn)
>
>  Wilcoxon rank sum test with continuity correction
>
> data:  FC by group
> W = 73, p-value = 0.05029
> alternative hypothesis: true location shift is not equal to 0
t.test(FC ~ group, data = crohn, var.equal = FALSE)
>
>  Welch Two Sample t-test
>
> data:  FC by group
> t = -2.8438, df = 34.098, p-value = 0.007483
> alternative hypothesis: true difference in means between group control
> and group FDR is not equal to 0
> 95 percent confidence interval:
>  -42.532111  -7.080916
> sample estimates:
> mean in group control     mean in group FDR
>              19.55556              44.36207
```

(28 to less than 32 weeks), and (iv) extremely preterm (less than 28 weeks).[9] We see that, compared to babies born at term, those born preterm have a lower threshold for SGA, with an increasingly bigger difference of up to -2441 g for the extremely preterm infants. We can also observe that the SGA threshold is lower for girls compared to boys. With the help of the function `quantreg::residuals`, we can obtain for each infant, conditionally on their gestational age and sex, the residuals $y - \hat{Q}_Y(0.1 \mid \boldsymbol{x})$ which we use to create the SGA indicator variable as defined in Equation (8.20). The NCHS provides several other variables related to the infants and their mothers. In the final part of R Code 8.10, odds ratios and 95% confidence intervals, estimated by means of logistic regression, are shown for infant mortality in relation to gestational age, sex, SGA, and maternal health lifestyle during pregnancy with four categories: (i) no alcohol use or tobacco smoking, (ii) alcohol use only, (iii) tobacco smoking only, and (iv) alcohol use and tobacco smoking. Naturally,

[9] It would be more efficient to include gestational age as a continuous variable. However, given its nonlinear relationship with birth weight, we refer the reader to Chapter 9 for more advanced models.

R Code 8.10: Small-for-gestational and mortality among singletons born in the United States, 2001.

```
# singletons only
dd <- subset(USbirths2001, USbirths2001$multiple == 0)
# some missing ga_group and tox
dd <- na.omit(dd[,c("bw", "ga_group", "female", "tox", "died", "cause")])
# about 3.3 million
nrow(dd)
> [1] 3341744
# quantile regression at p = 0.1 (note method Frisch-Newton for large size)
fit.rq <- rq(bw ~ ga_group + female, data = dd, method = "fn", tau = 0.1)
summary(fit.rq, se = "nid")
>
> Call: rq(formula = bw ~ ga_group + female, tau = 0.1, data = dd,
>          method = "fn")
>
> tau: [1] 0.1
>
> Coefficients:
>                    Value       Std. Error  t value     Pr(>|t|)
> (Intercept)        2891.00000     1.06508  2714.35901     0.00000
> ga_group(17,27] -2441.00000    12.06183  -202.37400     0.00000
> ga_group(27,31] -1864.00000     0.80681 -2310.34313     0.00000
> ga_group(31,36]  -793.00000     1.90840  -415.53187     0.00000
> female            -113.00000     1.12109  -100.79434     0.00000
# define sga as indicator of negative residuals
dd$sga <- as.numeric(residuals(fit.rq) < 0)
# logistic regression for mortality
fit.glm <- glm(died ~ ga_group + female + sga + tox,
   family = "binomial", data = dd)
round(exp(cbind(OR = coef(fit.glm), confint.default(fit.glm))), 3)
>                                OR   2.5 %  97.5 %
> (Intercept)                 0.002   0.002   0.002
> ga_group(17,27]           299.363 288.111 311.056
> ga_group(27,31]            20.425  19.325  21.588
> ga_group(31,36]             3.546   3.392   3.707
> female                      0.854   0.828   0.881
> sga                         4.965   4.790   5.146
> toxalcohol                  1.429   1.373   1.488
> toxsmoking                  1.102   0.830   1.462
> toxalcohol and smoking      1.438   1.203   1.718
```

extremely preterm infants are at the highest risk of death. Notably, SGA quintuplicates the odds of not reaching one year of age, independently from the gestational age and the other factors. The deleterious effects of alcohol use do not seem to be paralleled by those of smoking, while being a female is confirmed to be a protective factor.

In the previous examples, the choice of the quantile to estimate was rather immediate. When this is not the case, the preference goes often towards quartiles ($p \in \{0.25, 0.5, 0.75\}$),

R Code 8.11: Regression quantiles for birth weight as a function of cause of death, with ANOVA test for equality of slopes, among singletons born in the United States, 2001.

```
> Call:
> rq(formula = bw ~ cause, tau = c(0.05, 0.33, 0.66, 0.95), data = dd,
>     method = "fn")
>
> Coefficients:
>               tau= 0.05 tau= 0.33 tau= 0.66 tau= 0.95
> (Intercept)        2438      3147      3572      4196
> causelbw/pt       -2183     -2750     -3052     -3431
> causesids          -709      -284      -227      -170
> causeother        -2098     -2353      -912      -426
>
> Degrees of freedom: 3341744 total; 3341740 residual
> Quantile Regression Analysis of Deviance Table
>
> Model: bw ~ cause
> Tests of Equality of Distinct Slopes: tau in {  0.05 0.33 0.66 0.95  }
>
>             Df Resid Df   F value    Pr(>F)
> causelbw/pt  3 13366973 17220.090 < 2.2e-16 ***
> causesids    3 13366973    27.091 < 2.2e-16 ***
> causeother   3 13366973  9578.433 < 2.2e-16 ***
> ---
> Signif. codes:  0 '***' 0.001 '**' 0.01 '*' 0.05 '.' 0.1 ' ' 1
```

deciles ($p \in \{0.1, 0.2, \ldots, 0.9\}$) or a combination thereof. Whatever the choice, it is recommended to consider a good range of quantiles so as to be able to examine the location, scale and shape of the conditional distribution (e.g., $p \in \{0.1, 0.25, 0.5, 0.75, 0.9\}$). How far one should go on the tails is first and foremost a matter of the local density of the observations (e.g., see Section 6.3.3). Moreover, a test for the equality of the slopes (Section 8.3.3) is needed to support the conclusions of the analysis. An illustration is given in the next example.

Comparing birth weights of infants, children or even adults who developed specific conditions or died from certain causes to those in the general population may offer important insights into the aetiological role of factors at different stages of gestation (Birch et al., 2010). However, in general there are no predetermined quantiles to analyse. In R Code 8.11, we show a quantile regression analysis of the NCHS data discussed in the preceding example where five quantiles ($p \in \{0.1, 0.25, 0.5, 0.75, 0.9\}$) of birth weight are compared among infants who survived past one year of age and those who did not. The latter are grouped into 3 categories based on the underlying cause of death as given by the International Classification of Diseases, Tenth Revision (ICD-10) code, namely, extremely low birth weight or extreme immaturity (P07.0, P07.2), sudden infant death syndrome (SIDS, R95) and other causes of death.

Unsurprisingly, all the regression quantiles for infants who died of low birth weight or extreme immaturity are negative. This is also true for SIDS and other causes of death. If we look more specifically at the distribution of birth weights for the SIDS group, this appears

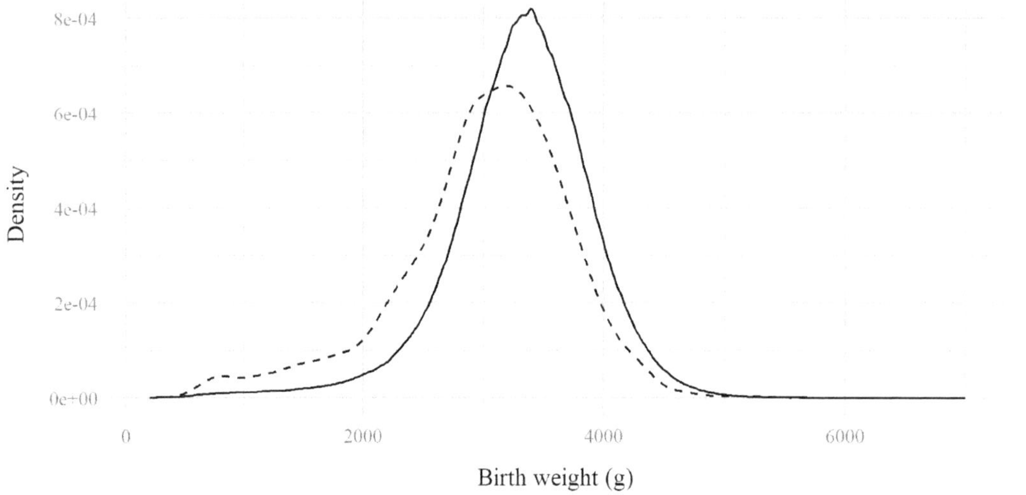

Figure 8.8
Estimated density of birth weights in infants who survived past one year of age (solid line) and infants who died of sudden infant death syndrome (dashed line) using the NCHS Births dataset.

left-shifted more heavily at lower quantiles than at higher quantiles (a visual aid is given in Figure 8.8). The ANOVA test unequivocally rejects the L-shift hypothesis for all groups.

We conclude this section with the analysis of the Childhood Respiratory Disease Study dataset introduced in Section 8.3.3. The model is specified as in Equation (8.19). Now suppose that there is a general interest in the entire quantile process to explore if and for which p any of age, height and sex has a particular association with FEV quantiles. In R Code 8.12, we run the Khmaladze test, restricted to the interval $[\epsilon, 1-\epsilon]$ with $\epsilon = 0.05$, for the null hypothesis that age, height, and sex have a L-shift effect (i.e., constant QTE) on FEV against the alternative hypothesis of a more complex effect. The joint test is significant at the 5% level, while the tests for individual slopes signals that only height is significant at the 10% level. In contrast, neither the joint nor the single-slope Khmaladze tests reject the LS-shift hypothesis at the 10% level. In summary, the tests point to a L-shift effect of age and sex, and to a LS-shift effect of height. These results are broadly consistent with the ANOVA findings in R Code 8.8.

The corresponding quantile processes, as estimated in the final part of R Code 8.12 with the command `rq(..., tau = -1)`, are shown in Figure 8.9. Overall, the algorithm yields 942 breakpoints. Starting from the intercept, we notice that its graph reminds us of the shape of a typical QF. Indeed, the predicted QF of FEV for the reference individual (a ten-year-old girl who is 61 inches tall) is nothing other than the intercept quantile process $\{\hat{Q}_Y(p \mid \boldsymbol{x} = \mathbf{0}) = \hat{\beta}_0(p) : p \in [0,1]\}$. A close inspection of the graph in the [0.9,1] interval, which is highlighted in the figure inset, reveals occasional dips (i.e., quantile crossing).[10] The quantile process for age, $\{\hat{\beta}_1(p) : p \in [0,1]\}$, moves up and down within a fairly narrow range for the most part of the p range, which is consistent with the L-shift hypothesis. The conspicuous jumps at the tails are a natural consequence of the local sparsity. In contrast, the quantile process for height, $\{\hat{\beta}_2(p) : p \in [0,1]\}$, does show a non-constant pattern that

[10]This is to be expected at such a fine grid of quantile levels and can be easily remedied with the rearrangement operator (see Section 8.3.5).

R Code 8.12: Estimation and testing of quantile regression processes in the Childhood Respiratory Disease Study dataset.

```
# test location-shift hypothesis
ff <- fev ~ I(age - 10) + I(height - 61) + sex
kt <- KhmaladzeTest(ff, tau = 1:19/20, nullH = "location", data = fev)
Qtools::KhmaladzeFormat(kt, epsilon = 0.05)
> Khmaladze test for the location-shift hypothesis
> Joint test is significant at 5% level
> Test(s) for individual slopes:
> I(age - 10) not significant at 10% level
> I(height - 61) significant at 10% level
> sexmale not significant at 10% level
# test location--scale-shift hypothesis
kt <- KhmaladzeTest(ff, tau = 1:19/20, nullH = "location-scale", data = fev)
Qtools::KhmaladzeFormat(kt, epsilon = 0.05)
> Khmaladze test for the location-scale-shift hypothesis
> Joint test is not significant at 10% level
> Test(s) for individual slopes:
> I(age - 10) not significant at 10% level
> I(height - 61) not significant at 10% level
> sexmale not significant at 10% level
# estimate the quantile process
fit.rqs <- rq(ff, tau = -1, data = fev) # use plot(fit.rqs) for plotting
# number of breakpoints
with(fit.rqs, ncol(sol))-1
> [1] 942
```

is statistically consistent with a LS-shift effect: at lower quantiles an increase in height of 1 inch has a smaller effect on FEV than at a higher quantile. Considering that lung volume correlates with the cube of height (Hepper et al., 1960; Cook and Hamann, 1961), it not surprising to see such an effect. Indeed, on the log scale, the Khmaladze test does not reject the L-shift hypothesis for height at the 10% significance level (R Code 8.13).

R Code 8.13: Testing of the quantile regression processes (log-response scale) in the Childhood Respiratory Disease Study dataset.

```
gg <- I(log(fev)) ~ I(age - 10) + I(height - 61) + sex
kt <- KhmaladzeTest(gg, tau = 1:19/20, nullH = "location", data = fev)
Qtools::KhmaladzeFormat(kt, epsilon = 0.05)
> Khmaladze test for the location-shift hypothesis
> Joint test is not significant at 10% level
> Test(s) for individual slopes:
> I(age - 10) not significant at 10% level
> I(height - 61) not significant at 10% level
> sexmale not significant at 10% level
```

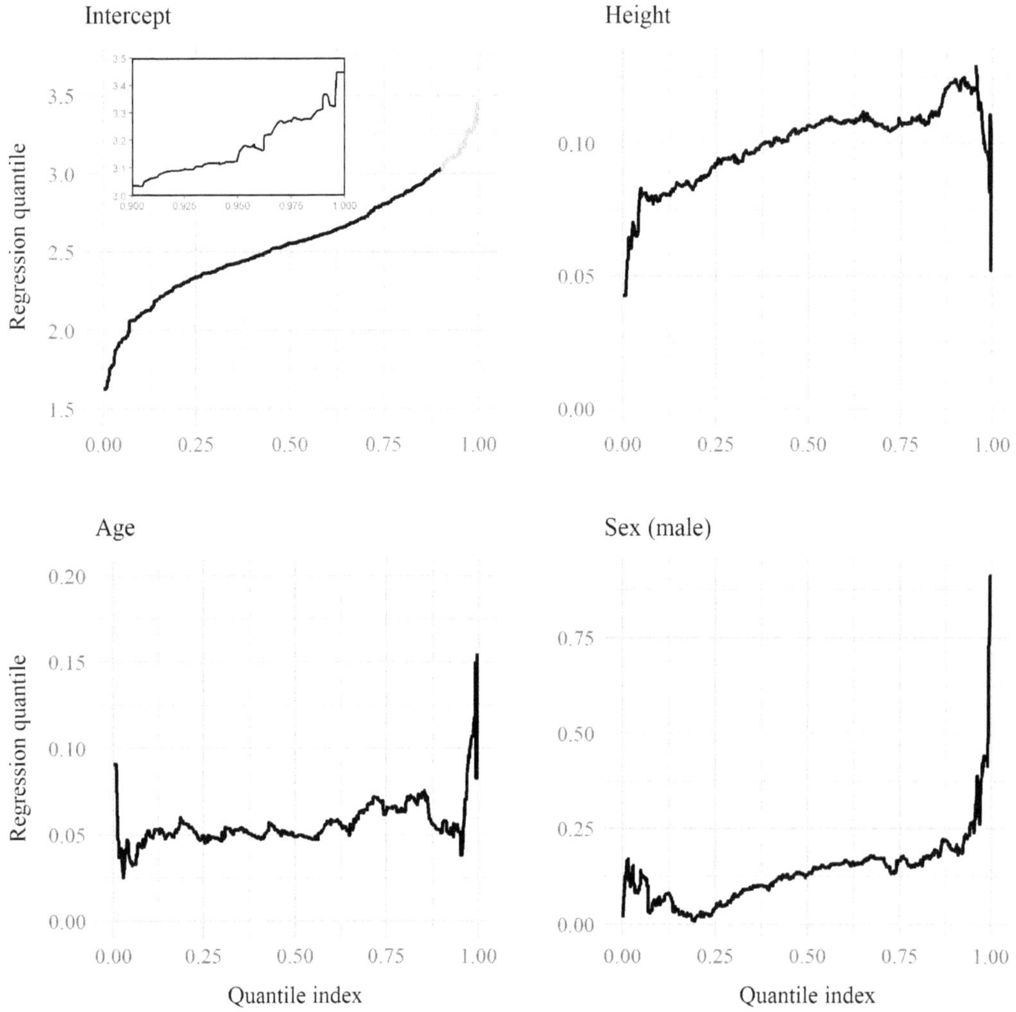

Figure 8.9
Quantile regression processes obtained from the analysis of forced expiratory volume as a function of age, height and sex using the Childhood Respiratory Disease Study dataset. The inset in the intercept panel highlights the behaviour of the estimated intercept coefficient for quantile indices between 0.9 and 1.

Lastly, concerning the quantile process $\{\hat{\beta}_3(p) : p \in [0,1]\}$ related to sex (Figure 8.9), there are hints that it increases over p, although not significantly in a statistical sense according to the Khmaladze test.

8.3.5 Potential pitfalls and remedies

This section deals with common difficulties that arise in quantile regression analyses, namely quantile crossing, sparsity, non-uniqueness of the solution, and rounding.

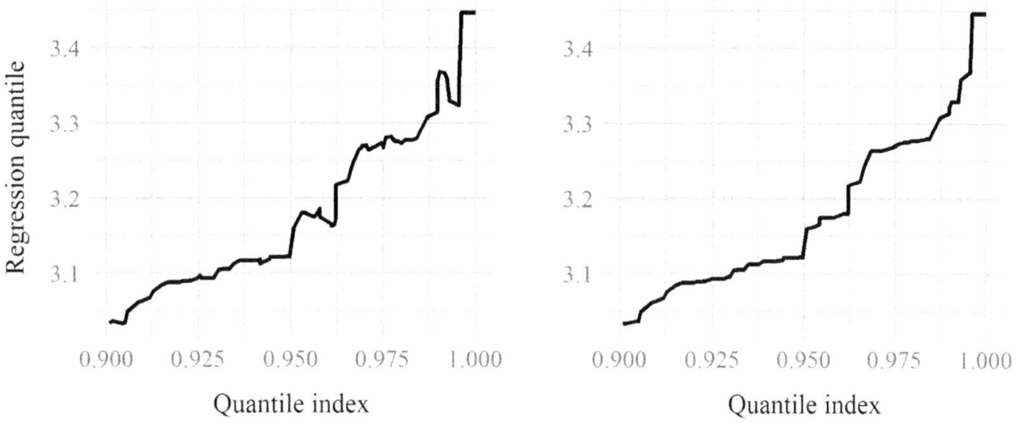

Figure 8.10
Quantile crossing (left) and rearranged quantiles (right) in the Childhood Respiratory Disease Study dataset.

Let's start with the quantile crossing issue. It is natural to ask that the conditional QF $Q_Y(p \mid \boldsymbol{x})$, like any QF, is nondecreasing with respect to p. However, it can happen that $\hat{Q}_Y(p_r \mid \boldsymbol{x}) > \hat{Q}_Y(p_s \mid \boldsymbol{x})$ for some $p_r < p_s$. As remarked in Section 8.3.1, if multiple quantiles are estimated *sequentially* without any constraint on monotonicity, then nothing prevents quantile crossing from happening. Some consolation comes from the observation that such a violation typically occurs in outlying regions of the design space (or even outside the convex hull of the design matrix) where sparsity is most common (Koenker, 2005, p. 56). This is not necessarily a bad thing if we look at it as a warning sign of a possible misspecification of the model, or as an excessive faith in the amount of information available in the data, or a combination of both. We now consider a few examples.

We had a first glimpse of quantile crossing in Figure 8.9 with the quantile process $\{\hat{\beta}_0(p)\}$. With such a fine grid of quantiles, it is almost inevitable to observe occasional dips over the range of p. Luckily, the rearrangement operator `quantreg::rearrange` (Chernozhukov et al., 2010) allows one to sort (rearrange) monotonically the original non-monotone predictions into a monotone rearranged curve. The latter is shown to be closer to the true quantile curve than the original curve in finite samples (Chernozhukov et al., 2010). Figure 8.10 compares the estimated QF $\hat{Q}_Y(p \mid \boldsymbol{x} = \mathbf{0})$, limited to the range $p \in [0.9, 1]$, and the rearranged function obtained with the code fragment shown below.

```
p <- fit.rqs$sol["tau",]
Q0hat <- fit.rqs$sol["(Intercept)",]
sf <- rearrange(stepfun(p, c(-Inf, Q0hat)))
Q0hat <- sf(p)
plot(p, Q0hat, type = "l", xlim = c(0.9,1), ylim = c(3, 3.5))
```

Now we consider cases when only a limited number of quantiles are estimated and wonder about the validity of the interpretation of the estimated coefficients in general. This situation is more insidious than the previous in some respects. Consider again the

R Code 8.14: Quantile crossing in the Childhood Respiratory Disease Study dataset.

```
p <- c(0.1, 1:3/4, 0.9)
fit <- rq(fev ~ I(age - 10) + I(height - 61) + sex, tau = p, data = fev)
Qhat <- predict(fit)
colnames(Qhat) <- p
print(sel <- which(t(apply(Qhat, 1, diff)) < 0, arr.ind = TRUE))
>       row col
> [1,] 222   2
fev[sel[1],]
>     age   fev height    sex
> 222   3 1.072     46 female
Qhat[sel[1],]
>       0.1      0.25       0.5      0.75       0.9
> 0.5512874 0.6168261 0.6140000 0.7462553 0.8073554
```

Childhood Respiratory Disease Study and say we fit model (8.19) for five quantiles, $p \in \{0.1, 0.25, 0.5, 0.75, 0.9\}$, as in R Code 8.8. Is there any crossing at some values of $\boldsymbol{x}$? This check is immediate for $\hat{\beta}_0(p)$ as these are the estimated quantiles of Y at $\boldsymbol{x} = \mathbf{0}$. On the other hand, we cannot immediately tell if there is a problem with the other coefficients as these are, in general, regression slopes.[11] Not to worry, because, as shown in R Code 8.14, it is sufficient to compute predictions of the conditional quantiles for the entire dataset using `quantreg::predict.rq`, and then check for violations of monotonicity in p by inspecting negative first differences of $\hat{Q}_Y(p \mid \boldsymbol{x}_i)$ conditional on $\boldsymbol{x}_i$. In our example, such violations occur for only one value of $\boldsymbol{x}$ between the first quartile and the median, and are confined to the lower tail of the age and height distribution, with a very small negative first difference, negligible in practice.

In the third and final act of the quantile crossing drama, we consider a situation in which, although quantile crossing may be considered more problematic, a simple solution is within reach. We use the `cars` dataset, containing 50 measurements of the speed (mph) of cars and the distances (ft) taken to stop. Suppose we estimate a simple linear model $Q_Y(p \mid x) = \beta_0(p) + \beta_1(p)x$ for 19 vigintiles of the distance Y as a function of the speed (x) as shown in R Code 8.15. The predicted conditional quantiles are depicted in the left plot of Figure 8.11. Quantile crossing appears to have gone berserk and, to be fair, not unexpectedly considering the observations are few and far between. When dealing with a linear model that in all likelihood is LS-shift at most (as it seems to be the case as suggested by the Khmaladze test in R Code 8.15), a remedy to the crossing of the quantile curves is offered by restricted quantile regression (He, 1997). Basically, it imposes a LS-shift regression model that satisfies monotonicity in a natural way. It should be stressed that, differently from the rearrangement operator discussed previously, restricted quantile operates on the $\hat{\boldsymbol{\beta}}(p)$ itself, not on the predictions $\hat{Q}_Y(p)$. Model fitting, which proceeds in three steps (see He, 1997, for details), is implemented in the package `Qtools`. Its application to the `cars` dataset is illustrated in R Code 8.15 while the results are given in the right plot of Figure 8.11. Asymptotic properties of the restricted estimator are discussed by Zhao (2000).

[11] If we fitted a model without an intercept, then the interpretation of the coefficients for the levels of a categorical variable would not be that of a contrast with the reference level. For example, the estimate of $\beta_1(p)$ one would obtain from `rq(fev ~ 0 + sex, data = fev)` is no longer the quantile treatment effect of male sex, but it is the actual estimated pth quantile of FEV for males, $\hat{Q}_Y(p \mid x = 1)$.

R Code 8.15: Quantile crossing in the `cars` dataset.

```
# standard quantile regression
fit.rq <- rq(dist ~ speed, tau = 1:19/20, data = cars)
Qhat <- predict(fit.rq)

# total number of quantile crossings across 19 vigintiles
sum(t(apply(Qhat, 1, diff)) < 0)
> [1] 23
# test location-scale-shift hypothesis
kt <- KhmaladzeTest(dist ~ speed, tau = 1:19/20,
nullH = "location-scale", data = cars)
Qtools::KhmaladzeFormat(kt, epsilon = 0.05)
> Khmaladze test for the location-scale-shift hypothesis
> Joint test is not significant at 10% level
> Test(s) for individual slopes:
>  not significant at 10% level
# restricted quantile regression
fit.rrq <- Qtools::rrq(dist ~ speed, tau = 1:19/20, data = cars)
```

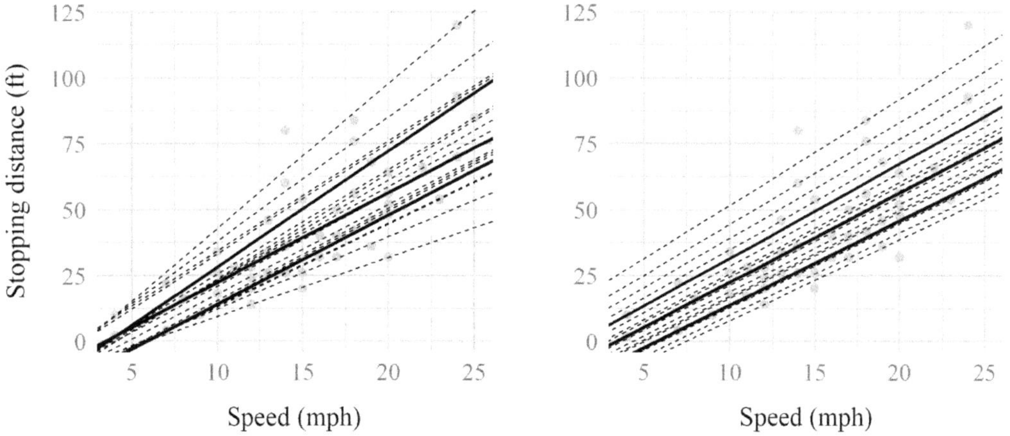

Figure 8.11
Quantile crossing (left) and restricted quantiles (right) in the `cars` dataset. Each line represents one of the 19 vigintiles with quartiles emphasised using solid lines.

R Code 8.16: Nonpositive estimates of sparsity using the `cars` dataset.

```
summary(rq(dist ~ speed, data = cars, tau = 0.975), se = "nid")

> Warning in summary.rq(rq(dist ~ speed, data = cars, tau = 0.975), se =
"nid"): 5 non-positive fis

>
> Call: rq(formula = dist ~ speed, tau = 0.975, data = cars)
>
> tau: [1] 0.975
>
> Coefficients:
>              Value     Std. Error t value   Pr(>|t|)
> (Intercept) -12.00000  36.63431   -0.32756   0.74467
> speed         5.50000   1.93077    2.84861   0.00645
h <- bandwidth.rq(0.975, nrow(cars), hs = TRUE)/2
Qhat <- predict(rq(dist ~ speed, data = cars, tau = c(0.975-h, 0.975+h)))
dQ <- apply(Qhat, 1, diff)
dQ[which(!(dQ > 0))]
> [1]  0.0  0.0  0.0  0.0 -1.5
cars[which(!(dQ > 0)),]
>    speed dist
> 46    24   70
> 47    24   92
> 48    24   93
> 49    24  120
> 50    25   85
```

Another issue related to quantile crossing concerns the situation when the estimate $\hat{q}(p) = 1/\hat{f}(Q(p))$, which is needed to calculate the standard errors for $\hat{\boldsymbol{\beta}}(p)$, is not positive. This occurs if $\boldsymbol{x}_i^\top\{\hat{\boldsymbol{\beta}}(p+h) - \hat{\boldsymbol{\beta}}(p-h)\} \leq 0$ in the application of the estimator in (8.13). One practical solution is to replace $\hat{f}$ by its positive part $\hat{f}^+ = \max\{0, \hat{f}\}$ (Koenker, 2005, p. 80).[12] Otherwise, one could try addressing quantile crossing directly. R Code 8.16 shows an example of what happens when calculating NID standard errors at a rather extreme quantile for which it is more likely to incur in `nonpositive fis` ($\hat{f}_i$s). In this example, 5 out of 50 estimates $\hat{f}_i$s are either zero or negative, all at the far end of the design space.

One aspect of quantile regression that oftentimes prompts concern and criticism is the non-uniqueness of the solution to the minimisation problem (8.6). This is a natural consequence of the absolute deviations loss. As opposed to the limiting loss function, which is smooth when the response is absolutely continuous, in finite samples the loss function may be piecewise constant for some intervals of p. Multiple solutions occur when, loosely speaking, the objective function is 'flat' at the minimum.[13] Non-uniqueness is a finite-sample artifact arising from the piecewise linear nature of the objective function; under standard regularity conditions, the population objective is strictly convex at the solution, ensuring

[12] This is the solution implemented in the function `quantreg::summary.rq`.

[13] More technically, the solution is unique when the LP objective function meets the polyhedral constraint set at only one vertex.

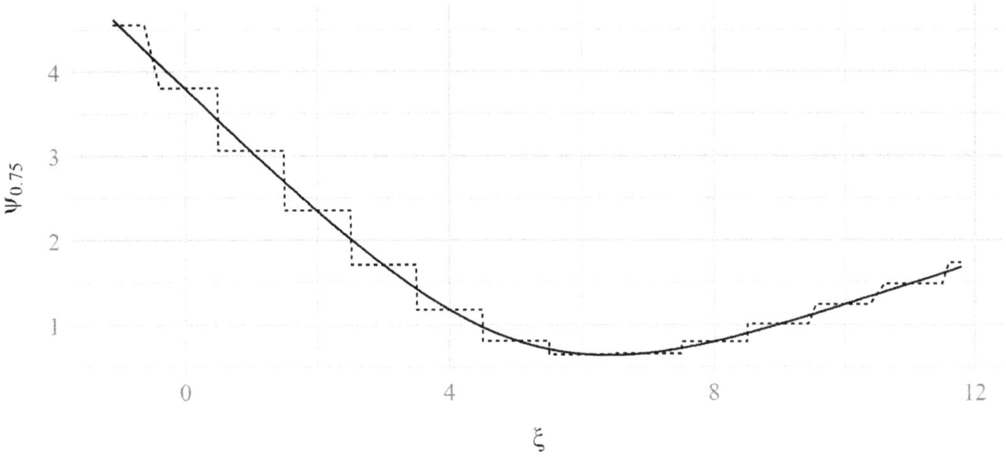

Figure 8.12
Quantile loss function with data randomly generated from a continuous distribution before (solid) and after (dashed) rounding.

asymptotic uniqueness provided that the corresponding population quantile is unique. If, on the other hand, the population quantile is an interval, then it is perfectly normal to look for set-valued estimates. See the discussion by Portnoy and Mizera (1998) in Ellis (1998) as well as Koenker (2005, p. 118).

The non-uniqueness of regression quantiles spontaneously arises when the response is continuous but heavily rounded.[14] Continuous measurements are necessarily discretised, but aggressive rounding may in fact force the loss function to be piecewise constant even with a large sample size (e.g., consider the gestational age data in Section 4.2). An illustration is given in Figure 8.12 where, given a sample of observations y_i, $i = 1, \ldots, 2000$, randomly generated from $\mathcal{N}(5, 4)$, the loss function $2000^{-1} \sum_i^{2000} \psi_{0.75}(y_i - \xi)$ is plotted against the values of the parameter ξ. The smooth appearance of the curve obtained with raw observations (solid line) is in stark contrast with the curve obtained after rounding the observations to the nearest integers (dashed line). No value of n can remedy the potential harm of rounding.

In this circumstance, one can artificially 'reconstruct' the original variable by adding random noise (jitter) to the discretised values. Let y_r denote the variable y rounded to the nearest integer. Independently generate R jittered variables $y^* = y_r + u$, where u is uniform between -0.5 and 0.5, and, for each variable, calculate the pth sample quantile $\tilde{\xi}_p^{(j)}$, $j = 1, \ldots, R$. Finally, average the noise out with $R^{-1} \sum_{j=1}^{R} \tilde{\xi}_p^{(j)}$. The latter will yield an estimate that, so to speak, recoups the lost continuity. A numerical example is given in R Code 8.17 with $n = 100$ observations sampled from $\mathcal{N}(5, 4)$, quantile $p = 0.75$ and $R = 1000$ jittered samples.

[14]All the more so if the response is genuinely discrete. Quantile regression with discrete responses is discussed in Chapter 10.

R Code 8.17: Jittering for quantile estimation of a rounded variable.

```
set.seed(123)
nn <- 100
p <- 0.75
R <- 1000
y <- sort(rnorm(nn, 5, 2))
yr <- round(y)
ystar <- replicate(R, yr + runif(nn, -0.5, 0.5))

# original
quantile(y, probs = p, names = FALSE)
> [1] 6.383638
# rounded
quantile(yr, probs = p, names = FALSE)
> [1] 6
# jittered
mean(apply(ystar, 2, quantile, probs = p))
> [1] 6.406889
```

9

Nonlinear quantile regression

9.1 Nonlinear parametric models

Nonlinear regression models are those in which the parameters enter the model nonlinearly. In nonlinear (parametric) quantile regression, the model assumes a relationship of the type

$$Q_{Y_i}(p \mid \boldsymbol{x}_i) = g\left(\boldsymbol{\beta}(p); \boldsymbol{x}_i\right), \qquad i = 1, \ldots, n, \tag{9.1}$$

where the $(k+1) \times 1$ vector $\boldsymbol{\beta}(p) = (\beta_0(p), \beta_1(p), \ldots, \beta_k(p))^\top$ collects the regression coefficients associated with $\boldsymbol{x}_i = (1, x_{i,1}, \ldots, x_{i,k})^\top$ (e.g., see Oberhofer, 1982; Koenker and Park, 1996). The function g is nonlinear with respect to $\boldsymbol{\beta}(p)$ or some of its components. The observations are assumed to be independent (though, of course, not necessarily identically distributed). Estimating $\boldsymbol{\beta}(p)$ in model (9.1) requires solving

$$\min_{\boldsymbol{\beta} \in \mathbb{R}^{k+1}} \frac{1}{n} \sum_{i=1}^{n} \psi_p\{y_i - g\left(\boldsymbol{\beta}; \boldsymbol{x}_i\right)\}, \tag{9.2}$$

which can be solved using a variety of techniques, most notably linear programming algorithms applied to local linear approximations of (9.2) (Koenker and Park, 1996; Koenker, 2005). Briefly, let $r_i(\boldsymbol{\beta}) = y_i - g\left(\boldsymbol{\beta}; \boldsymbol{x}_i\right)$, $i = 1, \ldots, n$, be functions continuously differentiable in $\boldsymbol{\beta}$. A practical approach is to solve $\min_{\boldsymbol{\beta} \in \mathbb{R}^{k+1}} n^{-1} \sum_{i=1}^{n} \psi_p(r_i(\boldsymbol{\beta}))$ via a sequence of locally linearised problems (akin to a Gauss–Newton scheme) in $\boldsymbol{\zeta}_\beta$, where the residual vector $\boldsymbol{r}(\boldsymbol{\beta}) = (r_1(\boldsymbol{\beta}), \ldots, r_n(\boldsymbol{\beta}))^\top$ is approximated locally as $\boldsymbol{r}(\boldsymbol{\beta} + \boldsymbol{\zeta}_\beta) \approx \boldsymbol{r}(\boldsymbol{\beta}) + \nabla \boldsymbol{r}(\boldsymbol{\beta})\, \boldsymbol{\zeta}_\beta$ and $\nabla \boldsymbol{r}(\boldsymbol{\beta})$ denotes the Jacobian matrix of $\boldsymbol{r}(\boldsymbol{\beta})$. The expression for the step $\boldsymbol{\zeta}_\beta$ is given in Koenker (2005, p. 212). Then, the algorithm updates $\boldsymbol{\beta}$ with $\boldsymbol{\beta} + \lambda \boldsymbol{\zeta}_\beta$, where the step length $\lambda \in [0, 1]$ minimises

$$\sum_{i=1}^{n} \psi_p \left\{r_i\left(\boldsymbol{\beta} + \lambda \boldsymbol{\zeta}_\beta\right)\right\}. \tag{9.3}$$

The algorithm proceeds until convergence (see Koenker and Park, 1996, for the conditions and criteria of convergence). The resulting estimator, $\hat{\boldsymbol{\beta}}_n(p)$, is consistent and asymptotically normal, with a limiting variance-covariance matrix analogous to that in (8.10). See Box 9.1 for details.

DOI: 10.1201/9781003326410-9

Box 9.1: Consistency and asymptotic normality of the nonlinear quantile regression estimator (regularity conditions)

Define the nonlinear quantile regression estimator as

$$\hat{\boldsymbol{\beta}}_n(p) \equiv \operatorname*{argmin}_{\boldsymbol{\beta}\in B} \frac{1}{n}\sum_{i=1}^{n} \psi_p\{y_i - g(\boldsymbol{\beta};\boldsymbol{x}_i)\},$$

where $B \subset \mathbb{R}^{k+1}$ is compact. Koenker (2005, p. 123) provides the conditions for consistency and asymptotic normality of $\hat{\boldsymbol{\beta}}_n(p)$, summarised as follows:

1. The regression function $g(\cdot;\cdot)$ varies in a controlled way with respect to $\boldsymbol{\beta}$, so that distinct parameter values yield correspondingly distinct fitted values. This guarantees identifiability and a well-behaved objective function.

2. There exist positive-definite matrices $\boldsymbol{H}(p)$ and $\boldsymbol{J}$ such that:

$$\boldsymbol{H}(p) = \lim_{n\to\infty} \boldsymbol{H}_n(p)$$
$$\boldsymbol{J} = \lim_{n\to\infty} \boldsymbol{J}_n,$$

where $\boldsymbol{H}_n(p) = n^{-1}\sum \dot{\boldsymbol{g}}_i\dot{\boldsymbol{g}}_i^\top f_i(\xi_i)$ and $\boldsymbol{J}_n = n^{-1}\sum \dot{\boldsymbol{g}}_i\dot{\boldsymbol{g}}_i^\top$, f_i is the conditional density of Y_i given $\boldsymbol{x}_i$ and evaluated at $\xi_i = Q_{Y_i}(p|\boldsymbol{x}_i)$ (assumed to exist and be positive), and $\dot{\boldsymbol{g}}_i = \nabla_{\boldsymbol{\beta}} g(\boldsymbol{\beta}_0;\boldsymbol{x}_i)$, the gradient of g with respect to $\boldsymbol{\beta}$, evaluated at the true value $\boldsymbol{\beta}_0(p)$, and

$$\max_{i=1,\ldots,n} \|\dot{\boldsymbol{g}}_i\|/\sqrt{n} \to 0.$$

Under these conditions, the following Bahadur representation is valid:

$$\sqrt{n}\left(\hat{\boldsymbol{\beta}}_n(p) - \boldsymbol{\beta}_0(p)\right) = \boldsymbol{H}^{-1}(p)\frac{1}{\sqrt{n}}\sum_{i=1}^{n} \dot{\boldsymbol{g}}_i\{p - I(r_i(p) < 0)\} + o_{\mathbb{P}}(1),$$

where $r_i(p) = y_i - g(\boldsymbol{\beta}_0(p);\boldsymbol{x}_i)$ and $p - I(r_i(p) < 0)$ is the derivative (in the subgradient sense) of $\psi_p(r_i(p))$. From this, it follows that

$$\sqrt{n}\left(\hat{\boldsymbol{\beta}}_n(p) - \boldsymbol{\beta}_0(p)\right) \xrightarrow{d} \mathcal{N}(\boldsymbol{0},\boldsymbol{\Omega}),$$

with limiting covariance matrix $\boldsymbol{\Omega} = p(1-p)\boldsymbol{H}^{-1}(p)\boldsymbol{J}\boldsymbol{H}^{-1}(p)$.

A nonlinear model depends on either prior knowledge of the phenomenon or the introduction of new, strong theory to explain the observed relationship with potential predictive power. Nonlinear modelling is more common in some fields than others. For instance, in kinesiology, pharmacokinetics, and enzyme kinetics, the study of the dynamics of an agent in a system involves the estimation of nonlinear models; phenomena like human growth, certain disease mechanisms and the effects of harmful environmental substances such as lead and mercury, may show strong nonlinearities over time. Typically, the preferred strategy in these applications is to find a parametric model with substantive interpretability, possibly parsimonious (in general more parsimonious than polynomials), and valid beyond the observed range of the data. Of course, in general not all nonlinear models possess easily interpretable parameters, which is why it may be more attractive to obtain the marginal effects (or QTEs) $\partial Q_Y(p \mid \boldsymbol{x})/\partial x_j$, $j = 1, \ldots, k$. This issue is analogous to the interpretation of regression coefficients in generalised linear models with nonlinear link functions.

For example, Geraci (2019c) considers a pharmacokinetics study with the biexponential model

$$Q_Y(p \mid x) = \beta_1(p) \exp\{-\beta_2(p)x\} + \beta_3(p) \exp\{-\beta_4(p)x\}, \tag{9.4}$$

where Y denotes the drug concentration and x the time of the measurement since injection. This model is linear in the multipliers (or amplitudes) $\beta_1(p)$ and $\beta_3(p)$, but nonlinear in the rate constants $\beta_2(p) > 0$ and $\beta_4(p) > 0$. In pharmacokinetic experiments, the drug concentration usually declines at different rates in different 'compartments' of the organism, rapidly in the initial phases and more slowly afterwards. To identify which phase is which, one may impose a constraint such as $\beta_2(p) > \beta_4(p)$ (Pinheiro and Bates, 2000). Another example in Geraci (2019c) pertains to growth curve modelling and the well-known logistic model

$$Q_Y(p \mid x) = \frac{\beta_1(p)}{1 + \exp\{(\beta_2(p) - x)/\beta_3(p)\}}, \tag{9.5}$$

where Y is a variable related to growth (e.g., weight, height, etc.) and x denotes time.

In R Code 9.1, we illustrate how to fit model (9.4) with the `nlrq` function in the `quantreg` package (Koenker, 2025). The data come from an old pharmacokinetics study (Kwan et al., 1976) in which indomethacin (an anti-inflammatory drug) was injected intravenously in 6 human subjects and 11 measurements of plasma concentrations of the drug (mcg/ml) were made on each individual[1] between 15 minutes and 8 hours post-injection. The `nlrq` function is akin to the `nls` function for nonlinear least squares estimation available in the `stats` package (R Core Team, 2026), which is also shown in R Code 9.1 for comparison. Indeed, the `formula` argument in `nlrq` requires the `nls` format and accepts self-starting models. The latter are special functions that compute starting values from the data, which otherwise should be provided by the user. Model (9.4) is specified by means of the self-starting biexponential function `SSbiexp`, which is parameterised in terms of the two amplitudes `A1` and `A2`, and the *natural logarithm* (to ensure positivity) of the rate constants `lrc1` and `lrc2`.[2] Estimation in `nlrq` is carried out using an interior point algorithm (Koenker and Park, 1996), while standard errors are obtained via bootstrap applied to the linearisation of the model evaluated at the estimated parameters.

Predicted trajectories of indomethacin concentration for $p \in \{0.1, 0.5, 0.9\}$ are plotted in Figure 9.1. The rapid initial decline during the distribution phase is governed by the first exponential component, whose median decay-rate parameter is estimated as $\hat{\beta}_2(0.5) = \exp(0.61) = 1.83$. The subsequent slowdown during the elimination phase is mainly determined by the second exponential component, with median decay-rate parameter estimated as $\hat{\beta}_4(0.5) = \exp(-1.31) = 0.27$. There is also evidence of decreasing variability over time, as suggested by the narrowing distance between the quantiles 0.9 and 0.1, together with a change in skewness from negative to approximately symmetric or positive, as suggested by the relative distances between the median and the upper and lower quantiles.

When the nonlinear quantile function has a relatively simple form, then one may want to just apply the standard machinery for linear quantile functions based on a linearisation of the model. The properties of quantiles (see Section 2.2.2) turn out to be very handy in this case. If the model is of the type

$$Q_Y(p \mid \boldsymbol{x}) = g\left(\boldsymbol{x}^\top \boldsymbol{\beta}(p)\right),$$

[1]In this analysis, we ignore the correlation induced by the repeated measures design and refer the reader to Chapter 13 for more details on this topic.

[2]The interested reader can find in R also the self-starting function `SSlogis` for fitting model (9.5). See also Section 13.1.2.

R Code 9.1: Nonlinear quantile regression analysis of drug concentration in the `Indometh` dataset.

```
library(quantreg)
data(Indometh)
# fit nonlinear median regression
fit.nlrq <- nlrq(conc ~ SSbiexp(time, A1, lrc1, A2, lrc2),
data = Indometh, tau = 0.5)
summary(fit.nlrq)
>
> Call: nlrq(formula = conc ~ SSbiexp(time, A1, lrc1, A2, lrc2), data =
>            Indometh, tau = 0.5, control = list(maxiter = 100, k = 2,
>            InitialStepSize = 1, big = 1e+20, eps = 1e-07, beta = 0.97),
>            trace = FALSE)
>
> tau: [1] 0.5
>
> Coefficients:
>        Value    Std. Error t value  Pr(>|t|)
> A1     2.51563  0.45243     5.56027  0.00000
> lrc1   0.60635  0.25357     2.39122  0.01984
> A2     0.47000  0.22996     2.04382  0.04523
> lrc2 -1.31317  0.34015    -3.86052  0.00027
# fit nonlinear mean regression
fit.nls <- nls(conc ~ SSbiexp(time, A1, lrc1, A2, lrc2), data = Indometh)
summary(fit.nls)
>
> Formula: conc ~ SSbiexp(time, A1, lrc1, A2, lrc2)
>
> Parameters:
>        Estimate Std. Error t value Pr(>|t|)
> A1       2.7734     0.2533  10.951 3.97e-16 ***
> lrc1     0.8864     0.2222   3.988 0.000178 ***
> A2       0.6067     0.2671   2.272 0.026598 *
> lrc2    -1.0919     0.4089  -2.671 0.009658 **
> ---
> Signif. codes:  0 '***' 0.001 '**' 0.01 '*' 0.05 '.' 0.1 ' ' 1
>
> Residual standard error: 0.1745 on 62 degrees of freedom
>
> Number of iterations to convergence: 0
> Achieved convergence tolerance: 3.302e-07
```

g is nondecreasing and g^{-1} exists in an analytically explicit form, then for the equivariance property we have

$$Q_{g^{-1}(Y)}(p \mid \boldsymbol{x}) = \boldsymbol{x}^\top \boldsymbol{\beta}(p).$$

The idea is illustrated in R Code 9.2 with a simple exponential model. The estimates for $Q_Y(0.9) = \exp\{\boldsymbol{x}^\top \boldsymbol{\beta}(0.9)\}$ are obtained with **`nlrq`**, while the equivalent log-linear model $Q_{\log Y}(0.9) = \boldsymbol{x}^\top \boldsymbol{\beta}(0.9)$ is fitted with the interior point algorithm in **`rq`**. Standard errors

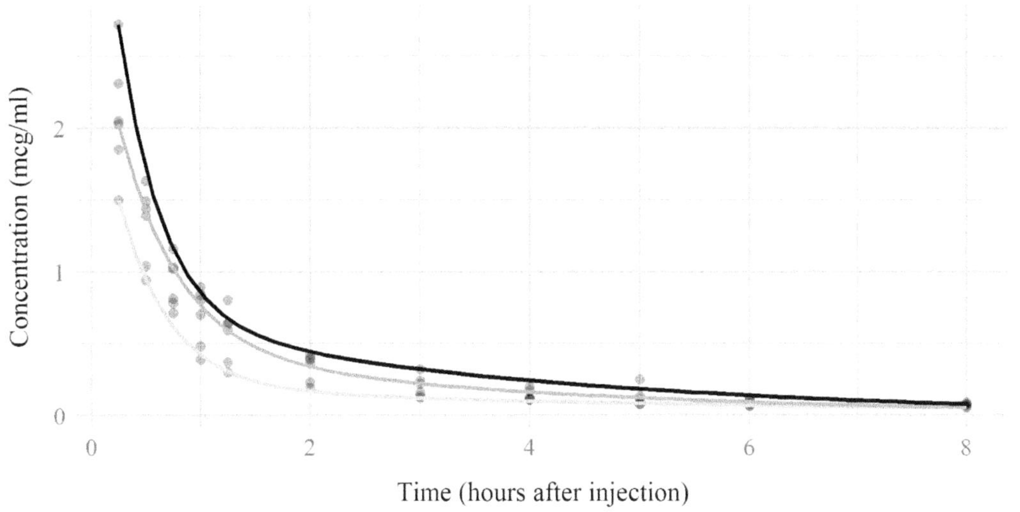

Figure 9.1
Predicted quantiles ($p \in \{0.1, 0.5, 0.9\}$) of indomethacin concentration conditional on time since injection in the Indomethacin dataset using the biexponential model (9.4).

are based on $R = 200$ bootstrap replications in both cases. As one might anticipate, point estimates and standard errors from the two estimation routines are similar. Of course, some discrepancies should be expected due to different approximations involved and the sampling variability of the bootstrap. The main advantage of working on the linear scale is arguably the speed of computation and the scalability of the fast algorithms (Section 8.3.1).

9.2 Transformation models

Unless there is a belief, grounded in theory or in one's own empirical experience, that the model has a particular parametric form, a more relaxed approach is to let the data inform the choice of the model among a *family* of parametric models. In this section, we consider flexible, parsimonious families of parametric transformations that are applied to the response in order to obtain approximate linearity on the transformed scale. We refer to these models as transformation models to stress that a *transformation parameter* (usually low-dimensional) drives the final functional form of the fitting.

The goal is to fit the model

$$Q_{h(Y_i;\lambda_p)}(p \mid \boldsymbol{x}_i) = \boldsymbol{x}_i^\top \boldsymbol{\beta}(p), \qquad i = 1, \ldots, n. \tag{9.6}$$

The assumption is that the transformation h is able to capture the most salient aspects of what is supposed to be g^{-1} in (9.1) thanks to the flexibility afforded by the parameter λ_p, so that the pth quantile function of the transformed response variable is (approximately) linear. The parameter λ_p, which is estimated from the data, may control different aspects of the shape of the transformation, such as inflection points, convexity and symmetry. It is indexed by p since the model assumes that quantile functions at different values of p may have different shapes.

R Code 9.2: Comparison of nonlinear quantile regression for an exponential model with log-linear quantile regression.

```
set.seed(123)

n <- 5000
x <- runif(n, 0, 4)
eta <- 2 + 0.5*x
y <- exp(eta)*rnorm(n,1,0.2)
all(y > 0)
> [1] TRUE
z <- log(y)

# fit nonlinear model
system.time(fit.nlrq <- nlrq(y ~ exp(a + b*x), tau = 0.9,
start = list(a = 0, b = 0)))
>    user  system elapsed
>    0.95    0.20    1.14
summary(fit.nlrq, R = 200)
>
> Call: nlrq(formula = y ~ exp(a + b * x), start = list(a = 0, b = 0),
>     tau = 0.9, control = list(maxiter = 100, k = 2, InitialStepSize = 1,
>         big = 1e+20, eps = 1e-07, beta = 0.97), trace = FALSE)
>
> tau: [1] 0.9
>
> Coefficients:
>   Value     Std. Error t value   Pr(>|t|)
> a   2.23625   0.00889  251.60562   0.00000
> b   0.49323   0.00354  139.50945   0.00000
# fit log-linear model
system.time(fit.rq <- rq(z ~ x, tau = 0.9, method = "fn"))
>    user  system elapsed
>    0.02    0.00    0.02
summary(fit.rq, se = "boot", R = 200)
>
> Call: rq(formula = z ~ x, tau = 0.9, method = "fn")
>
> tau: [1] 0.9
>
> Coefficients:
>              Value     Std. Error t value   Pr(>|t|)
> (Intercept)   2.23960   0.00776  288.50568   0.00000
> x             0.49214   0.00299  164.85225   0.00000
```

The Box–Cox transformation (Box and Cox, 1964) has been one of the first to be applied in quantile transformation models (Buchinsky, 1995; Chamberlain, 1994; Fitzenberger et al., 2010; Mu and He, 2007). Originally, Box and Cox (1964) proposed using power transformations to address lack of linearity, homoskedasticity and normality of the residuals in mean regression modelling. Sakia (1992, p. 175) reported that 'seldom does this transformation fulfil the basic assumptions of linearity, normality and homoskedasticity simultaneously as originally suggested by Box & Cox (1964). The Box–Cox transformation has found more practical utility in the empirical determination of functional relationships in a variety of fields, especially in econometrics'. This particular choice, applicable when Y has a lower bound (singly bounded), say $Y > 0$, yields the transformation

$$h_{BC}(Y; \lambda_p) = \begin{cases} \dfrac{Y^{\lambda_p} - 1}{\lambda_p} & \text{if } \lambda_p \neq 0 \\ \log Y & \text{if } \lambda_p = 0. \end{cases} \tag{9.7}$$

Dehbi et al. (2016) considered the Aranda-Ordaz transformations (Aranda-Ordaz, 1981) to model variables whose support is bounded from below and from above (doubly bounded), say $0 < Y < 1$. There are two families of such transformations, one symmetric[3]

$$h_{AOs}(Y; \lambda_p) = \begin{cases} \dfrac{2}{\lambda_p} \dfrac{Y^{\lambda_p} - (1-Y)^{\lambda_p}}{Y^{\lambda_p} + (1-Y)^{\lambda_p}} & \text{if } \lambda_p \neq 0, \\ \log\left(\dfrac{Y}{1-Y}\right) & \text{if } \lambda_p = 0, \end{cases} \tag{9.8}$$

and one asymmetric

$$h_{AOa}(Y; \lambda_p) = \begin{cases} \log\left\{\dfrac{(1-Y)^{-\lambda_p} - 1}{\lambda_p}\right\} & \text{if } \lambda_p \neq 0, \\ \log\{-\log(1-Y)\} & \text{if } \lambda_p = 0. \end{cases} \tag{9.9}$$

Notwithstanding their merits, the transformations in (9.7) and (9.8) are still bounded for all $\lambda_p \neq 0$: the range of h_{BC} is not $\mathbb{R}$, but is still a singly bounded interval, with lower (or upper) endpoint $-1/\lambda_p$ if $\lambda_p > 0$ (or $\lambda_p < 0$); the range of h_{AOs} is not $\mathbb{R}$, but $(-2/|\lambda|, 2/|\lambda|)$. To overcome range problems, which give rise to computational difficulties, Geraci and Jones (2015) proposed to use instead one-parameter transformations with range $\mathbb{R}$. For $Y > 0$, the transformation takes two forms, symmetric

$$h_{MCJs}(Y; \lambda_p) = \begin{cases} \dfrac{1}{2\lambda_p}\left(Y^{\lambda_p} - \dfrac{1}{Y^{\lambda_p}}\right) & \text{if } \lambda_p \neq 0 \\ \log Y & \text{if } \lambda_p = 0. \end{cases} \tag{9.10}$$

and asymmetric

$$h_{MCJa}(Y; \lambda_p) = \begin{cases} \dfrac{1}{2\lambda_p}\left\{\{\log(1+Y)\}^{\lambda_p} - \dfrac{1}{\{\log(1+Y)\}^{\lambda_p}}\right\} & \text{if } \lambda_p \neq 0 \\ \log\log(1+Y) & \text{if } \lambda_p = 0. \end{cases} \tag{9.11}$$

The symmetric and asymmetric transformations for $0 < Y < 1$ are, respectively,

$$h_{MCJs}(Y; \lambda_p) = \begin{cases} \dfrac{1}{2\lambda_p}\left\{\{Y/(1-Y)\}^{\lambda_p} - \dfrac{1}{\{Y/(1-Y)\}^{\lambda_p}}\right\} & \text{if } \lambda_p \neq 0 \\ \log(Y/(1-Y)) & \text{if } \lambda_p = 0. \end{cases} \tag{9.12}$$

[3]The symmetry here is that $h_{AOs}(y; \lambda_p) = -h_{AOs}(1-y; \lambda_p) = h_{AOs}(y; -\lambda_p)$.

and

$$h_{MCJa}(Y;\lambda_p) = \begin{cases} \dfrac{1}{2\lambda_p}\left\{\{-\log(1-Y)\}^{\lambda_p} - \dfrac{1}{\{-\log(1-Y)\}^{\lambda_p}}\right\} & \text{if } \lambda_p \neq 0 \\ \log - \log(1-Y) & \text{if } \lambda_p = 0. \end{cases} \quad (9.13)$$

Since the transformations in Equations (9.10)–(9.13) have range $\mathbb{R}$ for all λ_p, they admit explicit inverse transformations. Also, it is sufficient to restrict λ_p to $\lambda_p \geq 0$, given that for these transformations $h(Y; -\lambda) = h(Y; \lambda)$, so that positive and negative values of λ produce identical curves. In addition, in the case of a single covariate, every estimated quantile that results will be monotone increasing, decreasing or constant, although different estimated quantiles can have different shapes from this collection.[4]

The behaviour of these transformations is illustrated in Figure 9.2 for selected values of λ. The Box–Cox and symmetric Aranda–Ordaz transformations display the range restrictions discussed above when $\lambda \neq 0$. In contrast, the MCJ transformations map the support of Y onto the real line for all values of λ, while retaining flexibility in curvature. On the bounded support, the symmetric MCJ transformation is centred around $Y = 1/2$, whereas the asymmetric version allows different behaviour near the two endpoints.

If h is a nondecreasing function on its domain (as is the case for all transformations considered in this section), parameter estimation can be carried out on the scale of either h or h^{-1}. The latter entails a nonlinear optimisation as discussed in Section 9.1 where

$$\min_{\boldsymbol{\beta}\in\mathbb{R}^{k+1},\lambda\in\mathbb{R}} \frac{1}{n}\sum_{i=1}^{n} \psi_p\{y_i - h^{-1}(\boldsymbol{x}_i^\top\boldsymbol{\beta};\lambda)\} \quad (9.14)$$

can be solved via, for example, the interior point algorithm for nonlinear quantile functions (Koenker and Park, 1996). However, since this nonlinear problem may suffer from numerical instability and the advantage of assuming model (9.6) is precisely to avoid it, a two-stage estimator that operates on the scale of h is preferred (Chamberlain, 1994; Buchinsky, 1995). In practice, $\boldsymbol{\beta}(p)$ is estimated conditionally over a grid of values for the transformation parameter.[5] The estimation algorithm consists of the following two optimisation problems:

1. Obtain the pth regression quantile $\hat{\boldsymbol{\beta}}_p(\lambda)$ conditional on λ by minimising the linear quantile loss function, that is,

$$\hat{\boldsymbol{\beta}}_p(\lambda) = \underset{\boldsymbol{\beta}\in\mathbb{R}^{k+1}}{\operatorname{argmin}} \sum_{i=1}^{n} \psi_p\left\{h\left(y_i;\lambda\right) - \mathbf{x}_i^\top\boldsymbol{\beta}\right\}. \quad (9.15)$$

This can be efficiently solved by standard linear programming algorithms.

[4]Geraci and Jones (2015) proposed also a transformation that unifies the symmetric and asymmetric versions of h_{MCJ} into a single two-parameter transformation, namely

$$h^*_{MCJ}(Y;\lambda_p) = h_{MCJ}(Y^*;\lambda_p),$$

where h_{MCJ} is given in (9.10) and

$$Y^* = h_{BC}(1+Y;\delta_p)$$

if $Y > 0$, and

$$Y^* = h_{BC}(1/(1-Y);\delta_p)$$

if $Y \in (0,1)$. The additional parameter δ_p controls the asymmetry: symmetric forms of h_{MCJ} correspond to $\delta_p = 1$ while asymmetric forms of h_{MCJ} to $\delta_p = 0$.

[5]The nonlinear estimation in (9.14) has the advantage of allowing inference on λ_p, which is treated as a parameter to be estimated along with $\boldsymbol{\beta}(p)$. On the other hand, two-stage (linear) estimation treats λ_p as *given*. Of course, the strategy based on linear estimation is numerically more stable but at the cost of potentially underestimating the variability associated with the regression coefficients. The package `Qtools` provides functions for all the estimation procedures discussed in this section, namely `tsrq` and `tsrq2` for two-stage estimation, `nlrq1` and `nlrq2` for nonlinear estimation, and `rcrq` for residual CUSUM process estimation.

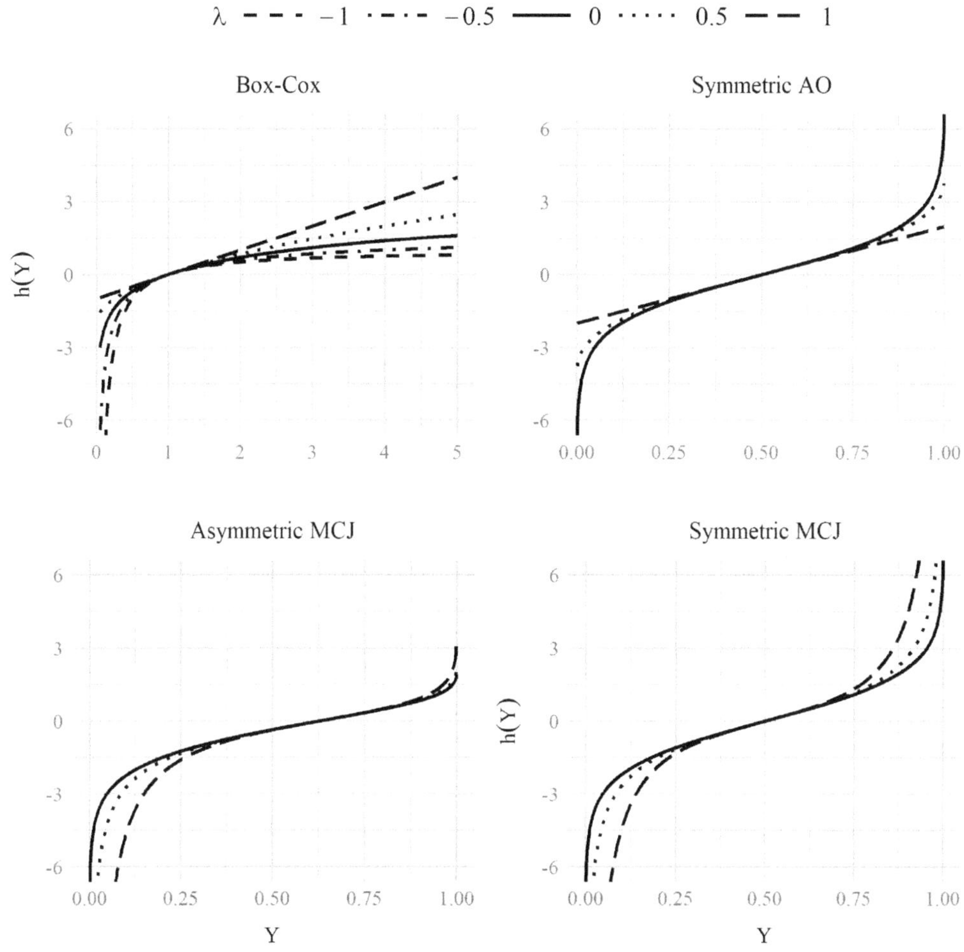

Figure 9.2
Transformation functions defined in Equations (9.7), (9.8), (9.12) and (9.13). The Box–Cox transformation is shown for $\lambda \in \{-1, -0.5, 0, 0.5, 1\}$. The symmetric Aranda–Ordaz, symmetric MCJ and asymmetric MCJ transformations are shown for $\lambda \in \{0, 0.5, 1\}$ only.

2. Estimate the parameter λ_p, conditional on $\hat{\boldsymbol{\beta}}_p(\lambda)$, as

$$\hat{\lambda}_p = \underset{\lambda \in \mathbb{R}}{\operatorname{argmin}} \sum_{i=1}^{n} \psi_p \left\{ y_i - h^{-1}\left(\mathbf{x}_i^\top \hat{\boldsymbol{\beta}}_p(\lambda); \lambda\right) \right\}, \tag{9.16}$$

which can be solved numerically using a grid search.

The inverse transformation in (9.16) is always well defined when using the transformations in (9.10)–(9.13). This cannot be said for either the Box–Cox transformation, for which h^{-1} in (9.16) is available only if $\lambda \boldsymbol{x}_i^\top \hat{\boldsymbol{\beta}}_p(\lambda) + 1 > 0$, or the symmetric Aranda-Ordaz transformation, for which h^{-1} in (9.16) is available only if $\left|\frac{1}{2}\lambda \mathbf{x}_i^\top \hat{\boldsymbol{\beta}}_p(\lambda)\right| < 1$. In the latter two cases, it might be tempting to truncate the image of h^{-1} by discarding observations that do not fall in the admissible set (Fitzenberger et al., 2010). Such crude expedients should be discouraged since they may lead to an unacceptable loss of information depending on the particular sample in a given application. An alternative estimation strategy that avoids the troublesome inversion of the Box–Cox and Aranda-Ordaz transformations is based on the residual CUSUM process (Mu and He, 2007). The resulting estimator is computationally more intensive than the two-stage estimator and its limiting joint distribution requires resampling since it is not amenable to analytic computation.

Predictions can be easily obtained from (9.6) by virtue of the equivariance property with

$$Q_Y(p \mid \boldsymbol{x}) = h^{-1}\left\{\boldsymbol{x}^\top \boldsymbol{\beta}(p); \lambda_p\right\}. \tag{9.17}$$

The interpretation of $\boldsymbol{\beta}(p)$, on the other hand, may not be as immediate. Hence, it may be preferable to compute the QTE of the covariate, say, x_j, $j = 1, \ldots, k$, on $Q_Y(p)$. That is, the derivative of the composition $Q \circ \eta$, that is,

$$\frac{\partial Q_{Y|\mathbf{X}}(p)}{\partial x_j} = \frac{\partial Q_{Y|\mathbf{X}}(p)}{\partial \eta(p)} \cdot \frac{\partial \eta(p)}{\partial x_j}, \tag{9.18}$$

provided the transformation is differentiable, where $\eta(p) = \boldsymbol{x}^\top \boldsymbol{\beta}(p)$. Once the estimates $\hat{\boldsymbol{\beta}}(p)$ and $\hat{\lambda}_p$ are obtained, these can be plugged in Equations (9.17) and (9.18).

The large-sample properties of the two-stage estimator for monotonic quantile regression models have been studied by Powell (1991) (see also Chamberlain, 1994; Machado and Mata, 2000). Under regularity conditions, it can be shown that the joint estimator $\left(\hat{\boldsymbol{\beta}}(p)^\top, \hat{\lambda}_p\right)$ is consistent and normally distributed with limiting variance-covariance matrix akin to that defined in Box 9.1 (see Machado and Mata, 2000; Fitzenberger et al., 2010). Joint inference on $\boldsymbol{\beta}_p$ and λ_p can be done also by means of the bootstrap. A naïve approach to inference on $\boldsymbol{\beta}_p$ only is to apply one of the several methods developed for linear quantile regression estimators (Section 8.3.2) by assuming that the transformation parameter is known. Unless λ_p is truly given and set by the analyst, conditional inference may produce substantially lower standard errors for the regression coefficients and thus lead to overstating their significance.

In R Code 9.3, we fit a linear conditional quantile regression model for ozone concentration on solar radiation using the New York Air Quality dataset (Geraci, 2016b), which contains 153 air quality measurements on 6 variables collected in New York between May and September 1973. The fitted model is subsequently assessed by means of the CUSUM lack-of-fit test introduced in Section 8.3.3. The test applied to the `rq` fit signals that, at $p = 0.9$, a linear relationship between ozone and solar radiation is not satisfactory.

In R Code 9.4, we fit instead the transformation model (9.6) using the function `tsrq` in the R package `Qtools`. The `formula` argument specifies the model for the linear predictor, while the argument `tsf` selects the transformation h: `bc` for the Box–Cox model, `ao` for the Aranda–Ordaz families, and `mcjI` for Geraci and Jones's (2015) transformations. Additional arguments include `symm`, a logical flag indicating whether the symmetric or asymmetric version of `ao` and `mcjI` should be used; `dbounded`, a logical flag indicating whether the response variable is doubly bounded (the default assumes a strictly positive response); `lambda`, a numerical vector defining the grid of candidate values for estimating λ_p; and `conditional`, a logical flag indicating whether λ_p is treated as known (in which case the argument `lambda` provides the fixed value) (Geraci, 2016b).

The CUSUM lack-of-fit test applied to the transformation model does not reject the fitted model (9.6) with transformation (9.10). The transformation parameter estimate $\hat{\lambda}_{0.9} = 1.5$ suggests that the 90th percentile of Y (hence, on the h^{-1} scale) grows at a below-linear rate.

In Figure 9.3, predictions (9.17) and QTEs (9.18) obtained with the function `predict.rqt` are shown for the quantiles $p \in \{0.1, 0.5, 0.9\}$. The effect of solar radiation on

R Code 9.3: Estimation and lack-of-fit testing for a linear quantile regression model using the New York Air Quality dataset.

```
library(Qtools)
library(quantreg)
data(airquality)

# fit linear model
fit.rq <- rq(Ozone ~ Solar.R, data = airquality, tau = 0.9)
summary(fit.rq, se = "nid")
>
> Call: rq(formula = Ozone ~ Solar.R, tau = 0.9, data = airquality)
>
> tau: [1] 0.9
>
> Coefficients:
>             Value    Std. Error t value  Pr(>|t|)
> (Intercept) 17.97727  8.27201    2.17327  0.03192
> Solar.R      0.38636  0.05425    7.12205  0.00000
# CUSUM lack-of-fit test
LOFTest(fit.rq, alpha = 0.05, B = 1000, seed = 987)
> Lack-of-fit test for quantile regression based on the cusum process
> A large test statistic (small p-value) is evidence of lack of fit
> Quantile 0.9: Test statistic = 0.0457; p-value = 0.02
```

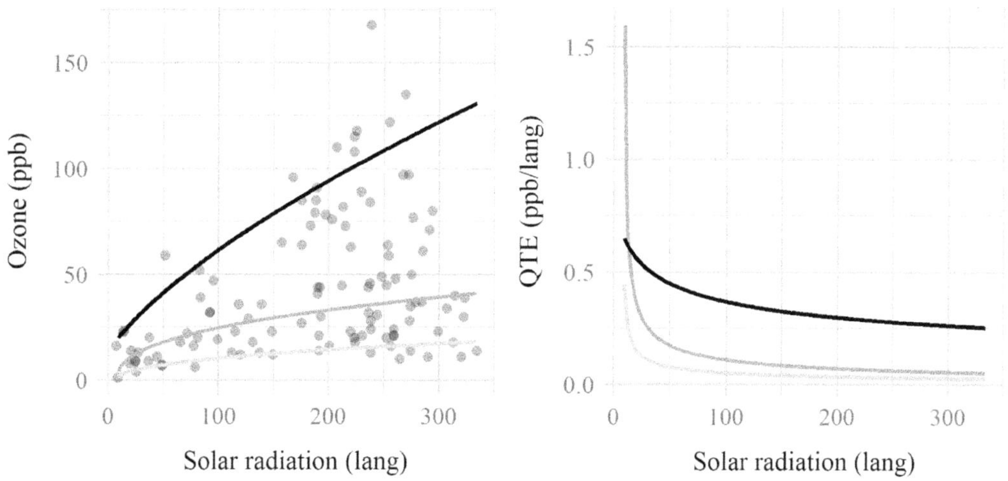

Figure 9.3
Predicted quantiles ($p \in \{0.1, 0.5, 0.9\}$) of ozone conditional on solar radiation (left) and corresponding quantile treatment effects (right) in the New York Air Quality dataset using the transformation model (9.10).

R Code 9.4: Estimation and lack-of-fit testing for a transformation-based quantile regression model using the New York Air Quality dataset.

```
# fit transformation model
fit.rqt <- tsrq(Ozone ~ Solar.R, data = airquality, tsf = "mcjI",
symm = TRUE, dbounded = FALSE, lambda = seq(1, 3, by = 0.005),
conditional = FALSE, tau = .9)
summary(fit.rqt, conditional = FALSE, se = "nid")
> call:
> summary.rqt(object = fit.rqt, se = "nid", conditional = FALSE)
>
> Proposal I symmetric transformation (singly bounded response)
>
> Summary for unconditional inference
>
> tau =  0.9
>
> Optimal transformation parameter:
>       Value  Std. Error Lower bound Upper bound
>   1.5000000   0.7047198   0.1187745   2.8812255
>
> Coefficients linear model (transformed scale):
>                 Value Std. Error Lower bound Upper bound
> (Intercept) 16.557327  22.692722  -27.919590   61.034245
> Solar.R      1.443407   4.202759   -6.793849    9.680663
>
> Degrees of freedom: 111 total; 109 residual
# CUSUM lack-of-fit test
LOFTest(fit.rqt, alpha = 0.05, B = 1000, seed = 987)
> Lack-of-fit test for quantile regression based on the cusum process
> A large test statistic (small p-value) is evidence of lack of fit
> Quantile 0.9: Test statistic = 0.0212; p-value = 0.134
```

different quantiles of ozone levels shows a nonlinear behaviour, especially at lower ranges of radiation (below 50 Ly).

Transformation models are a relatively easy-to-implement alternative for nonlinear modelling. In some applications (Boghossian et al., 2016), they may be competitive as compared to methods based on smoothing, which is introduced in the next section.

9.3 Nonparametric models

Nonparametric regression is a generic label for regression methods where the functional form of the relationship between the response and the predictor is unknown. A nonparametric model is essentially data-driven. A location-shift model for a continuous response with a univariate covariate x is generally presented in the following form:

$$Y_i = g\left(x_i\right) + \varepsilon_i, \qquad i = 1, \ldots, n, \tag{9.19}$$

where $\mathrm{E}(\varepsilon_i) = 0$. The function g is an infinite-dimensional parameter whose estimation requires some minimal assumptions such as smoothness. On the other hand, the distribution of the error ε may or may not be specified. In a distribution-free quantile regression framework, we assume that $Q_{\varepsilon_i}(p) = 0$, for the quantile level p under consideration, and

$$Q_{Y_i}(p \mid x_i) = g_p(x_i), \qquad i = 1, \ldots, n, \tag{9.20}$$

where g_p reflects the heterogeneous effect of x on the conditional distribution of Y. In this section, we consider two important and popular methods to estimate nonparametric functions, namely (penalised) smoothing splines and kernel smoothers. In the following, we focus on models with just one covariate x. In the presence of multiple linear and nonlinear terms, the literature on additive quantile regression offers several cues (e.g., see Yu and Lu, 2004; Horowitz and Lee, 2005; Li and Racine, 2008; Koenker, 2011; Fasiolo et al., 2021).

In smoothing splines, the complexity of g is approximated by a sequence of locally linear polynomials (a naïve global polynomial trend can be considered to be a special case). Estimation typically requires the introduction of a penalty term to control the trade-off between bias (smoothness) and variability of the fit. The choice of the penalty is arguably a key aspect as this is directly connected with the computation of the splines.[6] Consider the minimisation problem

$$\min_{g\in\mathcal{G}} \underbrace{\sum_{i=1}^{n} \psi_p\left\{y_i - g(x_i)\right\}}_{\text{fidelity}} + \lambda \underbrace{\int \left\{g''(x)\right\}^2 \mathrm{d}x}_{\text{penalty}}, \tag{9.21}$$

where the integral is taken over the covariate space. While the metric of the first term in Equation (9.21), sometimes termed 'fidelity' (as an indication of the fidelity of the fit to the observations), is L_1, the metric of the second term, called 'penalty', is L_2. The L_2-norm penalty has a long tradition in nonparametric regression, especially for computational reasons as it is tied to least squares fitting. Early references to the combination of least absolute minimisation and the L_2-norm penalty include Bloomfield and Steiger (1983), Eubank (1988) and Jones (1988) in the discussion of the paper by Cole (1988). The solution to (9.21) is a natural cubic spline with knots $\{x_1, \ldots, x_n\}$, and natural boundary conditions (Bosch et al., 1995; Nychka et al., 1995). Problem (9.21) can be reduced to a quadratic programming problem for which Bosch et al. (1995) developed an easily implementable algorithm. Another algorithm proposed by Nychka et al. (1995) starts from smoothing the kink of ψ_p at 0 (e.g., see Section 5.4) and then iteratively applies the weighted least squares machinery. The parameter $\lambda \geq 0$ controls the degree of smoothing, where smaller (larger) values correspond to less (more) smoothing.

In the proposal for conditional quantile inference based on smoothing splines of Koenker et al. (1994) (see also Koenker, 2005, Chapter 7), the L_2 penalty is replaced by an L_1 penalty

$$\min_{g\in\mathcal{G}} \sum_{i=1}^{n} \psi_p\left\{y_i - g(x_i)\right\} + \lambda \int \left|g''(x)\right| \mathrm{d}x. \tag{9.22}$$

The advantage is that the optimisation problem in (9.22) retains the usual linear programming structure of quantile regression. On the other hand, the ensuing smoothed curves are piecewise linear, which may or may not be consistent with the underlying assumed model or with the intended aesthetic goal of the fitting. An excellent overview is given by Mizera (2018).

[6]Given that the literature on this topic is vast, the reader is referred to other publications for the underlying theory (see, e.g., Eubank, 1999; Green and Silverman, 1994; Ruppert et al., 2003; Wang, 2011).

R Code 9.5: Nonparametric quantile regression via penalised smoothing splines.

```
set.seed(123)
n <- 100
p <- 0.5
x <- sort(runif(n, 0, 2*pi))
y <- sin(x) + (1 + x)*rnorm(n, 0, 0.1)

# lambda set at an arbitrary value
fit.L2 <- Qtools::qsmspline(x = x, y = y, p = p, lambda = 0.1)
fit.L1 <- quantreg::rqss(y ~ qss(x, constraint= "N", lambda = 0.1), tau = p)
```

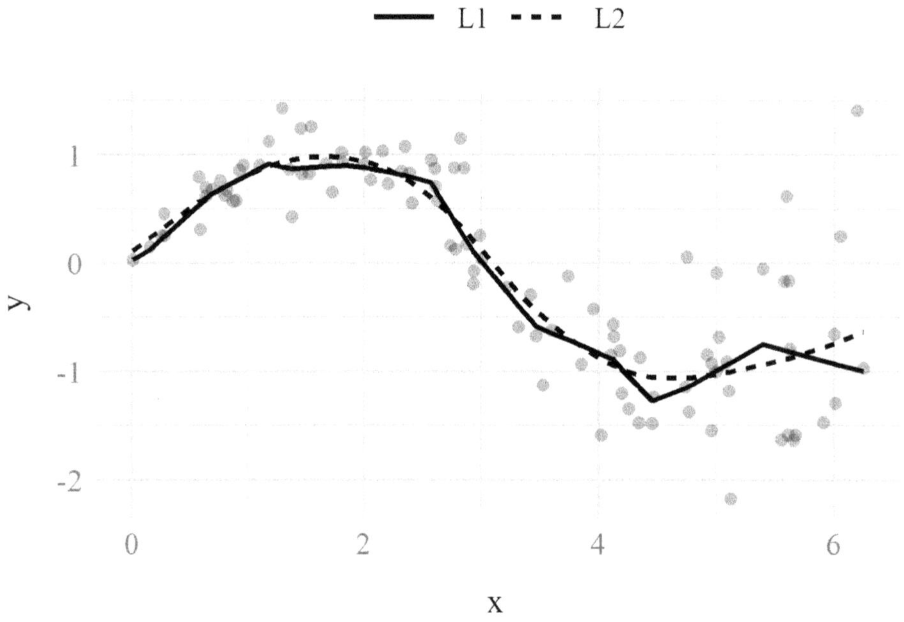

Figure 9.4
Data generated with R Code 9.5 together with the conditional median predicted by penalised smoothing splines.

R Code 9.5 provides a simulated example with data generated from the model $y = \sin(x) + (1+x)\varepsilon$, where $\varepsilon \sim \mathcal{N}(0, 0.1)$ and x is uniform on $(0, 2\pi)$. Median smoothing splines are fitted on a sample of $n = 100$ observations using Bosch et al.'s (1995) algorithm with L_2 penalty as implemented in the `Qtools::qsmspline` function and the linear program with L_1 penalty implemented in the `quantreg::rqss` function. Both estimated splines describe reasonably well the underlying true function (Figure 9.4). As mentioned above, the L_1-penalised spline is characterised by sharp corners at the x_is where the solution bends. In neither case, we concerned ourselves with finding the optimal level of smoothing. The small value of λ chosen arbitrarily for this particular example simply yields a stronger contrast between the two fits. Methods for λ-selection are usually based on cross-validation, which can be time-consuming, or on information criteria. An example of the latter is the modified

R Code 9.6: Selection of the penalty parameter in nonparametric quantile regression using the Schwarz Information Criterion.

```
# define grid - must be wide enough to include the minimum
lambda <- seq(0.1, 2, by = 0.1)
# selection of lambda for rqss based on SIC
sic <- rep(NA, length(lambda))
for(j in 1:length(lambda)){
    fit <- quantreg::rqss(y ~ qss(x, constraint= "N", lambda = lambda[j]),
    tau = p)
    sic[j] <- log(fit$fidelity/n) + 0.5/n*fit$edf*log(n)
}

# value of lambda that minimises SIC
lambda[which.min(sic)]
> [1] 1.2
```

Schwarz Information Criterion (SIC) of Koenker et al. (1994), namely

$$\mathrm{SIC}\left(m_{\lambda}\right) = \log\left[\frac{1}{n}\sum_{i=1}^{n}\psi_{p}\left\{y_{i} - \hat{g}_{p}\left(x_{i}\right)\right\}\right] + \frac{1}{2n}m_{\lambda}\log n,$$

where m_λ denotes the number of interpolated y_is, taken as the effective dimension of the fit with penalty parameter λ, and $\hat{g}_p$ is the estimate of g_p conditional on λ. A sketch of the idea is given in R Code 9.6 based on the data generated in R Code 9.5.

A different nonparametric estimation strategy makes use of the familiar kernel smoothing techniques. Yu and Jones (1998), for example, considered the loss

$$\sum_{i=1}^{n}\psi_{p}\left\{y_{i} - \beta_{0} - \beta_{1}\left(x_{i} - x\right)\right\} K\left(\frac{x_{i} - x}{h}\right), \tag{9.23}$$

which basically estimates the unknown function $g_p(x)$ via a locally linear model weighted by a kernel function centred at x. The latter can be suitably chosen among common density functions as those discussed in Section 7.1.2, while the choice of the bandwidth h requires further consideration as discussed in the example below. Fitting (9.23) becomes straightforward with linear programming algorithms that allow for the use of observation weights as with `quantreg::rq(..., weights)`.

R Code 9.7 shows Yu and Jones's (1998) single kernel estimation[7] of the conditional median using the simulated data of the previous example. The kernel K is the usual Gaussian density whereas the value of the bandwidth, $h = 0.20$, as implemented in the `R` package `KernSmooth` (Wand, 2025), results from the application of Yu and Jones's (1998) rule-of-thumb in combination with Ruppert et al.'s (1995) bandwidth as recommended by the same authors (Yu and Jones, 1998, p. 231). The estimated curve behaves sufficiently well except at the end of the x range where it becomes more wiggly due to the increased local variability (Figure 9.5).

[7]As opposed to a single kernel in the x-direction, one can apply a kernel also in the y-direction (double kernel estimation). See Yu and Jones (1998) and Li and Racine (2008) for more details.

R Code 9.7: Nonparametric quantile regression via a kernel weighted local linear regression.

```
set.seed(123)
n <- 100
p <- 0.5
x <- sort(runif(n, 0, 2*pi))
y <- sin(x) + (1 + x)*rnorm(n, 0, .1)
h_mean <- KernSmooth::dpill(x, y)
h <- h_mean * (p * (1 - p)/(dnorm(qnorm(p)))^2)^0.2
yhat <- rep(NA, n)
for(i in 1:n){
    x.c <- x[i] - x
    w <- dnorm(x.c/h)
    yhat[i] <- quantreg::rq(y ~ x.c, weights = w, tau = p)$coef[1]
}
```

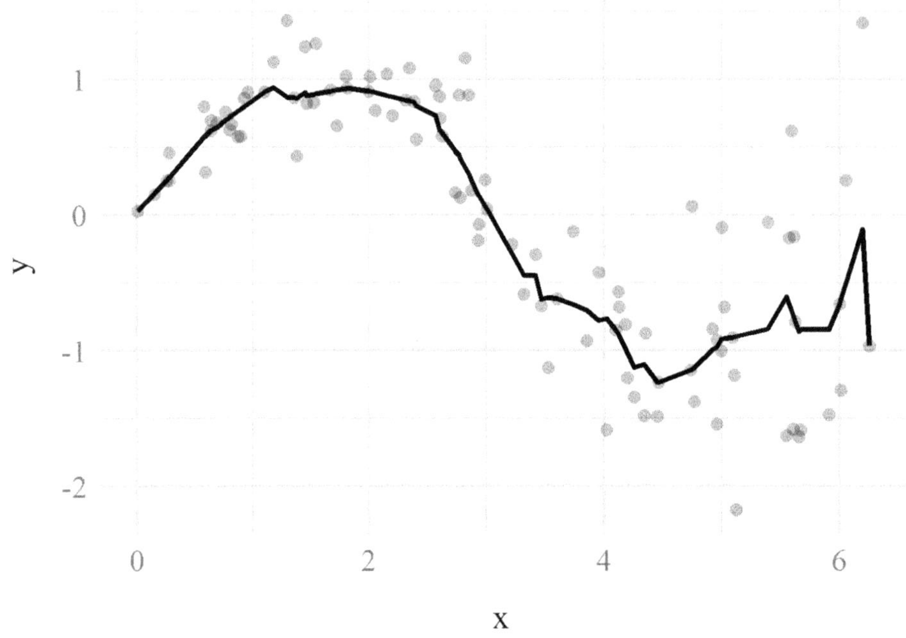

Figure 9.5
Data generated with R Code 9.7 together with the conditional median predicted by kernel weighted local linear regression.

9.4 Distribution-based quantile regression

The use of parametric distributions in conditional quantile modelling has progressively made its way into the statistical literature. In its original formulation (Koenker and Bassett, 1978), quantile regression responded to the need of modelling complex location, scale and shape effects in a robust manner, that is, without having to make stringent assumptions about the underlying distribution. If taken at face value, distribution-based quantile regression may seem like a step backward. However, it has certain merits, particularly when the model provides flexibility. As is often the case, this flexibility is needed to accommodate nonlinear relationships (which is why this topic is being discussed in this chapter). Arguably, models of this kind have appeared in the literature in two variants: (i) some or all the parameters of the selected distribution are defined as functions of the covariates and conditional quantiles of any order are obtained from the fitted distribution (hence, the parameters are not quantile-specific); and (ii) the distribution is parameterised in terms of a conditional quantile of fixed order so that the newly defined parameter is quantile-specific. While the former is a natural application of maximum likelihood estimation, the latter has questionable utility.

As far as the modelling type (i) goes, this is the conditional extension of what we termed 'maximum likelihood for quantiles' (MLQ) in Section 6.2. The work by Gilchrist (2000, 2008) takes a refreshing look at parametric modelling via quantile functions of common distributions. Perhaps the most appealing aspects of this modelling strategy are (i) the availability of a large number of flexible and interpretable models derived from the literature of density-based modelling and (ii) the malleability of quantile functions that can be combined and morphed into new models thanks to the desirable properties of quantiles (Section 2.2.2).

For didactic purposes, let's begin with a basic, though not completely innocuous, model. Suppose $Y \sim \mathcal{N}\left(\mu_x, \sigma_x^2\right)$, where $\mu_x = \beta_0 + \beta_1 x$ and $\sigma_x = \exp(\gamma_0 + \gamma_1 x)$, with the proviso that γ_0 is finite. This means that the conditional quantiles are nonlinear, that is,

$$Q_{Y|X}(p) = \beta_0 + \beta_1 x + \sigma \exp(\gamma_1 x)\Phi^{-1}(p),$$

where $\sigma = \exp(\gamma_0)$. The QTE associated with x is therefore

$$\frac{\partial Q_{Y|X}(p)}{\partial x} = \beta_1 + \gamma_1 \sigma \exp(\gamma_1 x)\Phi^{-1}(p),$$

which depends exponentially on x only when $p \neq 0.5$ and $\gamma_1 \neq 0$. If $p = 0.5$, then the QTE is constant and equal to β_1.

Directly fitting the linear-exponential model above requires some care. An unsuccessful attempt to fit three quantiles using nonlinear estimation via `quantreg::nlrq` is shown in R Code 9.8. The algorithm either throws an error or it remains trapped at the initial values. On the other hand, estimation is quite straightforward with generalised least squares (GLS) regression (Pinheiro and Bates, 2000) as shown in R Code 9.9. Predicting conditional quantiles as well as estimating QTEs requires only a few lines of code. These are plotted in Figure 9.6 for five values of p. Of course, here we know that the model is correctly specified but in real situations the modelling process, together with estimation, can be a time-consuming step. The main takeaway in this instance is that there could be distribution-based solutions that, from the modelling point of view, are essentially equivalent to distribution-free ones.

In the second, and more sophisticated, example of the MLQ method, we reproduce some of the results presented by Noufaily and Jones (2013). In their application, they assume that the strictly positive response variable Y follows a generalised gamma distribution (Stacy, 1962), parameterised in such a way that $\mu \in \mathbb{R}$, $\sigma > 0$ and $\kappa > 0$ represent, respectively, location, scale and shape parameters. For interpretation and computation, Noufaily and

R Code 9.8: Fitting linear-exponential models with nonlinear quantile regression.

```
set.seed(123)
n <- 500
X <- cbind(1, x <- runif(n, 0, 5))
b <- c(0,1)
g <- c(0.5,0.5)
y <- X%*%b + exp(X%*%g)*rnorm(n)

ss <- list(a = 0.1, b = 0.1, sigma = 0.1, g = 0.1)
try(nlrq(y ~ a + b*x + sigma*exp(g*x), tau = 0.1, start = ss))
> Error in numericDeriv(form[[3]], names(ind), env) :
>   Missing value or an infinity produced when evaluating the model
> Error in nlrq.calc(m, ctrl, trace) : optim unable to find valid step size
try(nlrq(y ~ a + b*x + sigma*exp(g*x), tau = 0.5, start = ss))
> Nonlinear quantile regression
>    model:  y ~ a + b * x + sigma * exp(g * x)
>     data:  parent.frame
>      tau:  0.5
> deviance:  1553.987
>      a      b sigma      g
>    0.1    0.1    0.1    0.1
try(nlrq(y ~ a + b*x + sigma*exp(g*x), tau = 0.9, start = ss))
> Nonlinear quantile regression
>    model:  y ~ a + b * x + sigma * exp(g * x)
>     data:  parent.frame
>      tau:  0.9
> deviance:  2000.803
>      a      b sigma      g
>    0.1    0.1    0.1    0.1
```

Jones (2013) deal with the transformed variable $Z = \{\log(Y) - \mu\}/\sigma$ (Lawless, 1980). The resulting PDF is given by

$$f_Z(z;\kappa) = \frac{\kappa^{\kappa-1/2}}{\Gamma(\kappa)} \exp\left\{\sqrt{\kappa}z - \kappa \exp\left(\frac{z}{\sqrt{\kappa}}\right)\right\},$$

while the quantile function associated with Y is obtained with

$$Q_Y(p) = \exp(\mu)\left\{\frac{Q_0(p)}{\kappa}\right\}^{\sigma\sqrt{\kappa}},$$

where $Q_0(p)$ is the quantile function of a gamma distribution with shape κ and unit scale (see Section A.3).

In the same spirit as Rigby and Stasinopoulos (2005), the distribution of Y is regressed on the covariate x so that

$$\mu_x = \beta_0 + \beta_1 x, \qquad \sigma_x = \exp(\gamma_0 + \gamma_1 x), \qquad \kappa_x = \exp(\delta_0 + \delta_1 x).$$

Differently from the previous example, the QTE associated with x has a rather complex

R Code 9.9: Predicting nonlinear conditional quantiles with generalised least squares regression.

```
library(nlme)
# fit a generalised least squares regression
# with exponential var function
fit <- gls(y ~ x, weights = varExp(form = ~ x))
# the fitted scale parameter sigma estimates
# exp(g[1]) from the data-generating model
gamma1 <- as.numeric(coef(fit$modelStruct))
print(thetahat <- c(coef(fit), gamma0 = log(fit$sigma),
gamma1 = gamma1))
> (Intercept)           x      gamma0      gamma1
>   0.1369662   0.9398373   0.4977479   0.5034510
# predict quantiles
probs <- c(0.10, 0.25, 0.50, 0.75, 0.90)
xx <- seq(0, 5, by = 0.05)
mu <- cbind(1, xx)%*%thetahat[c(1:2)]
sigma <- exp(cbind(1, xx)%*%thetahat[c(3:4)])
qhat <- sapply(probs, function(p) qnorm(p, mean = mu, sd = sigma))
# estimate QTE
qte <- gamma1*fit$sigma*exp(gamma1*xx)
qte <- coef(fit)[2] + outer(qte, qnorm(probs), "*")
```

form

$$\frac{\partial Q_{Y|X}(p)}{\partial x} = Q_{Y|X}(p)\frac{\partial \log Q_{Y|X}(p)}{\partial x},$$

where

$$\frac{\partial \log Q_{Y|X}(p)}{\partial x} = \beta_1 + \left\{\gamma_1\sigma_x\sqrt{\kappa_x} + \frac{\delta_1\sigma_x\kappa_x}{2\sqrt{\kappa_x}}\right\}\{\log Q_0(p) - \log\kappa_x\} - \delta_1\sigma_x\sqrt{\kappa_x}.$$

While the interpretation of marginal effects in nonlinear models is delicate in general, the one implied by the generalised gamma distribution is arguably unworkable.

In R Code 9.10, we fit the generalised gamma distribution to 298 immunoglobulin G (IgG) measurements (g/l) from healthy children aged six months to six years (Isaacs et al., 1983), with age as covariate. The value of the maximised log-likelihood and the parameter estimates are very close to those produced by Noufaily and Jones (2013). Predicted trajectories of IgG concentration for $p \in \{0.1, 0.25, 0.5, 0.75, 0.9\}$ are plotted in Figure 9.7. These are obtained with the MLQ method by plugging the parameter estimates directly into the quantile function, that is,

$$Q_{Y|X}(p) = \exp(\hat{\mu}_x)\left\{\frac{Q_0(p)}{\hat{\kappa}_x}\right\}^{\hat{\sigma}_x\sqrt{\hat{\kappa}_x}},$$

where $\hat{\mu}_x$, $\hat{\sigma}_x$ and $\hat{\kappa}_x$ are the ML estimates for a given value of $x \in [0, 6]$. Note that predictions extend to age 0 (newborns) although there are no observations available before age 0.5 years (analogously to Fig. 3 in Noufaily and Jones, 2013). One of the alleged advantages of having at one's disposal a 'global' model based on a parametric distribution (as opposed to a 'local' model such as a distribution-free quantile model) is the potential to make predictions beyond the range of the observed x. While this might be reasonable in principle,

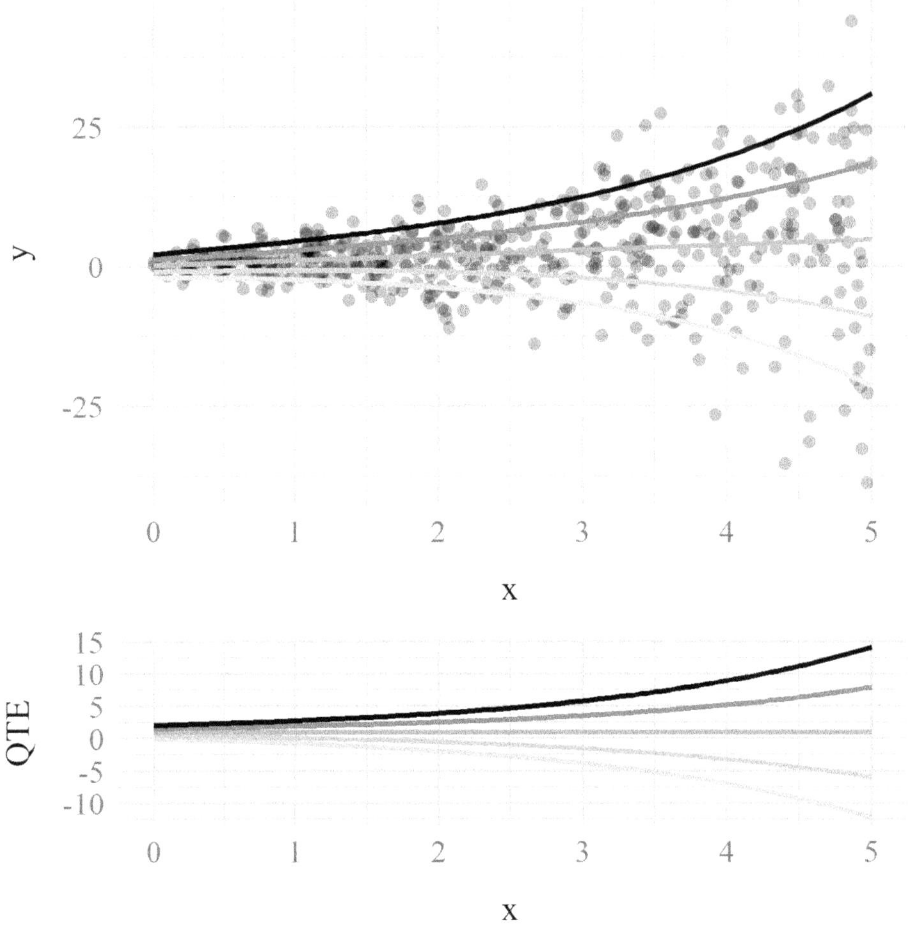

Figure 9.6
Data generated with R Code 9.9 together with five conditional quantiles (top) and quantile treatment effects (bottom) obtained from generalised least squares regression.

it needs careful scrutiny in reality. IgG concentrations in newborns are notoriously elevated due to maternal-foetal transfer during pregnancy. They decline in the first few months of age and increase afterwards, reaching the lowest levels at around 3-4 months (physiologic hypogammaglobulinaemia). IgG values 48–72 hours postnatally have been recorded in a range of approximately 4 to 20 g/l in term infants (Alkan Ozdemir et al., 2016). These considerations make the predictions below 6 months as shown in Figure 9.7 very unconvincing. For completeness, Figure 9.7 shows the estimated QTEs. It is fair to say that their graphical representation is much more intuitive than their analytical form. R Code 9.11 shows how to numerically calculate quantile treatment effects at different values of p using the parameter estimates obtained in R Code 9.10.

Concerning modelling type (ii), referred to at the beginning of this section, it is worth delving into something that goes under the label of *parametric quantile regression* (e.g., see Mazucheli et al., 2022; Stülp et al., 2025, and references therein) that is different from the MLQ method discussed above. Suppose $Y \sim F_Y(y;\boldsymbol{\theta})$, $Y \in (a,b) \subseteq \mathbb{R}$, is a random variable with CDF indexed by $\boldsymbol{\theta} \equiv (\gamma,\boldsymbol{\delta}) \in \boldsymbol{\Theta} \subseteq \mathbb{R}\times\mathbb{R}^q$, and $Q_Y(p;\boldsymbol{\theta}) = F_Y^{-1}(p;\boldsymbol{\theta})$ is given in explicit

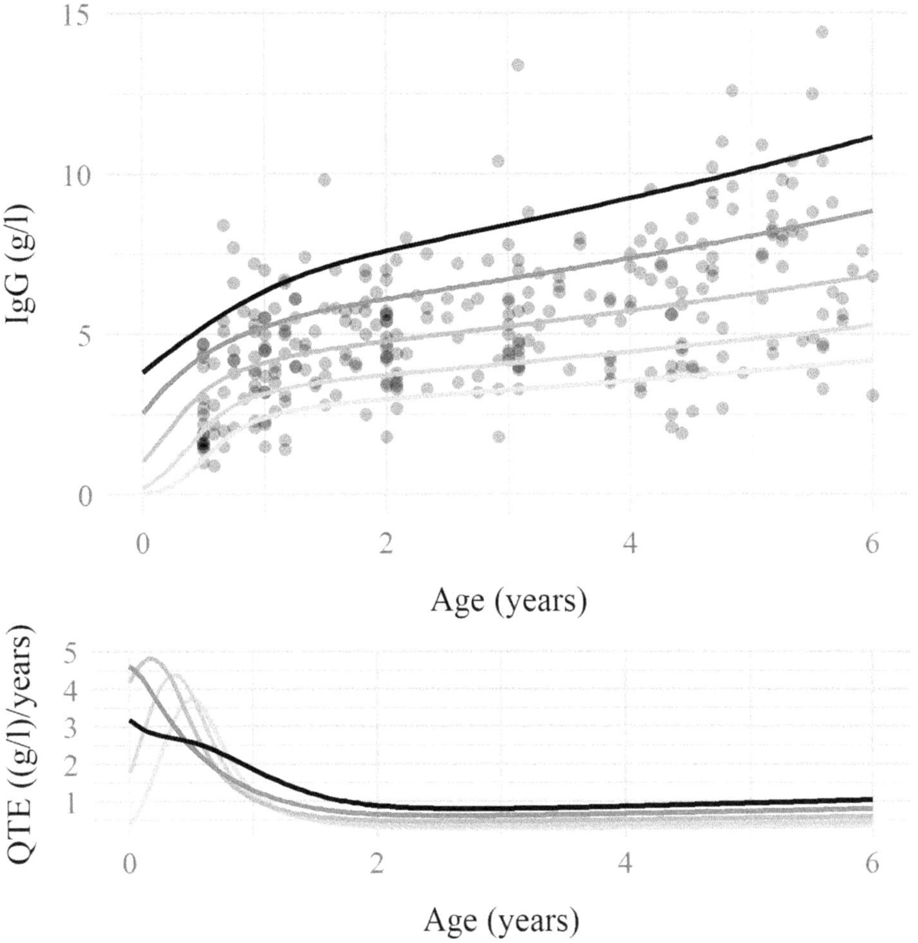

Figure 9.7
Predicted quantiles ($p \in \{0.1, 0.25, 0.5, 0.75, 0.9\}$) of immunoglobulin G concentration conditional on age (top) and estimated quantile treatment effects (bottom) in the IgG dataset using the generalised gamma model.

form. The scalar parameter γ is elected to be the main target of conditional inference. Define $Q_Y(p; \gamma, \boldsymbol{\delta}) = g_p(\gamma; \boldsymbol{\delta}) \equiv \mu_p$ for $p \in (0,1)$. The CDF is reparameterised in terms of μ_{p_0}, i.e. $F_Y(y; g_{p_0}^{-1}(\mu_{p_0}; \boldsymbol{\delta}))$. We use the notation p_0 to stress that this value is fixed and to distinguish it from p, the variable that generically indexes a quantile function. The composition $F_Y \circ g_{p_0}^{-1}$ is such that the p_0th quantile of Y is μ_{p_0}, i.e. $F_Y(\mu_{p_0}; g_{p_0}^{-1}(\mu_{p_0}; \boldsymbol{\delta}), \boldsymbol{\delta}) = p_0$. So, while γ may or may not have a straightforward interpretation, μ_{p_0} always does. The latter is subsequently modelled as a function of covariates and regression coefficients, i.e. $h(\mu_{p_0}) = \boldsymbol{x}^\top \boldsymbol{\beta}(p_0)$ where $h : (a, b) \rightarrow \mathbb{R}$ is the 'link' function. The estimation proceeds by maximum likelihood using the density

$$f_{Y|X}(y; g_{p_0}^{-1}\left[h^{-1}\{\boldsymbol{x}^\top \boldsymbol{\beta}(p_0)\}; \boldsymbol{\delta}\right], \boldsymbol{\delta}). \tag{9.24}$$

If, at this point, one finds the expression in (9.24) somewhat obscure and wonders how it serves the ultimate purpose of doing quantile regression, then they are not to blame. By anchoring the distribution of Y to its p_0th conditional quantile, the coefficients $\boldsymbol{\beta}(p_0)$ will reflect the quantile structure commanded by h *and* g at that particular quantile, which

R Code 9.10: Predicting nonlinear conditional quantiles in the IgG dataset using maximum likelihood for quantiles.

```
qgg <- function(p, mu, sigma, kappa){
 qp <- qgamma(p, shape = kappa, scale = 1)/kappa
 exp(mu)*qp^(sigma*sqrt(kappa))
}

dlgg <- function(x, mu, sigma, kappa, log = FALSE){
 z <- (x - mu)/sigma
 val1 <- (kappa - 0.5)*log(kappa) - log(sigma) - lgamma(kappa)
 val2 <- z*sqrt(kappa) - kappa*exp(z/sqrt(kappa))
 val <- val1 + val2
 if(!log) val <- exp(val)
 return(val)
}

loglik <- function(theta, x, y){
 p <- ncol(x)
 mu <- x %*% theta[1:p]
 sigma <- exp(x %*% theta[(p+1):(2*p)])
 kappa <- exp(x %*% theta[(2*p+1):(3*p)])
 val <- -sum(dlgg(y, mu, sigma, kappa, log = TRUE))
 return(val)
}

# obtain the MLE using the GG likelihood
data(ImmunogG, package = "Brq")
fit <- do.call(optim, args = list(par = rep(1, 6), fn = loglik, x =
cbind(1,ImmunogG$Age), y = log(ImmunogG$IgG), method = "BFGS"))
-fit$value # loglik
> [1] -138.5069
round(fit$par, 4) # estimates
> [1]  1.3931  0.0881 -1.0292  0.0109 -3.6458  4.9438
# predict quantiles using the GG distribution
xx <- seq(0, 6, by = 0.05)
muhat <- cbind(1, xx)%*%fit$par[c(1:2)]
sigmahat <- exp(cbind(1, xx)%*%fit$par[c(3:4)])
kappahat <- exp(cbind(1, xx)%*%fit$par[c(5:6)])
probs <- c(0.10, 0.25, 0.50, 0.75, 0.90)
qhat <- sapply(probs, function(p) qgg(p, mu = muhat,
sigma = sigmahat, kappa = kappahat))
```

will ultimately determine its role and interpretation. For example, Mazucheli et al. (2022) considered the unit-interval bounded variable $Y = [1 + \exp\{-(Z - \gamma)/\delta\}]^{-1}$, $\gamma \in \mathbb{R}$, $\delta > 0$, where $Z \sim \text{Logistic}(0, 1)$ follows the standard logistic distribution (see definition in Appendix A.3). This distribution is called *logit-logistic* in Smithson and Shou (2017). The

R Code 9.11: Numerical approximation of the nonlinear quantile treatment effects using the results in R Code 9.10.

```
# define quantile as a function of x, given the parameter theta and
# probability p
Q <- function(x, theta, p){
    mu <- sum(c(1,x)*theta[1:2])
    sigma <- exp(sum(c(1,x)*theta[3:4]))
    kappa <- exp(sum(c(1,x)*theta[5:6]))
    qgg(p, mu, sigma, kappa)
}
# estimate the QTE for the quantile, e.g., 0.1 along a sequence xx
library(numDeriv)
qte <- sapply(xx, function(x) grad(func = Q, x, theta = fit$par, p = 0.1))
```

density and quantile functions of Y are given by

$$f(y;\gamma,\delta) = \frac{\delta\exp(\gamma)\left(\dfrac{y}{1-y}\right)^{\delta}}{y(1-y)\left\{1+\exp(\gamma)\left(\dfrac{y}{1-y}\right)^{\delta}\right\}^{2}} \tag{9.25}$$

and

$$Q(p;\gamma,\delta) \equiv \mu_p = \frac{\exp\left(-\frac{\gamma}{\delta}\right)\left(\dfrac{p}{1-p}\right)^{\frac{1}{\delta}}}{1+\exp\left(-\frac{\gamma}{\delta}\right)\left(\dfrac{p}{1-p}\right)^{\frac{1}{\delta}}}, \tag{9.26}$$

respectively. We need to define the parameter γ in (9.26) as a function of the pth quantile μ_p. We first take the logit transformation of μ_p

$$\log\left(\frac{\mu_p}{1-\mu_p}\right) = \frac{1}{\delta}\log\left(\frac{p}{1-p}\right) - \frac{\gamma}{\delta}$$

and then solve for γ, that is,

$$\gamma = g_p^{-1}(\mu_p;\delta) = \log\left(\frac{p}{1-p}\right) - \delta\log\left(\frac{\mu_p}{1-\mu_p}\right). \tag{9.27}$$

Expression (9.27) is plugged back into the PDF (9.25) for fixed $p = p_0$. As a result, we have

$$f(y; g_{p_0}^{-1}(\mu_{p_0};\delta),\delta) = \frac{\delta\left(\dfrac{p_0}{1-p_0}\right)\left(\dfrac{1-\mu_{p_0}}{\mu_{p_0}}\right)^{\delta}\left(\dfrac{y}{1-y}\right)^{\delta}}{y(1-y)\left\{1+\left(\dfrac{p_0}{1-p_0}\right)\left(\dfrac{1-\mu_{p_0}}{\mu_{p_0}}\right)^{\delta}\left(\dfrac{y}{1-y}\right)^{\delta}\right\}^{2}}. \tag{9.28}$$

Next, one can take the regression model to be

$$\log\left(\frac{\mu_{p_0}}{1-\mu_{p_0}}\right) = \boldsymbol{x}^{\top}\boldsymbol{\beta}(p_0), \tag{9.29}$$

and estimate $\boldsymbol{\beta}(p_0)$ by maximising the (log)likelihood based on (9.28).

R Code 9.12: Simulation study comparing a misspecified distribution-based model and nonlinear quantile regression in terms of coefficient estimation and conditional quantile prediction at the quantile level $p = 0.75$.

```
set.seed(765)
R <- 500
p <- 0.75
b <- c(0, -2)
n <- 1000
res <- array(NA, dim = c(3,2,R,2))
nr <- array(NA, dim = c(R,2))
colnames(nr) <- dimnames(res)[[4]] <- c("unitquantreg","nlrq")

for(i in 1:R){
    x <- runif(n, -2, 2)
    y <- rbeta(n, shape1 = exp(x), shape2 = exp(2*x))
    y[y == 0] <- 0.001
    y[y == 1] <- 0.999
    dd <- data.frame(x = x, y = y)

    fit.uq <- try(unitquantreg::unitquantreg(formula = y ~ x,
    data = dd, family = "ulogistic", tau = p, link = "logit"),
    silent = TRUE)
    fit.nlrq <- try(nlrq(y ~ SSlogis(x, Asym, xmid, scal),
    data = dd, tau = p), silent = TRUE)

    if(!any(inherits(fit.uq, "try-error"), inherits(fit.nlrq, "try-error")))
    {res[,,i,1] <- summary(fit.uq)[[1]][,1:2]
    res[,,i,2] <- summary(fit.nlrq)$coefficients[,1:2]
    nr[i,1] <- mean(y - predict(fit.uq, type = "quantile") < 0)
    nr[i,2] <- mean(y - predict(fit.nlrq) < 0)}
}
```

If the data truly came from the logit-logistic density in (9.28), then of course inference would be correct in terms of both point and interval estimation. Otherwise, the model would be misspecified. This is exemplified in R Code 9.12 with a small simulation based on 1000 replications, where the data are generated from a beta distribution and fitted to the quantile 0.75 with the logistic model in (9.29) and the density in (9.28) using the R package `unitquantreg` (Menezes and Mazucheli, 2023). The results are compared with those obtained from an analogous logistic model fitted with `quantreg::nlrq`. The contrast between the mean standard error and the standard deviation of the estimates obtained from the logit-logistic likelihood indicates, as expected, that the uncertainty is grossly underestimated (Table 9.1). Also, the proportion of negative residuals is severely off the target, hence the conditional quantile model itself, as determined by the structure of the whole model (9.28), is incorrect. In contrast, the nonlinear quantile model is free from any additional structure besides the logistic functional form and yields correct standard errors and proportion of negative residuals.

The final verdict on model (9.28) boils down to answering the question: is there any advantage in going through the hassle of reparameterising the density when, in fact, the data

Table 9.1
Simulation results from R Code 9.12 comparing `unitquantreg` and `nlrq` at the quantile level $p = 0.75$. Reported are the mean estimated standard error (SE), the empirical standard deviation (SD) of the parameter estimates, and the proportion of negative residuals (PNR).

	unitquantreg	nlrq
Mean SE	0.3289	0.0506
SD of estimates	0.1372	0.0380
PNR	0.8681	0.7401

are compatible with the underlying distribution? The answer is that there is none because, on the logit scale, the model is just location-shift.[8] By replacing the logit $\log\{\mu_p/(1-\mu_p)\}$ in (9.27) with the expression in (9.29), one immediately realises that the slopes are constant with respect to p. The apparent flexibility of this parametrisation is illusory: once the distributional structure is fixed, the quantile process is fully determined and cannot exhibit the heterogeneous effects that motivate quantile regression in the first place. To put it simply, we can fit (only once!) the likelihood based on the original density in (9.25) with $\gamma = \boldsymbol{x}^\top \boldsymbol{b}$ and obtain $\boldsymbol{\beta}(p_0)$ in (9.29), for any value of p_0, with a simple transformation. Indeed, let

$$\hat{\boldsymbol{\theta}} = \begin{pmatrix} \hat{\boldsymbol{b}} \\ \hat{\lambda} \end{pmatrix}, \qquad \hat{\delta} = \exp(\hat{\lambda}),$$

where $\hat{\boldsymbol{b}} = (\hat{b}_0, \hat{b}_1)^\top$ and $\hat{\lambda}$ is the estimate of $\log \delta$. The 'quantile-specific' parameter estimate implied by model (9.28) is therefore equivalently obtained through the transformation

$$\hat{\boldsymbol{\beta}}(p_0) = g(\hat{\boldsymbol{\theta}}) = -\hat{\delta}^{-1}\left(\hat{\boldsymbol{b}} + \begin{bmatrix} \log\{(1-p_0)/p_0\} \\ 0 \end{bmatrix}\right).$$

Naturally, one needs to account for such a transformation when calculating the corresponding standard errors. By the delta method,

$$\mathrm{var}\{\hat{\boldsymbol{\beta}}(p_0)\} = \boldsymbol{G}\boldsymbol{V}\boldsymbol{G}^\top,$$

where $\boldsymbol{V}$ is the variance-covariance of $\hat{\boldsymbol{\theta}}$ estimated via the observed inverse Hessian, and

$$\boldsymbol{G} = \left.\frac{\partial g(\boldsymbol{\theta})}{\partial \boldsymbol{\theta}^\top}\right|_{\boldsymbol{\theta}=\hat{\boldsymbol{\theta}}} = \begin{bmatrix} -\hat{\delta}^{-1} & 0 & -\hat{\beta}_0(p_0) \\ 0 & -\hat{\delta}^{-1} & -\hat{\beta}_1(p_0) \end{bmatrix}.$$

This procedure is shown in R Code 9.13 where we used a single covariate and $p_0 = 0.75$. The estimated coefficients and standard errors from `unitquantreg` and the standard MLE, nearly identical to the fourth decimal, are reported in Table 9.2.

We conclude this example by noting that the MLQ estimation in R Code 9.13 could also be carried out using the logit-logistic family that is implemented in the R package

[8]We found that these issues, as well as other more serious ones, were not limited to the logistic model. In general, a word of caution must be given to the reader who is encouraged to critically examine the implications of this form of parametrisation when applied to a specific model.

R Code 9.13: Maximum likelihood estimation of quantile regression coefficients using different parameterisations of the logit-logistic distribution.

```
dlogis2 <- function(y, gamma, delta, log = FALSE){
 num <- delta*exp(gamma)*(y/(1-y))^delta
 den <- y*(1-y)*(1 + exp(gamma)*(y/(1-y))^delta)^2
 val <- num/den
 if(log) val <- log(val)
 return(val)
}

loglik_logis <- function(theta, x, y){
 k <- ncol(x)
 gamma <- x%*%theta[1:k]
 delta <- exp(theta[k+1])
 -sum(log(dlogis2(y, gamma, delta)))
}
# generate data from (unit) logistic with given p0
set.seed(765)
p0 <- 0.75
n <- 1000; delta <- 2; beta_p0 <- c(0, -2)
x <- runif(n, -2, 2)
eta <- cbind(1,x)%*%beta_p0
mu_p0 <- exp(eta)/(1 + exp(eta))
y <- unitquantreg::rulogistic(n, mu = mu_p0, theta = delta, tau = p0)
dd <- data.frame(x = x, y = y)

# fit (true) density with quantile-based parameterisation
fit.uq <- unitquantreg::unitquantreg(formula = y ~ x, data = dd,
family = "ulogistic", tau = p0, link = "logit")

# fit original density (i.e, not reparameterised)
fit.mle <- optim(par = c(0,0,0), loglik_logis, x = cbind(1,x), y = y,
hessian = TRUE)
deltahat <- exp(fit.mle$par[3])
# obtain the same estimates obtained with unitquantreg
betahat <- -(fit.mle$par[1:2] + c(log((1-p0)/p0), 0))/deltahat
# obtain standard errors with delta method
V_thetahat <- solve(fit.mle$hessian)
G <- matrix(0, nrow = 2, ncol = 3)
G[1,1] <- -1/deltahat; G[2,2] <- -1/deltahat; G[,3] <- -betahat
V_betahat <- G %*% V_thetahat %*% t(G)
se_betahat <- sqrt(diag(V_betahat))
```

`cdfquantreg` (Shou and Smithson, 2019), as illustrated in R Code 9.14. The term 'CDF-quantile regression' used by Shou and Smithson (2019) should *not* be interpreted as referring to the usual QTE $\beta(p)$ considered in this book; rather, these methods are more naturally viewed within the general MLQ framework, as they follow the same underlying logic.

Table 9.2
Estimates and standard errors (SE) of the p_0-specific regression coefficients in R Code 9.13 using **unitquantreg** and standard maximum likelihood estimation (MLE).

	unitquantreg		MLE	
Intercept	0.0249	0.0311	0.0250	0.0311
Slope	−1.9878	0.0236	−1.9878	0.0236

R Code 9.14: Maximum likelihood estimation for the logit-logistic distribution with the R package **cdfquantreg**.

```
fit.cdfq <- cdfquantreg::cdfquantreg(y ~ x | 1, data = dd,
family = "logit-logistic")
# account for parameterisation used in Smithson and Shou (2017)
d <- exp(-cdfquantreg:::coef.cdfqr(fit.cdfq)[3])
as.numeric(c(d*(-cdfquantreg:::coef.cdfqr(fit.cdfq)[1:2]), d))
> [1] 1.048678 3.971588 1.998017
# compare with MLE in previous R Code
c(fit.mle$par[1:2], exp(fit.mle$par[3]))
> [1] 1.048671 3.971571 1.998008
```

10

Quantile regression for discrete responses

10.1 Introduction

In Chapter 4, we introduced the issues surrounding quantile inference with discrete distributions. In particular, it was noted that sample quantiles may lack consistency and asymptotic normality. In this chapter, we discuss methods for conditional quantiles when the response is discrete.

Generalised linear models (GLMs) have a special place in the analysis of discrete responses. They are arguably *the* regression models for conditional discrete distributions. The reasons are manifold and include convenient interpretation of the regression coefficients (e.g., as (log) odds ratios in logistic regression or (log) rate ratios in Poisson regression), universal availability in statistical software, and the benefits of a well-developed, unifying maximum likelihood theory. Recent developments in quantile regression for discrete responses have shown that more flexibility can be achieved using a distribution-free approach.

Discrete outcomes provide a natural stress test for the quantile-based perspective. In continuous settings, quantiles inherit many convenient properties from the underlying distribution, whereas in discrete settings they are no longer uniquely defined and may exhibit flat regions. This apparent limitation is, however, informative: it reflects intrinsic features of the data-generating process that are often obscured by mean-based models. While generalised linear models focus on the conditional mean and require specific distributional assumptions, quantiles remain well-defined as order-based functionals and retain a direct interpretation in terms of outcome levels. Extending quantile methods to discrete responses is therefore not merely a technical exercise, but a necessary step towards a unified distributional framework that applies across both continuous and discrete domains.

We begin with a familiar regression model. Suppose Y is a count variable such as, for example, the number of patients entering the emergency department in a week or the number of particles recorded in a single-bin analysis, and let $\boldsymbol{x} = (1, x_1, x_2, \ldots, x_k)^\top$ denote a realisation of the covariate vector $\boldsymbol{X} = (1, X_1, X_2, \ldots, X_k)^\top$. A typical assumption for the distribution of Y conditional on $\boldsymbol{X} = \boldsymbol{x}$ is the Poisson model with parameter

$$\lambda_x \equiv \mathrm{E}(Y \mid \boldsymbol{x}) = \exp\left(\boldsymbol{x}^\top \boldsymbol{\beta}\right) \tag{10.1}$$

and logarithmic link function. The conditional variance also has the same form as the conditional mean, that is, $\lambda_x = \mathrm{var}(Y \mid \boldsymbol{x})$, while moments of order higher than 2 depend on the same parameter (e.g., see Section 3.1). If, on the one hand, a Poisson model is more enticing than a location-shift regression, on the other, every component of the conditional location-scale-shape (LSS) structure is controlled by the same parameter.

Conditional quantiles of discrete distributions follow the general definition $Q_Y(p \mid \boldsymbol{x}) = \inf\{y \in S_{Y|\boldsymbol{X}} : F_Y(y \mid \boldsymbol{x}) \geq p\}$, where $S_{Y|\boldsymbol{X}}$ denotes the support of Y conditional on $\boldsymbol{X}$. Because the conditional quantile function of a discrete response is stepwise in the support of Y, quantile effects are generally described through finite differences rather than derivatives, that is, $\Delta(p) = Q_Y(p \mid x_1, \tilde{\boldsymbol{x}}) - Q_Y(p \mid x_0, \tilde{\boldsymbol{x}})$, for a variation of the regressor, say, x, from

DOI: 10.1201/9781003326410-10

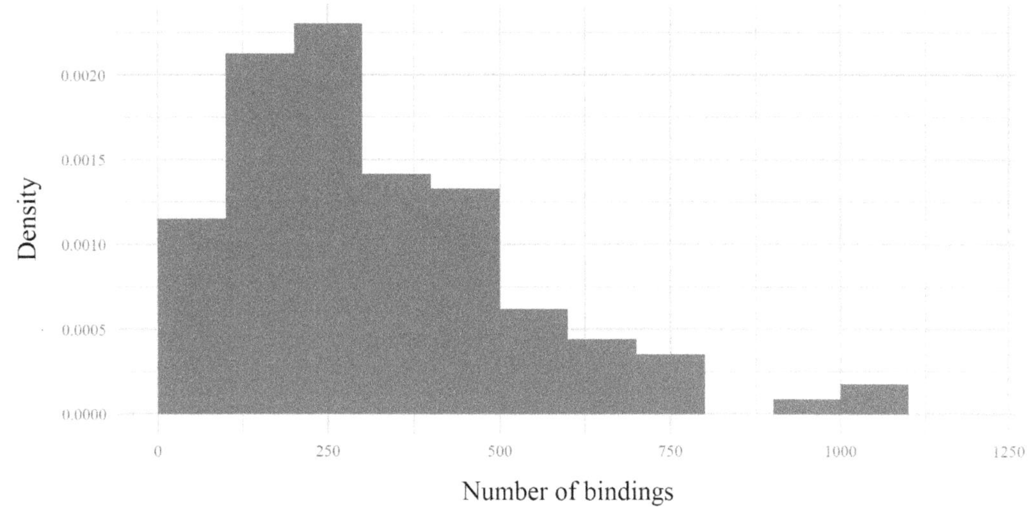

Figure 10.1
Histogram of the number of bindings in the Esterase Assay dataset.

x_0 to x_1, while keeping the covariates that do not depend on x, $\tilde{\boldsymbol{x}}$, fixed at some value. In many (non-trivial) cases, these calculations must be carried out numerically (rather than symbolically) since finding Q requires summing probabilities until F reaches or exceeds the quantile threshold p. In the Poisson regression model, for example, $Q_Y(p \mid \boldsymbol{x}) = \inf\{y \in \mathbb{N} : F_Y(y; \lambda_x) \geq p\}$ where $F_Y(y; \lambda_x) = \Gamma_r(\lfloor y+1 \rfloor, \lambda_x)$ (see definition in Appendix A.2). Although there is no closed formula, calculations can be easily carried out with commonly available software.

We now introduce an example to motivate the methods that will be presented in the next section. The Esterase Assay dataset, analysed by Carroll and Ruppert (1988, p. 48), provides $n = 113$ assay concentrations of an enzyme esterase and the counts of bindings in a binding experiment at those esterase levels. As explained by Carroll and Ruppert (1988), this is an example of a calibration experiment where the goal is to take observed counts in a new experiment and infer the concentration of esterase, especially for smaller values of the latter, by inverting the relationship found in the calibration (training) dataset. The histogram of the observed counts (Figure 10.1) suggests a right-skewed response distribution, which is qualitatively consistent with a Poisson-type count model. In contrast, the mean and variance of the counts are very different and equal to, respectively, 322 and 45,437. Still, we need to examine the regression fit before drawing any conclusions.

Based on the first specification of the model, with $\log(\lambda_x) = \beta_0 + \beta_1 x$, where x represents the esterase concentration, two problems are evident in the left plot of Figure 10.2. First, the log-linear specification in the original scale of esterase does not capture well the shape of the relationship, especially at low concentrations. Second, the model substantially underestimates the conditional variability (something Carroll and Ruppert (1988) expected based on their experience of analysing assay data), as shown by the 10th and 90th fitted quantiles lying too close to the median. Indeed, the predicted 10th and 90th centiles are very close to the median, with a proportion of negative residuals equal to 42% and 59%, respectively (Table 10.1), indicating that the fitted conditional spread is too narrow. The residual deviance, equal to 4606 on 111 degrees of freedom, confirms the presence of overdispersion, as does the p-value of the related test (Cameron and Trivedi, 1990). The overdispersion test is available in the R package `AER` (Kleiber and Zeileis, 2026, 2008).

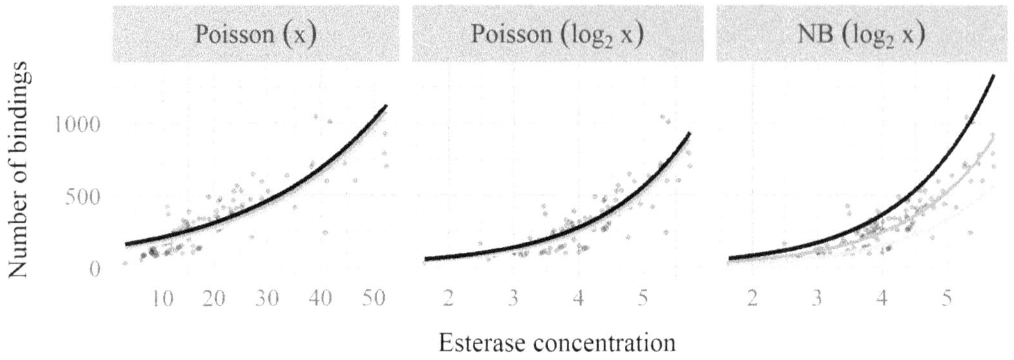

Figure 10.2
Predicted quantiles ($p \in \{0.1, 0.5, 0.9\}$) of number of bindings conditional on esterase concentration in the Esterase Assay dataset using: Poisson regression with x on the original scale (left); Poisson regression with x on the $\log_2$-scale (centre); negative binomial (NB) regression with x on the $\log_2$-scale (right).

Table 10.1
Summary statistics for the generalised linear models fitted in the Esterase Assay dataset: residual deviance (degrees of freedom), Akaike information criterion (AIC), overdispersion test p-value, negative binomial (NB) overdispersion parameter, proportion of negative residuals (PNR, %) at three conditional quantiles ($p \in \{0.1, 0.5, 0.9\}$).

Model	Poisson	Poisson ($\log_2 x$)	NB ($\log_2 x$)
Residual deviance (df)	4606 (111)	3340 (111)	115 (110)
AIC	5445	4179	1336
Dispersion test p-value	< 0.001	< 0.001	–
NB overdispersion	–	–	9.5
PNR	42, 49, 59	39, 50, 65	12, 42, 94

In R Code 10.1, we show the steps taken to address the issues noted above. By fitting the Poisson model[1] $\log(\lambda_x) = \beta_0 + \beta_1 \log_2 x$, we improve substantially the fit as it better aligns with the data points (centre plot of Figure 10.2), with the resulting residual deviance now considerably lower and equal to 3340 (Table 10.1). To address the issue of overdispersion, we fit a negative binomial regression with log link (again, with log-transformed x) by means of the R package `MASS` (Ripley and Venables, 2025; Venables and Ripley, 2002). While the model for the conditional mean is the same as in the Poisson discussed above, the conditional variance of the negative binomial allows for extra-variation.[2]

This distribution provides a superior fit compared to the Poisson, in terms of both AIC and proportions of negative residuals (Table 10.1). Still, the latter are not yet ideal, especially the value of 42% at the median. A simulation-based test comparing the observed

[1]In such a model, the interpretation of β_1 is the change in the log-expected count of bindings associated with a doubling of the esterase concentration.

[2]The negative binomial can be derived from a continuous mixture of Poisson distributions with mixing gamma distribution, that is, if $Y \mid U \sim \text{Pois}(U\lambda)$ and $U \sim \mathcal{G}(\theta, \theta)$, then marginally Y is negative binomial with mean λ and variance $\lambda + \lambda^2/\theta$. It follows that the Poisson is a limiting case of the negative binomial. The parameter θ can be fixed to a known value in `glm(..., family = negative.binomial(theta))` or estimated with `MASS::glm.nb`.

R Code 10.1: Estimating Poisson and negative binomial regressions in the Esterase Assay dataset.

```
library(MASS)
data("esterase", package = "Qtools")
dd <- data.frame(logx = log(esterase$Esterase, base = 2),
y = esterase$Count)
fit.pois <- glm(y ~ logx, data = dd, family = poisson)
fit.pois
>
> Call:  glm(formula = y ~ logx, family = poisson, data = dd)
>
> Coefficients:
> (Intercept)         logx
>      2.6685       0.7217
>
> Degrees of Freedom: 112 Total (i.e. Null);  111 Residual
> Null Deviance:      14710
> Residual Deviance: 3340   AIC: 4179
AER::dispersiontest(fit.pois)
>
>  Overdispersion test
>
> data:  fit.pois
> z = 5.4801, p-value = 2.126e-08
> alternative hypothesis: true dispersion is greater than 1
> sample estimates:
> dispersion
>   29.38335
summary(fit.nb <- glm.nb(y ~ logx, data = dd))
>
> Call:
> glm.nb(formula = y ~ logx, data = dd, init.theta = 9.506854602,
>     link = log)
>
> Coefficients:
>             Estimate Std. Error z value Pr(>|z|)
> (Intercept)   2.5317     0.1657   15.28   <2e-16 ***
> logx          0.7535     0.0397   18.98   <2e-16 ***
> ---
> Signif. codes:  0 '***' 0.001 '**' 0.01 '*' 0.05 '.' 0.1 ' ' 1
>
> (Dispersion parameter for Negative Binomial(9.5069) family taken to be 1)
>
>     Null deviance: 462.22  on 112  degrees of freedom
> Residual deviance: 115.31  on 111  degrees of freedom
> AIC: 1336
>
> Number of Fisher Scoring iterations: 1
>
>
>               Theta:  9.51
>           Std. Err.:  1.30
>
>  2 x log-likelihood:  -1330.002
```

R Code 10.2: Testing the proportion of negative residuals from the negative binomial regression for the Esterase Assay dataset.

```
# predicted quantiles of fitted negative binomial
probs <- c(0.10, 0.50, 0.90)
mu <- fitted(fit.nb)
qhat <- sapply(probs, function(p) qnbinom(p, size = fit.nb$theta, mu = mu))
# number of negative residuals
print(nr <- apply(sweep(-qhat, 1, dd$y, "+") < 0, 2, sum))
> [1]  14  47 106
# select the median
j <- 2
# exact probabilities under the fitted negative binomial model
pi_hat <- pnbinom(qhat[, j] - 1, size = fit.nb$theta, mu = mu)
# simulation-based exact p-value for the Poisson-binomial null
set.seed(123)
B <- 100000
Tb <- rowSums(matrix(runif(B * length(pi_hat)), B) < rep(pi_hat, each = B))
2 * min(mean(Tb <= nr[j]), mean(Tb >= nr[j]))
> [1] 0.09924
```

number of negative residuals with its distribution under the fitted negative binomial model provides some evidence, albeit limited, against the adequacy of the fitted median, as shown in R Code 10.2.

These findings highlight the limitations of mean-based models in capturing distributional features, motivating the quantile-based approaches developed in the next section.

10.2 Jittering

At the end of Chapter 8, we briefly considered the problems associated with (heavily) rounded continuous measurements. We also discussed jittering as an *ad hoc* solution. This form of smoothing, going back to Stevens (1950) (see also Koenker, 2005, p. 260), has been used in conditional quantile modelling of count responses.

Let Y be a random variable resulting from a count process with PMF

$$f_Y(k) = \begin{cases} \Pr(Y = k) & \text{if } k \in \mathbb{N} \\ 0 & \text{otherwise.} \end{cases}$$

Define $Z = Y + U$ as the variable obtained by jittering Y with a $[0, 1)$-uniform noise U, independent of Y. The PDF of this convolution is $f_Z(z) = \sum_{k=0}^{\infty} f_Y(k) I(k \leq z < k+1)$, for $z \geq 0$, while the CDF and QF are, respectively,

$$F_Z(z) = F_Y(k-1) + f_Y(k)(z-k), \quad k \leq z < k+1 \tag{10.2}$$

and

$$Q_Z(p) = Q_Y(p) + \frac{p - F_Y(Q_Y(p) - 1)}{f_Y\{Q_Y(p)\}}. \tag{10.3}$$

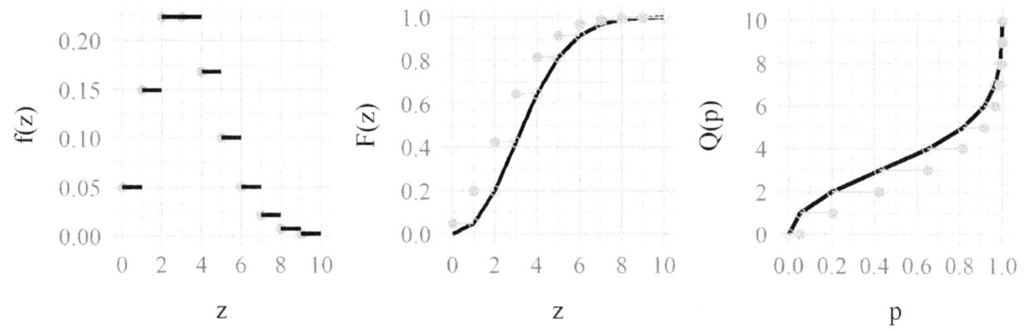

Figure 10.3
Probability (left), cumulative distribution (centre), and quantile (right) functions of a Poisson random variable (grey lines and circles) and those of its convolution with a uniform random variable (black lines).

An example is provided in Figure 10.3 using $Y \sim \text{Pois}(3)$. The PDF of Z retains the shape of the Poisson but it is stepwise-constant between integers. As a result, the CDF of Z is piecewise linear with slope $\Pr(Y = k)$ for $z \in [k, k+1)$. One can verify that F_Z dominates F_Y with the comparison quantile function $\kappa(z; F_Y, F_Z) = F_Z^{-1}(F_Y(z)) > z$. In particular, $Q_Z(p)$ is bounded from below by p (i.e., the pth quantile of U). The quantile of Y can be recovered, apart from boundary cases, with the relation $Q_Y(p) = \lceil Q_Z(p) - 1 \rceil$ for the present uniform-jitter construction.

By extending Equation (10.3) to the conditional case, Machado and Santos Silva (2005) considered the model

$$Q_Z(p \mid \boldsymbol{x}) = p + \exp\left\{\boldsymbol{x}^\top \boldsymbol{\beta}(p)\right\}, \tag{10.4}$$

where the offset p is the lower bound mentioned above. The exponential relationship follows from the tradition of assuming log-linear models for count responses. However, no parametric assumption is made on the distribution of Y. At this point, the linear transformation

$$Q_{h(Z;p)}(p \mid \boldsymbol{x}) = \boldsymbol{x}^\top \boldsymbol{\beta}(p), \tag{10.5}$$

where

$$h(Z; p) = \begin{cases} \log(Z - p) & \text{for } Z > p \\ \log \zeta & \text{for } Z \leq p, \end{cases} \tag{10.6}$$

completes the modelling step. The role of the small constant $\zeta \in (0, p)$ is to avoid taking the logarithm of a zero. Since quantiles are invariant to censoring from below up to the quantile of interest, this adjustment bears no consequences for estimation. As far as the interpretation of the regression slopes goes, we can refer to the QTE of the jittered (smoothed) model (assuming $\boldsymbol{x} = (1, x)^\top$ for simplicity)

$$\frac{\partial}{\partial x} Q_Z(p \mid x) = \beta_1(p) \exp\{\beta_0(p) + \beta_1(p)x\}.$$

This is not, however, the actual QTE on Y, but, loosely speaking, a smooth approximation of it.

Because jittering produces a continuous surrogate response, standard linear quantile regression machinery (Koenker and Bassett, 1978) can be applied to each jittered sample by fitting the model in (10.4). In practice, a sample of M jittered responses Z is taken to estimate $\hat{\boldsymbol{\beta}}_m(p)$, $m = 1, \ldots, M$, and the noise is then averaged out, $\hat{\boldsymbol{\beta}}(p) = \frac{1}{M}\sum_m \hat{\boldsymbol{\beta}}_m(p)$ (similarly to bootstrap point estimation), as exemplified by the following `R` pseudo-code:

1. generate the noise `u <- runif(n, 0, 0.999)` and create the vector `z <- y + u`, where `y` is the vector with `n` observed counts;
2. transform `z` $\longrightarrow$ `tz` using the transformation in (10.6) with an arbitrary small value of ζ (e.g., 10^{-5}) and fit the pth quantile regression model `fit <- rq(tz ~ x, tau = p)` conditional on the covariate `x`; record the regression quantiles `coef(fit)`;
3. repeat steps 1 and 2 above for M times, and then take the average of the M sets of estimated coefficients.

Of course, the seed for random number generation should be set to ensure replicable results. Calculating the variance of the estimator is more laborious than doing a simple average of variances (one could draw a parallel with Rubin's rule for bootstrap variance estimation). We refer the reader to Machado and Santos Silva (2005) for details on computation of the large-n approximation of the standard errors.

We now fit model (10.4) to the Esterase Assay dataset using the function `Qtools::rq.counts`. Estimation is carried out using $M = 50$ jittered responses (R Code 10.3). Figure 10.4 shows the resulting fit for three quantiles, which confirms the extra-Poisson variation already seen in the negative binomial fit (Figure 10.2). As compared to the latter, the discrete quantile regression model yields proportions of negative residuals that are closer to the target values: 12% ($p = 0.1$), 50% ($p = 0.5$), and 91% ($p = 0.9$). Unfortunately, the lack-of-fit test (see Section 8.3.3) applied to the `rq.counts` fits rejects the exponential model for $p = 0.1$ and, particularly, $p = 0.5$. Thus, further modelling would be required.[3]

Besides comparing models in terms of predictions, one may also wonder whether directly estimating a nearby conditional quantile leads to comparable or smaller estimation uncertainty. A direct comparison between standard errors of the estimates from a GLM with a nonlinear link function and those from a quantile regression model is not easily made because, in general, the conditional mean of the GLM does not correspond exactly to a fixed conditional quantile. However, provided the functional forms of the GLM mean and of the quantile model are the same, one can make a crude comparison by identifying a probability level p^* at which the fitted conditional mean is close to the fitted conditional quantile under the parametric model. This is shown in R Code 10.4. In this case, $p^* = 0.54$, so the fitted conditional mean lies close to the centre of the fitted negative binomial distribution. The maximum relative absolute error incurred by using the conditional mean as an approximation of $Q_{Y|X}(p^*)$ under the negative binomial model is small. Also small is the difference between the distribution-free predictions at p^* and the negative binomial mean. A comparison of the standard errors is therefore not unreasonable, although it should be interpreted only as an informal benchmark. In this example, the standard errors from the negative binomial fit are larger than those from the corresponding distribution-free quantile model.

[3] The interested reader may wish to explore, informally, a transformation-based model using `Qtools::tsrq(y ~logx, ...)`. They will find that the optimal transformation parameter gives support to the exponential model for $p = 0.1$ and $p = 0.9$, but to a cubic transformation for $p = 0.5$.

R Code 10.3: Regression quantiles for number of bindings as a function of log-esterase using the Esterase Assay dataset.

```
set.seed(456)
probs <- c(0.10, 0.50, 0.90)
fit.rq <- lapply(probs, function(p) rq.counts(y ~ logx, data = dd,
tau = p, M = 50))
names(fit.rq) <- probs
lapply(fit.rq, function(x) x$tTable)
> $`0.1`
>                 Value  Std. Error lower bound upper bound      Pr(>|t|)
> (Intercept) 1.7007797 0.043219829   1.6151368   1.7864227  4.997691e-67
> logx        0.8430162 0.008233246   0.8267014   0.8593309 1.006295e-111
>
> $`0.5`
>                Value Std. Error lower bound upper bound     Pr(>|t|)
> (Intercept) 2.433019 0.29269747   1.8530192   3.0130183 2.601964e-13
> logx        0.773822 0.06571307   0.6436072   0.9040368 2.915272e-21
>
> $`0.9`
>                 Value Std. Error lower bound upper bound     Pr(>|t|)
> (Intercept) 2.9288558 0.05540885   2.8190594   3.0386521 1.573341e-80
> logx        0.7334701 0.01368598   0.7063504   0.7605898 3.605990e-81
# PNR
qhat <- sapply(fit.rq, function(x) predict(x, type = "response"))
print(pnr <- apply(sweep(-qhat, 1, dd$y, "+") < 0, 2, mean)*100)
>      0.1      0.5      0.9
> 12.38938 50.44248 91.15044
# lack-of-fit test
lapply(fit.rq, LOFTest, alpha = 0.05, B = 1000, seed = 987)
> $`0.1`
> Lack-of-fit test for quantile regression based on the cusum process
> A large test statistic (small p-value) is evidence of lack of fit
> Quantile 0.1: Test statistic = 0.0501; p-value = 0.025
>
> $`0.5`
> Lack-of-fit test for quantile regression based on the cusum process
> A large test statistic (small p-value) is evidence of lack of fit
> Quantile 0.5: Test statistic = 0.1778; p-value = 0.007
>
> $`0.9`
> Lack-of-fit test for quantile regression based on the cusum process
> A large test statistic (small p-value) is evidence of lack of fit
> Quantile 0.9: Test statistic = 0.0142; p-value = 0.289
```

10.3 Conditional mid-quantiles for discrete responses

In Chapter 4, we illustrated mid-quantiles as a bridge between discrete and continuous variables. Sample mid-quantiles, like estimators based on jittering, are a form of

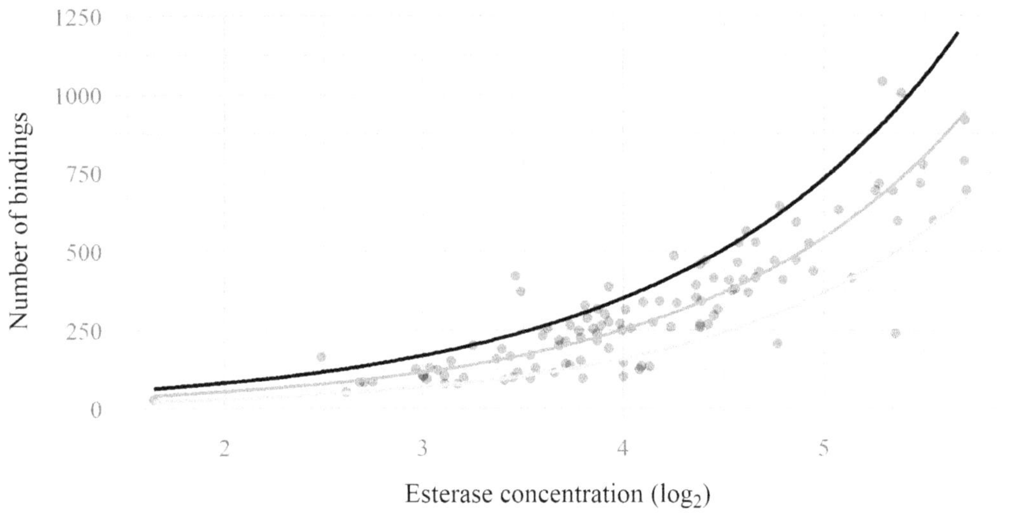

Figure 10.4
Predicted quantiles ($p \in \{0.1, 0.5, 0.9\}$) of number of bindings conditional on esterase concentration in the Esterase Assay dataset using jittered quantile regression.

R Code 10.4: Comparing standard errors of coefficient estimates from a negative binomial regression and an analogous quantile regression for counts.

```
# find a model-implied p such that Q(p) is close to the NB mean
mu <- fitted(fit.nb)
pstar <- mean(pnbinom(floor(mu), size = fit.nb$theta, mu = mu))
pstar
> [1] 0.5432082
# error when using mean as quantile pstar of NB
max(abs(mu - qnbinom(pstar, size = fit.nb$theta, mu = mu))/mu)
> [1] 0.004396209
# fit quantile model for p = pstar
set.seed(123)
fit.rq_pstar <- rq.counts(y ~ logx, data = dd, tau = pstar)
# error when using the NB mean as an approximation to Q(pstar)
max(abs(mu - predict(fit.rq_pstar))/mu)
> [1] 0.04472188
# compare standard errors of coefficients estimates
fit.rq_pstar$tTable[,1:2]
>                 Value Std. Error
> (Intercept) 2.4633308 0.07253134
> logx        0.7730618 0.01582709
summary(fit.nb)$coef[,1:2]
>              Estimate Std. Error
> (Intercept) 2.5317393 0.16570421
> logx        0.7535223 0.03970366
```

smoothing (interpolation) that removes some of the obstacles presented by discrete distributions. Geraci and Farcomeni (2022) have developed conditional mid-quantile methods for discrete responses. These methods are general as they can be applied to a large variety of discrete responses, including ordinal and count variables.

First, let's define the conditional mid-distribution function

$$\mathcal{F}_Y(y \mid \boldsymbol{x}) \equiv F_Y(y \mid \boldsymbol{x}) - 0.5 \cdot m_Y(y \mid \boldsymbol{x}), \tag{10.7}$$

where Y is a random variable with support $\mathcal{S}_Y \subset \mathbb{R}$, $F_Y(y \mid \boldsymbol{x}) = \Pr(Y \leq y \mid \boldsymbol{X} = \boldsymbol{x})$, and $m_Y(y \mid \boldsymbol{x}) = \Pr(Y = y \mid \boldsymbol{X} = \boldsymbol{x})$. In particular, $\mathcal{S}_Y = \{y_1, \ldots, y_m\}$ is a finite or countably infinite ($m = \infty$) subset of $\mathbb{R}$. The response Y can be binary, ordinal, or count. More generally, its support may contain positive or negative values and need not be equally spaced. Values need not be integers either. However, we do exclude nominal variables with more than two categories from the application of (10.7) as they do not have a natural ordering.

Let $\pi_j = \mathcal{F}_Y(y_j \mid \boldsymbol{x})$ (although π_j depends on $\boldsymbol{x}$, the notation has been simplified). The piecewise linear function that connects the points $(\pi_j, \mathcal{F}_Y^{-1}(\pi_j \mid \boldsymbol{x}))$, $j = 1, \ldots, m$, for given $\boldsymbol{x}$, is defined as the *conditional* mid-QF, denoted $\mathcal{Q}_Y(\cdot \mid \boldsymbol{x})$. Geraci and Farcomeni (2022) assume a quantile-specific regression model that is linear on the scale of h, that is,

$$\mathcal{Q}_{h(Y)}(p \mid \boldsymbol{x}) = \boldsymbol{x}^\top \boldsymbol{\beta}(p), \tag{10.8}$$

where h is a known monotone and differentiable 'link' function, and $\boldsymbol{\beta}(p)$ is a vector of q unknown regression coefficients for a given $p \in (0, 1)$. Equivalently, $\mathcal{Q}_Y(p \mid \boldsymbol{x}) = h^{-1}\{\boldsymbol{x}^\top \boldsymbol{\beta}(p)\}$. There are several possible choices for h, including the transformations discussed in Section 9.2. As in the unconditional case, conditional mid-quantiles cannot be obtained by inverting $\mathcal{F}_Y(y \mid \boldsymbol{x})$ except for points $y \in \mathcal{S}_Y$. Therefore, we introduce the piecewise linear function $\mathcal{F}_Y^c(y \mid \boldsymbol{x})$ that connects the points (y_j, π_j), $j = 1, \ldots, m$, and satisfies $\mathcal{F}_Y^c(y \mid \boldsymbol{x}) \equiv \mathcal{Q}_Y^{-1}(y \mid \boldsymbol{x})$, for all $y \in \mathbb{R}$.

To exemplify, consider the heteroscedastic model $Y = \lfloor x \rfloor + \lfloor x + 1 \rfloor \epsilon$, for $x \geq 0$, where ϵ is discrete uniform between 1 and 6 (die rolling). For fixed $\boldsymbol{x}$, the mid-quantiles are a linear function[4] of p between π_1 and π_6 (otherwise, they are constant), as shown in the left plot of Figure 10.5. If we assume the identity link $h(u) = u$, the conditional mid-quantile model (10.8) has equation $\mathcal{Q}_Y(p \mid x) = (0.5 + 6p) + (1.5 + 6p)x$. The first and third mid-quartiles with equations $\mathcal{Q}_Y(0.25 \mid x) = 2 + 3x$ and $\mathcal{Q}_Y(0.75 \mid x) = 5 + 6x$, respectively, are shown in the right plot of Figure 10.5.

Estimation is carried out in a two-step approach (Geraci and Farcomeni, 2022). In the first step, one needs to obtain the mid-probabilities as in the unconditional case (see Equation (4.12)), except now they need to be conditional on the covariates. It is relatively straightforward to obtain a sample estimate of $\mathcal{F}$ in (10.7). One choice, computationally fast, is distributional regression (Section 12.2). Let $F_{ij} = \Pr\{Y_i \leq z_j \mid \boldsymbol{x}_i\}$, where z_j, $j = 1, \ldots, s$, is the jth distinct observation of Y that occurs in the sample, with $z_j < z_{j+1}$ for all $j = 1, \ldots, s - 1$. Then it is just a matter of running s binomial regressions and predicting $\hat{F}_{ij}$, for $i = 1, \ldots, n$ and $j = 1, \ldots, s$. The latter are used to build an $n \times s$ matrix $\boldsymbol{F}$, whose ijth generic element is $\hat{F}_{ij}$. As a by-product, the row-wise iterated differences of $\boldsymbol{F}$ will give estimates of $m_{ij} = \Pr\{Y_i = z_j \mid \boldsymbol{x}_i\}$, $j = 2, \ldots, s$. The $n \times s$ matrices $\boldsymbol{M}$, whose ijth generic element is $\hat{m}_{ij}$ for $j = 2, \ldots, s$ and $\hat{F}_{ij}$ for $j = 1$, and $\boldsymbol{F}$ are used to obtain the quantity of interest $\boldsymbol{G} = \boldsymbol{F} - 0.5 \cdot \boldsymbol{M}$, whose ijth generic element is $\hat{\pi}_{ij} = \hat{\mathcal{F}}_Y(y_j \mid \boldsymbol{x}_i)$. This is shown in R Code 10.5 using a logistic regression model.

[4] For example, conditional on $x = 0$, the mid-probabilities are given by $\pi_j = (j - 0.5)/6$, $j = 1, \ldots, 6$. If one applies Definition 14 in Section 4.4, then $\mathcal{Q}_Y(p) = (1 - \gamma)j + \gamma(j + 1)$ if $p = (1 - \gamma)\pi_j + \gamma\pi_{j+1}$. After some algebra, one finds $\gamma = 0.5 - j + 6p$, hence $\mathcal{Q}_Y(p) = 0.5 + 6p$. Conditional on $x = 1$, $\mathcal{Q}_Y(p) = 2 + 12p$. And so on.

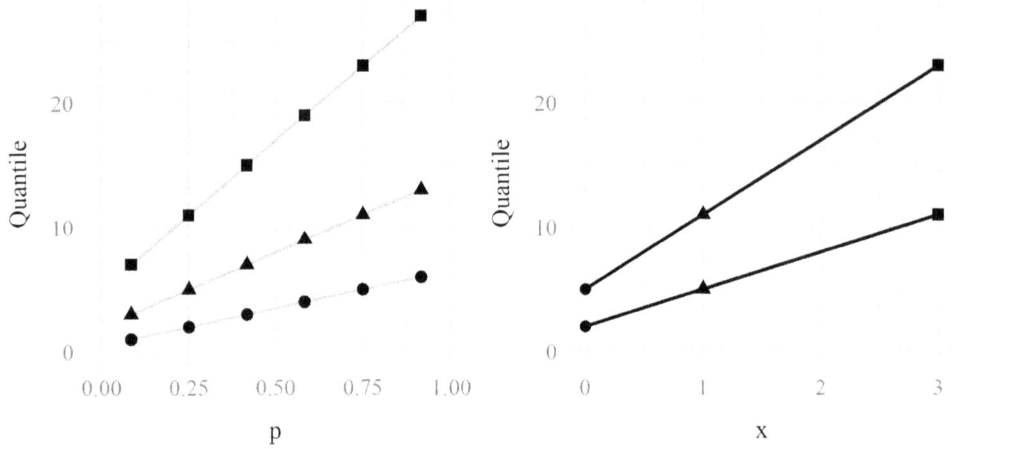

Figure 10.5
Mid-quantiles as a function of p by values of $x \in \{0, 1, 3\}$ (left), and as a function of x by values of $p \in \{0.25, 0.75\}$ (right).

These calculations are already implemented in the function `Qtools::cmidecdf`, whose argument `ecdf_est` provides some alternative estimators of the conditional cumulative distribution $F_{Y|X}$, including the nonparametric estimator of Li and Racine (2008) from the package `np` (Racine and Hayfield, 2026; Hayfield and Racine, 2008)

$$\hat{F}_Y(y \mid \boldsymbol{x}) = \frac{n^{-1}\sum_{i=1}^{n} I(y_i \leq y)K_{\nu}(\boldsymbol{x}_i, \boldsymbol{x})}{\hat{\rho}_x(\boldsymbol{x})}, \tag{10.9}$$

where K_{ν} is a product kernel for the covariates, with bandwidth vector $\boldsymbol{\nu} = (\nu_1, \ldots, \nu_k)^{\top}$, and $\hat{\rho}_x(\boldsymbol{x}) = n^{-1}\sum_{i=1}^{n} K_{\nu}(\boldsymbol{x}_i, \boldsymbol{x})$.

[!t]

In the second step of the estimation algorithm, the goal is to minimise the loss[5]

$$\varphi_n(\boldsymbol{\beta}; p) = n^{-1}\sum_{i=1}^{n}\left\{p - \hat{\mathcal{F}}_Y^{c}(q_i \mid \boldsymbol{x}_i)\right\}^2, \tag{10.10}$$

where $q_i = h^{-1}\{\boldsymbol{x}_i^{\top}\boldsymbol{\beta}\}$, with respect to $\boldsymbol{\beta}$. The objective function (10.10) can be optimised directly or solved explicitly as follows. Consider the linear interpolating function

$$\hat{\mathcal{F}}_Y^{c}(q_i \mid \boldsymbol{x}_i) = b_{j_i}(q_i - z_{j_i}) + \hat{\pi}_{j_i}, \qquad z_{j_i} \leq q_i \leq z_{j_i+1}, \tag{10.11}$$

where $b_{j_i} = \dfrac{\hat{\pi}_{j_i+1} - \hat{\pi}_{j_i}}{z_{j_i+1} - z_{j_i}}$ and $\hat{\pi}_{j_i} = \hat{\mathcal{F}}_Y(z_{j_i} \mid \boldsymbol{x}_i)$. The index $j_i = 1, \ldots, s-1$ identifies, for a given $i = 1, \ldots, n$, the value z_{j_i} among the z's such that $\hat{\mathcal{F}}_Y(z_{j_i} \mid \boldsymbol{x}_i) \leq p \leq \hat{\mathcal{F}}_Y(z_{j_i+1} \mid \boldsymbol{x}_i)$.

[5]Mid-quantiles can be characterised as minimisers of the classical quantile check-loss objective, provided that the loss is integrated with respect to the continuous mid-CDF rather than the ordinary CDF. Specifically, for a discrete response Y with continuous mid-CDF $\mathcal{F}_Y^{c}$, the population pth mid-quantile may be written as

$$\theta_p = \underset{\theta \in \mathbb{R}}{\operatorname{argmin}} \int \psi_p(y - \theta)\,\mathrm{d}\mathcal{F}_Y^{c}(y), \qquad \psi_p(u) = u\{p - I(u < 0)\}.$$

Thus, the usual check loss remains applicable, but the reference distribution is the continuous mid-distribution induced by the mid-CDF.

R Code 10.5: Computation of fitted cumulative probabilities and corresponding mid-distribution functions for a discrete response using binary logistic regression.

```
# generate data
set.seed(123)
n <- 200
x <- sample(0:3, n, replace = TRUE)
y <- floor(x) + floor(x + 1)*sample(1:6, n, replace = TRUE)
# unique values
print(z <- sort(unique(y)))
> [1] 1 2 3 4 5 6 7 8 9 11 13 14 15 17 19 20 23 27
# fit a logistic regression for each threshold
X <- cbind(1,x)
b <- sapply(z, function(z) glm.fit(X, 1*(y <= z),
family = binomial())$coefficients)
# estimate CDF
F <- 1/(1 + exp(-X%*%b))
# estimate PMF
M <- t(diff(t(F)))
# estimate mid-CDF
G <- F - 0.5 * cbind(F[,1], M)
```

Conditionally on $\hat{\boldsymbol{\pi}} = (\hat{\pi}_{1_1}, \ldots, \hat{\pi}_{s_1}, \ldots, \hat{\pi}_{1_n}, \ldots, \hat{\pi}_{s_n})^\top$, it is easy to verify that the minimiser of (10.10) when using (10.11) has the computationally convenient form

$$\hat{\boldsymbol{\beta}}(p; \hat{\boldsymbol{\pi}}) = (\boldsymbol{X}^\top \boldsymbol{X})^{-1} \boldsymbol{X}^\top \boldsymbol{u}, \tag{10.12}$$

where $\boldsymbol{X}$ is the matrix with ith row $\boldsymbol{x}_i^\top$ and $\boldsymbol{u}$ is a $n \times 1$ vector with ith element $h\left(\frac{p-\hat{\pi}_{j_i}}{b_{j_i}} + z_{j_i}\right)$.

We need to consider that, for some values of $\boldsymbol{x}_i$, it can occur that $p < \hat{\pi}_{i1}$ or $p > \hat{\pi}_{is}$ (i.e., the conditional mid-quantile lies on the flat ends of the function). We have seen, in Section 4.4, that the sampling behaviour of the mid-quantiles is dependent on whether p is outside or inside the mid-probability range. For this reason, Geraci and Farcomeni (2022) restrict p to be in the interval $\mathcal{I} = \left[\max_i \hat{\mathcal{F}}_Y(z_1 \mid \boldsymbol{x}_i), \min_i \hat{\mathcal{F}}_Y(z_s \mid \boldsymbol{x}_i)\right]$. Of course, one can still obtain an estimate of $\boldsymbol{\beta}(p)$ using (10.12), regardless of whether p is within the interval $\mathcal{I}$ for all or for some of the observations. However, all the elements u_i of $\boldsymbol{u}$ for which p is outside the mid-probability range are 'censored', that is, constrained to be equal to $h(z_1)$ (if $p < \hat{\pi}_{i1}$) or $h(z_s)$ (if $p > \hat{\pi}_{is}$). As a result, the linear predictor of the mid-quantile model will flatten out if p approaches 0 or 1, with slopes tending to 0 and the intercept tending to the sample minimum or maximum, at a speed that depends on the censoring rate. This is a reasonable behaviour since it reflects all we can say about the relationship between Y and $\boldsymbol{X}$ when we have little or no information.

An example of the application of the function `midrq` in `Qtools` is illustrated in R Code 10.6. The data are generated from the L-shift model $Y = \lfloor 1 + 2x \rfloor + \epsilon$, for $x \geq 0$, with discrete uniform error. In this instance, the estimation of the conditional mid-probabilities is carried out using the logistic model (`ecdf_est = "logit"`). Choices for the link function in (10.8) include the identity link (if `lambda = NULL`) and the Box–Cox transformation in (9.7) with parameter λ_p specified via `lambda`. The results in R Code 10.6 show that the estimated QTEs are all close to 2 (as expected) for p within the interval $\mathcal{I} = [0.165, 0.848]$. For mid-quantiles outside this range, the function returns warnings indicating the number

R Code 10.6: Estimating conditional mid-quantiles with the R package `Qtools`.

```
library(Qtools)
set.seed(123)
n <- 1000
p <- c(1e-5, 0.15, 0.25, 0.5, 0.75, 0.85, 1-1e-5)
x <- runif(n, 0, 3)
y <- floor(1 + 2*x) + sample(1:6, n, replace = TRUE)
range(y)
> [1]  2 12
fit <- midrq(y ~ x, tau = p, lambda = NULL,
control = midrqControl(ecdf_est = "logit"))

> Warning in midrq.fit(x = x, y = y, offset = offset, lambda = lambda, binary =
binary, : tau = 1e-05 is outside mid-probabilities range [0.165, 0.848] for 914 out
of 1000 observations. See details for ?midrq
> Warning in midrq.fit(x = x, y = y, offset = offset, lambda = lambda, binary =
binary, : tau = 0.15 is outside mid-probabilities range [0.165, 0.848] for 13 out
of 1000 observations. See details for ?midrq
> Warning in midrq.fit(x = x, y = y, offset = offset, lambda = lambda, binary =
binary, : tau = 0.85 is outside mid-probabilities range [0.165, 0.848] for 1 out of
1000 observations. See details for ?midrq
> Warning in midrq.fit(x = x, y = y, offset = offset, lambda = lambda, binary =
binary, : tau = 0.99999 is outside mid-probabilities range [0.165, 0.848] for 648
out of 1000 observations. See details for ?midrq

t(coef(fit))
>           (Intercept)           x
> 1e-05        1.995213 0.004940506
> 0.15         2.137433 1.940587634
> 0.25         2.711182 2.000783632
> 0.5          3.959115 2.049314622
> 0.75         5.374261 1.977113850
> 0.85         6.186943 1.877852234
> 0.99999     11.969556 0.014488135
```

of observations that do not contribute to estimation – this number increases as p approaches 0 or 1. At rather extreme values of p, which are used purely to illustrate the concept discussed above, the slope is 'squashed' towards 0 and the intercepts approach the minimum and maximum of the response, which in this case are, respectively, 2 and 12.

Regarding the large-n behaviour of the estimator in (10.12), Geraci and Farcomeni (2022) show the consistency and approximate normality (see Box 10.1 for a summary). The variance-covariance matrix is derived with the total variance law as follows:

$$\operatorname{var}\left\{\hat{\boldsymbol{\beta}}(p;\hat{\boldsymbol{\pi}})\right\} = \mathrm{E}_{\hat{\pi}}\left\{\operatorname{var}_{\hat{\boldsymbol{\beta}}}(\hat{\boldsymbol{\beta}} \mid \hat{\boldsymbol{\pi}})\right\} + \operatorname{var}_{\hat{\pi}}\left\{\mathrm{E}_{\hat{\boldsymbol{\beta}}}(\hat{\boldsymbol{\beta}} \mid \hat{\boldsymbol{\pi}})\right\}. \tag{10.13}$$

The first term reflects second-stage estimation variability conditional on the estimated mid-distribution, whereas the second term captures the additional uncertainty propagated from the first-stage estimation of the conditional mid-distribution.

The first term in the right-hand side of (10.13) is obtained via a Huber–White variance-covariance estimator, which is given by $(\boldsymbol{X}^\top\boldsymbol{X})^{-1}\boldsymbol{X}^\top\boldsymbol{D}\boldsymbol{X}(\boldsymbol{X}^\top\boldsymbol{X})^{-1}$, where $\boldsymbol{D} = \operatorname{diag}(\hat{e}_1^2, \ldots, \hat{e}_n^2)$ and $\hat{e}_i = h(y_i) - \boldsymbol{x}_i^\top\hat{\boldsymbol{\beta}}(p;\hat{\boldsymbol{\pi}})$, $i = 1, \ldots, n$. To obtain an estimate of the second term, we have, by the delta method,

$$\text{var}_{\hat{\pi}} \left\{ \text{E}_{\hat{\beta}}(\hat{\beta} \mid \hat{\pi}) \right\} \approx \nabla_{\hat{\pi}} \hat{\beta}(p; \hat{\pi})^\top \text{var}(\hat{\pi}) \nabla_{\hat{\pi}} \hat{\beta}(p; \hat{\pi}). \tag{10.14}$$

The expression for $\text{var}(\hat{\pi})$ depends on the variance of the estimator $\hat{F}_{Y|\boldsymbol{X}}$ used in the first step of the estimation algorithm (e.g., the nonparametric estimator in (10.9) or the semiparametric logistic model). The calculation of (10.13) is implemented in the function `Qtools::vcov.midrq`. The latter also provides an alternative estimation (with the argument `numerical = TRUE`) based on the (generalised) inverse of the numerical Hessian, whose computation tends to be faster than the calculation of (10.13), already at $n > 500$. This is due to the slower evaluation of the Jacobian $\nabla_{\hat{\pi}} \hat{\beta}(p; \hat{\pi})$, which is obtained via numerical differentiation.

Box 10.1: Consistency and asymptotic normality of the conditional mid-quantile regression estimator

The large-n distribution of conditional mid-quantiles is approximately normal. Consistency and asymptotic normality are stated in the following theorems.

Theorem 5. *Generate n independent conditional responses from a discrete distribution with parameters satisfying* (10.8). *Assume that the marginal density of the continuous covariates in the $k \times 1$ vector $\boldsymbol{x}$ is strictly positive and that $0 < F_Y(y \mid \boldsymbol{x}) < 1$. Assume also that the kernel K_ν in* (10.9) *is symmetric, bounded, and compactly supported, and that $n \prod_j \nu_j \to \infty$, while $\nu_j \to 0$ for all $j = 1, \ldots, k$. For a fixed $p \in (0,1)$, let $\hat{\beta}(p)$ denote the solution of* (10.10) *and let $\beta^*(p)$ be its population counterpart.*
Then, as $n \to \infty$, $\sup_z \left| \hat{\mathcal{F}}_Y^c(z \mid \boldsymbol{x}) - \mathcal{F}_Y^c(z \mid \boldsymbol{x}) \right| \to 0$. Additionally, $\|\hat{\beta}(p) - \beta^*(p)\| \to 0$ and $\left| \hat{\mathcal{Q}}_{h(Y)}(p \mid \boldsymbol{x}) - \mathcal{Q}_{h(Y)}(p \mid \boldsymbol{x}) \right| \to 0$, for all $p \in (0,1)$.

Theorem 6. *In addition to the assumptions in Theorem 5, assume that $\hat{\mathcal{F}}_Y^c \left(h^{-1}(\boldsymbol{x}^\top \beta) \mid \boldsymbol{x} \right)$ is differentiable with respect to β. Assume also that the $n \times (k+1)$ design matrix $\boldsymbol{X}$ is full rank, that $\lim_n n^{-1} \boldsymbol{X}\boldsymbol{X}^\top$ exists and is a positive-definite matrix. Finally, assume that $\sqrt{n \prod_j \nu_j (\sum_j \nu_j)^2} = O_{\mathbb{P}}(1)$. Then,*

$$V(\beta^*(p))^{-1/2} \sqrt{n \prod_j \nu_j} (\hat{\beta}(p) - \beta^*(p)) \xrightarrow{d} \mathcal{N}(\boldsymbol{0}, \boldsymbol{I}_{k+1}),$$

where $V(\beta^*(p)) = J(\beta^*(p))^{-1} D(\beta^*(p)) J(\beta^*(p))^{-1}$,
$J(\beta^*) = \text{E}\left\{ \nabla_\beta^2 \varphi_n(\beta; p) \Big|_{\beta=\beta^*} \right\}$, *and* $D(\beta^*) = \text{var}\left\{ \sqrt{n \prod_j \nu_j} \nabla_\beta \varphi_n(\beta^*; p) \Big|_{\beta=\beta^*} \right\}$.

For a proof of Theorems 5 and 6, see Geraci and Farcomeni (2022).

Remark 8. *The approximate sampling distribution of $\hat{\mathcal{Q}}_Y(p \mid \boldsymbol{x}) = h^{-1}\left(\boldsymbol{x}^\top \hat{\beta}(p) \right)$ follows from the results in Theorems 5 and 6. Let $\boldsymbol{\Sigma}_p = V(\beta^*(p))(n \prod_j \nu_j)^{-1}$ and let $\delta(\boldsymbol{b}) = \nabla_\beta h^{-1}(\boldsymbol{x}^\top \beta) \Big|_{\beta=\boldsymbol{b}}$. Then for large n, the distribution of $\hat{\mathcal{Q}}_Y(p \mid \boldsymbol{x})$ is approximately normal with mean $h^{-1}(\boldsymbol{x}^\top \beta^*(p))$ and variance $\delta(\beta^*(p))^\top \boldsymbol{\Sigma}_p \delta(\beta^*(p))$.*

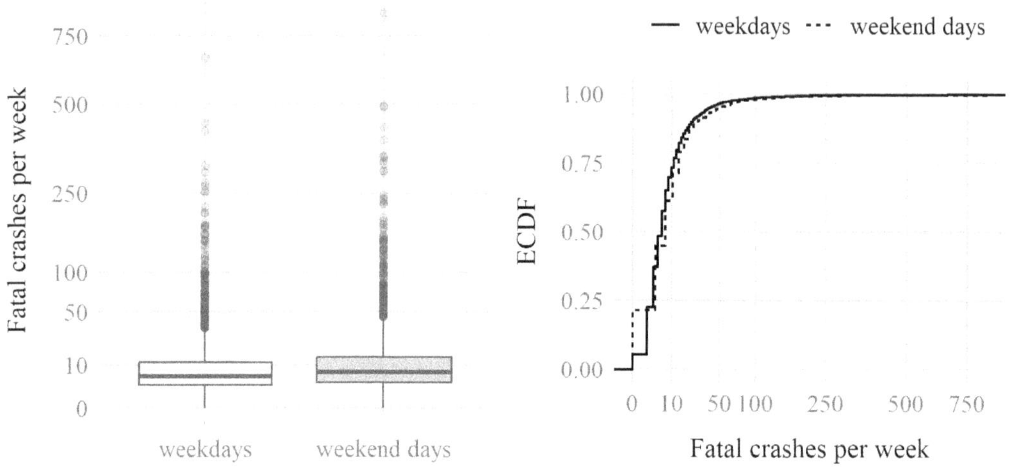

Figure 10.6
Boxplot (left) and cumulative distribution of weekly number of vehicle crashes (right) by day of the week in counties of the conterminous United States, 2020. The weekly number of crashes is shown on the square-root scale.

We now conduct an analysis of fatal motor vehicle crashes in the conterminous United States (US) during 2020 using the `fars` dataset already introduced in Section 7.1.5. The response Y is the total number of fatal crashes recorded in a US county for a given type of day of the week, either a weekday or a weekend day. The distribution of the weekly number of crashes[6] stretches from 0 to 843.5, and presents positive skewness with a long, sparse tail (left plot in Figure 10.6) for both weekdays and weekend days. Moreover, the ECDF for weekdays tends to lie above that for weekend days (right plot in Figure 10.6), suggesting stochastic dominance of the weekday distribution by the weekend-day distribution. For example, the sample quantiles for $p \in \{0.25, 0.5, 0.75, 0.9\}$ are, respectively, $2.8, 5.6, 11.2$, and 23.8 for weekdays and $3.5, 7, 14$, and 28 for weekend days. Hence, we should expect a positive weekend effect on the number of accidents.

Let θ and ϕ denote, respectively, latitude and longitude in radians of the counties centroids. The model for the pth conditional mid-quantile of Y is specified as follows:

$$\mathcal{Q}_{\log(Y)}(p \mid \boldsymbol{x}) = \log(d) + \beta_0(p) + \beta_1(p)x_1 + \beta_2(p)x_2 + \beta_3(p)x_3 + \beta_4(p)x_4 + \beta_5(p)x_5, \quad (10.15)$$

where d is the offset for length of exposure (either 5 for weekdays or 2 for weekends), x_1 is the indicator for weekend day, $x_2 = \cos(\theta) - \cos(\theta_c)$, $x_3 = \sin(\theta) - \sin(\theta_c)$, $x_4 = \cos(\phi) - \cos(\phi_c)$, and $x_5 = \sin(\phi) - \sin(\phi_c)$. To facilitate the interpretation of the intercept, cosines and sines are centred about those of the reference location with coordinates $\theta_c = 0.65$ and $\phi_c = -1.68$, that is, 37.5° N and 96.5° W (approximately the centre of the contiguous US). The transformed intercept $\exp(\beta_0(p))$ is therefore the predicted pth mid-quantile of the number of vehicle crashes per day that occur in a weekday at the centre location. In the implementation, zero counts are replaced by 0.1 to avoid taking logarithms of zero. The

[6]The count of crashes is divided by 5 for weekdays or by 2 for weekend days to obtain a daily rate, and then multiplied by 7 to obtain a weekly number.

R Code 10.7: Preparation of variables for modelling fatal motor vehicle crashes in counties of the conterminous United States, 2020.

```
library(Qtools)
data("fars", package = "Qtools")
dd <- fars
# offset weekend/weekday
dd$offset <- log(ifelse(dd$weekend == 1, 2, 5))
# replace zero counts with small constant
dd$y <- ifelse(dd$count == 0, 0.1, dd$count)
# transform longitude and latitude degrees to radians
theta_c <- 37.5*pi/180
phi_c <- -96.5*pi/180
lat_rad <- dd$latitude*pi/180
lon_rad <- dd$longitude*pi/180
# centre coordinates around reference location
dd$lat_cos <- cos(lat_rad) - cos(theta_c)
dd$lat_sin <- sin(lat_rad) - sin(theta_c)
dd$lon_cos <- cos(lon_rad) - cos(phi_c)
dd$lon_sin <- sin(lon_rad) - sin(phi_c)
```

data preparation steps, including the construction of the offset term and the transformation of geographical coordinates, are shown in R Code 10.7.

The results of the fitted model for $p \in \{0.75, 0.9\}$ are given in R Code 10.8. The predictions for the centre location are $\exp(\hat{\beta}_0(0.75)) = 1.0$ and $\exp(\hat{\beta}_0(0.9)) = 1.9$ fatal crashes per day between Monday and Friday. Relative to this location, predictions slope upward towards the southeast and towards the west at faster rates when $p = 0.9$ than when $p = 0.75$. Predictions for $p = 0.9$ across the conterminous US are shown in Figure 10.7, obtained using the package `usmap` (Di Lorenzo, 2025). The statistical significance of these spatial trends is confirmed by the small standard errors. Also significant is the weekend effect, which corresponds to approximately a 30% increase in the conditional quantiles of county-level fatal crash counts on weekend days relative to weekdays. Although the model adjusts for the differing lengths of the weekday and weekend observation periods, it does not account for county-level exposure measures such as population size, vehicle registrations, traffic volume, or vehicle miles travelled.

One may wonder if the weekend effect is the same also at lower quantiles. The last line in R Code 10.8 gives the mid-probability range, which is 0.22 to 0.93. As we do not want to push it too far, we consider eight deciles starting from the second (i.e., $p = 0.2$). The estimated effect of weekend days relative to weekdays, adjusted for spatial trends, does not appear to be constant across quantiles (Figure 10.8). Thus, the data do not support a simple location-shift model for $\log(Y)$. There is a negative or null effect at the lowest deciles and a positive effect at higher deciles, suggesting that the distribution of county-level crash counts is more dispersed on weekend days. In other words, relative to weekdays, weekend days are associated with both a greater prevalence of very small counts and a greater prevalence of large counts. This pattern may reflect increased heterogeneity across counties during weekends, possibly arising from differences in travel, recreation, tourism, nightlife, or other factors that vary geographically and temporally.

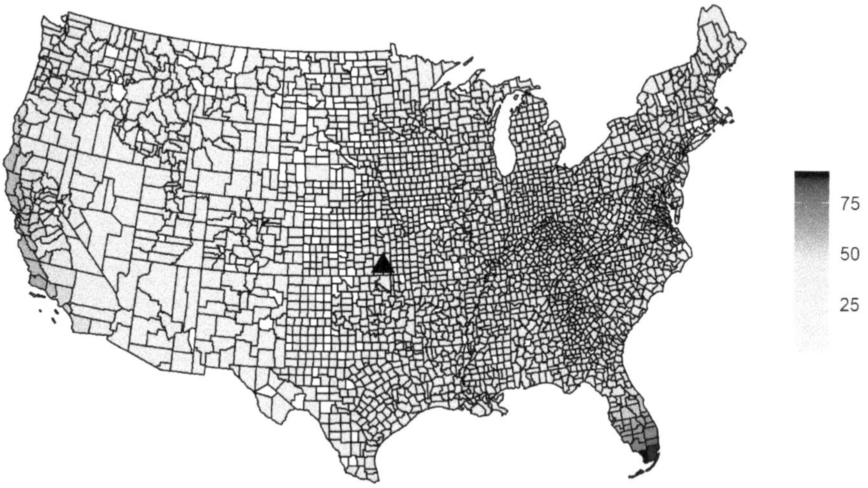

Figure 10.7
Predicted conditional mid-quantile ($p = 0.9$) of the number of car crashes during weekdays in counties of the conterminous United States, 2020. Counties with no recorded accidents are shown in white. The black triangle marks the location with coordinates (37.5° N, 96.5° W).

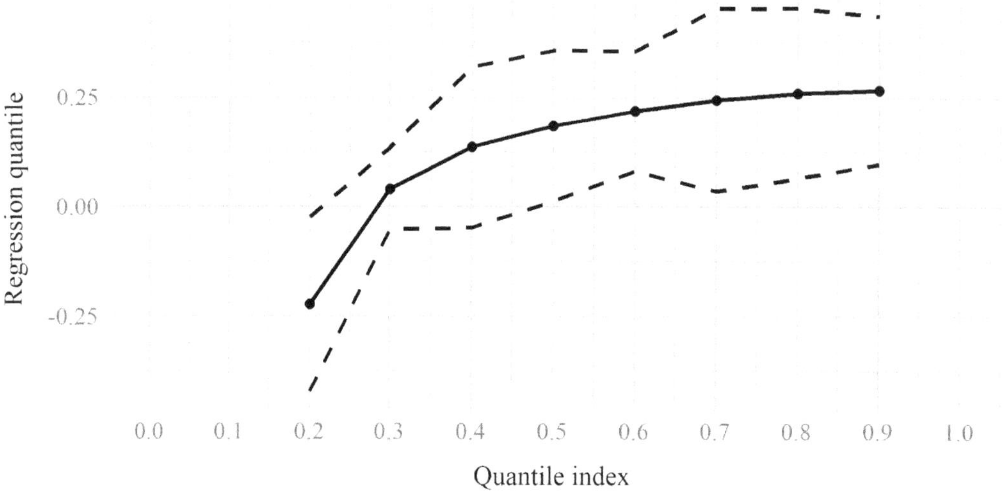

Figure 10.8
Estimated mid-quantile effect of weekend days relative to weekdays on 8 deciles ($p \in \{0.2, 0.3, \ldots, 0.9\}$) of the number of car crashes in counties of the conterminous United States along with 95% pointwise confidence intervals (dashed lines) using the log-linear model (10.15).

R Code 10.8: Mid-quantile regression analysis of weekly number of vehicle crashes by day of the week in counties of the conterminous United States, 2020.

```
# fit mid-quantile regression with log-transformation (Box-Cox lambda = 0)
# mid-probabilities estimated via logistic regression (ecdf_est = "logit")
fit.midrq <- midrq(y ~ weekend + lat_cos + lat_sin + lon_cos + lon_sin,
tau = c(0.75, 0.9), data = dd, offset = dd$offset, lambda = 0,
control = midrqControl(ecdf_est = "logit"))
# summary with numerical approximation of variance-covariance
summary(fit.midrq, numerical = TRUE)
> call:
> midrq(formula = y ~ weekend + lat_cos + lat_sin + lon_cos + lon_sin,
>     data = dd, tau = c(0.75, 0.9), lambda = 0, offset = dd$offset,
>     control = midrqControl(ecdf_est = "logit"))
>
>
> Coefficients linear predictor:
> $`0.75`
>                 Estimate   Std.Err       Lower      Upper
> (Intercept)   0.03319257 0.07300705  -0.1098986  0.1762838
> weekend       0.25264591 0.07723297   0.1012721  0.4040198
> lat_cos      -9.47221862 0.35476115 -10.1675377 -8.7768995
> lat_sin     -12.61027612 0.41248055 -13.4187231 -11.8018291
> lon_cos       1.76166776 0.33189280   1.1111698  2.4121657
> lon_sin      15.15694098 1.96175478  11.3119723 19.0019097
>
> $`0.9`
>                Estimate    Std.Err        Lower      Upper
> (Intercept)   0.6418873 0.07963822   0.48579927  0.7979754
> weekend       0.2662994 0.08611407   0.09751897  0.4350799
> lat_cos     -14.7716604 0.51924327 -15.78935852 -13.7539623
> lat_sin     -18.1971416 0.53631609 -19.24830180 -17.1459813
> lon_cos       1.7349710 0.35847484   1.03237326  2.4375688
> lon_sin      17.3189139 2.37886801  12.65641831 21.9814096
>
>
> Degrees of freedom: 5662 total; 5656 residual
# mid-probability range
attr(fit.midrq$midFit$G, "range")
> [1] 0.2178268 0.9349392
# weekend effect
100*(exp(coef(fit.midrq)["weekend",])-1)
>     0.75      0.9
> 28.74273 30.51258
```

10.4 Latent response

The latent variable interpretation of the classical GLM for a binary response $Y \sim \text{Bin}(1, \pi)$ posits that Y is the result of the dichotomisation $Y = I(Y^* > 0)$, where

$$Y^* = \boldsymbol{x}^\top \boldsymbol{\beta} + \epsilon \tag{10.16}$$

and $\epsilon \sim F$. The unobserved, continuous response Y^* is typically interpreted as an underlying utility or propensity that drives the observed binary outcome (e.g., decision to make a purchase). The popular logit, probit and complementary log-log link functions in GLMs, which arise from specific parametric assumptions on the distribution F of the latent error term, establish a relationship between the *probability* $\Pr(Y = 1) = \pi$ and the covariates $\boldsymbol{x}$.

One line of work on quantile regression for binary responses relies on the same continuous latent variable formulation in (10.16). Maximum score estimation (Manski, 1975, 1985) consists in finding the vector $\boldsymbol{\beta}$ that maximises the number of *correct* predictions of the binary outcome, based on the sign of $\boldsymbol{x}^\top \boldsymbol{\beta}$, that is,

$$\hat{\boldsymbol{\beta}} = \underset{\boldsymbol{\beta}:\|\boldsymbol{\beta}\|=1}{\text{argmax}} \sum_{i=1}^{n} I\left\{Y_i = I(\boldsymbol{x}_i^\top \boldsymbol{\beta} > 0)\right\}. \tag{10.17}$$

The condition $\|\boldsymbol{\beta}\| = 1$ is used to ensure identifiability since $I(\sigma \boldsymbol{x}^\top \boldsymbol{\beta} > 0) = I(\boldsymbol{x}^\top \boldsymbol{\beta} > 0)$ for all $\sigma > 0$. The estimator in (10.17) is equivalent to minimising the number of *incorrect* predictions, that is,

$$\hat{\boldsymbol{\beta}} = \left\{ \underset{\boldsymbol{\beta}:\|\boldsymbol{\beta}\|=1}{\text{argmin}} \sum_{i=1}^{n} \left| Y_i - I(\boldsymbol{x}_i^\top \boldsymbol{\beta} > 0) \right| \right\} \equiv \hat{\boldsymbol{\beta}}(0.5), \tag{10.18}$$

which can be viewed as a median-type regression criterion for the transformed variable $Y = I(Y^* > 0)$. In fact, since the indicator function is monotone, the equivariance $Q_{I(Y^*>0)}(0.5) = I\{Q_{Y^*}(0.5) > 0\}$ applies. To obtain the regression quantiles for model (10.16), it is just a matter of extending (10.18) to any $p \in (0, 1)$ and solve

$$\min_{\boldsymbol{\beta}:\|\boldsymbol{\beta}\|=1} \sum_{i=1}^{n} \psi_p \left\{Y_i - I(\boldsymbol{x}_i^\top \boldsymbol{\beta} > 0)\right\}. \tag{10.19}$$

The maximum score estimator does not require full specification of the distribution of ϵ, is consistent under mild conditions, and can be extended to ordered discrete responses (Lee, 1992). Unfortunately, the optimisation problem poses numerical challenges due to the discontinuity of the indicator function and the non-convexity of the loss function. Moreover, the asymptotic distribution is nonstandard. To improve its behaviour, Horowitz (1992) and Kordas (2006) considered smoothed variants of (10.17) obtained by replacing the indicator function with a continuous approximation (see Section 5.4).

Although the formulation in (10.19) resembles quantile regression, its interpretation differs fundamentally from that of models for continuous responses. When the outcome is binary, the conditional quantiles of Y are fully determined by the success probability $\Pr(Y = 1 \mid \boldsymbol{x})$ and therefore do not convey distributional information beyond this single quantity. The criterion in (10.19) is thus more appropriately interpreted as a cost-sensitive classification rule rather than as a model for conditional quantiles of the observed response. In particular, the parameter p governs the relative weight assigned to false positives and false negatives, rather than indexing different parts of the distribution of $Y \mid \boldsymbol{X}$. This perspective becomes clearer when contrasted with the models discussed in Chapter 12, where binomial regression and distributional regression explicitly target the conditional distribution function and its evaluation at fixed thresholds.

These difficulties with both interpretation and computation (see also Scharf et al., 2022) motivate continuing methodological development in this area.

Part IV

Further topics

11

Quantile-based distributional comparisons

11.1 Quantile differences and ratios

11.1.1 Quantile differences

This chapter focuses on comparisons between distributions based on quantiles and related functionals. The building blocks are quantile differences and quantile ratios, which provide simple, interpretable comparisons between parts of a distribution and form the basis of many measures of location, dispersion, and asymmetry.

Consider the difference $Q_Y(p_s) - Q_Y(p_r)$, where, without loss of generality, $0 < p_r < p_s < 1$. This can be viewed as a special case of a quantile regression model with intercept only, that is,

$$Q_Y(p) = \beta_0(p),$$

so that

$$Q_Y(p_s) - Q_Y(p_r) = \beta_0(p_s) - \beta_0(p_r).$$

For the particular choice of $p_r = 1 - p_s = p < 0.5$, we obtain the inter-quantile range $IPR(p) = Q_Y(1-p) - Q_Y(p)$ introduced in Chapter 3. Now consider a linear model of the type

$$Q_{Y|\mathbf{X}}(p_s) - Q_{Y|\mathbf{X}}(p_r) = \boldsymbol{x}^\top \boldsymbol{\delta}. \tag{11.1}$$

The $(k+1) \times 1$ vector $\boldsymbol{\delta} = (\delta_0, \delta_1, \ldots, \delta_k)^\top$ collects the regression coefficients associated with $\boldsymbol{x} = (1, x_1, \ldots, x_k)^\top$. The intercept δ_0 is interpreted as the quantile difference conditional on $x_j = 0$, $j = 1, \ldots, k$, while δ_j represents the *quantile difference treatment effect* associated with x_j. Although $\boldsymbol{\delta}$ depends on p_r and p_s, we suppress this dependence for notational simplicity. If we assume a linear relationship $Q_{Y|\mathbf{X}}(p) = \boldsymbol{x}^\top \boldsymbol{\beta}(p)$ as in (8.1), then Model (11.1) follows from

$$Q_{Y|\mathbf{X}}(p_s) - Q_{Y|\mathbf{X}}(p_r) = \boldsymbol{x}^\top \boldsymbol{\beta}(p_s) - \boldsymbol{x}^\top \boldsymbol{\beta}(p_r) = \boldsymbol{x}^\top \left(\boldsymbol{\beta}(p_s) - \boldsymbol{\beta}(p_r)\right). \tag{11.2}$$

Let

$$\hat{\boldsymbol{\theta}} = \begin{pmatrix} \hat{\boldsymbol{\beta}}(p_r) \\ \hat{\boldsymbol{\beta}}(p_s) \end{pmatrix} \tag{11.3}$$

be the estimator of $\boldsymbol{\theta} = (\boldsymbol{\beta}(p_r)^\top, \boldsymbol{\beta}(p_s)^\top)^\top$, the stacked vector of quantile regression coefficients. From the joint asymptotic normality of quantile regression estimators for multiple quantile levels, as seen in Section 8.3.2, we have

$$\sqrt{n}\left(\hat{\boldsymbol{\theta}} - \boldsymbol{\theta}\right) \dot{\sim} \mathcal{N}\left(\mathbf{0}, \boldsymbol{\Sigma}\right), \tag{11.4}$$

where $\boldsymbol{\Sigma}$ is the $2(k+1) \times 2(k+1)$ block covariance matrix whose generic (l, m) block is

$$(\min(p_l, p_m) - p_l p_m)\, \boldsymbol{H}(p_l)^{-1} \boldsymbol{J} \boldsymbol{H}(p_m)^{-1}.$$

DOI: 10.1201/9781003326410-11

Now define

$$\boldsymbol{\delta} = \boldsymbol{\beta}(p_s) - \boldsymbol{\beta}(p_r).$$

This is a linear transformation

$$\boldsymbol{\delta} = \boldsymbol{C}\boldsymbol{\theta}, \qquad \boldsymbol{C} = \begin{pmatrix} -\boldsymbol{I}_{k+1} & \boldsymbol{I}_{k+1} \end{pmatrix}.$$

Therefore, by the delta method,

$$\sqrt{n}\left(\hat{\boldsymbol{\delta}} - \boldsymbol{\delta}\right) \dot{\sim} \mathcal{N}\left(\boldsymbol{0}, \boldsymbol{C}\boldsymbol{\Sigma}\boldsymbol{C}^\top\right). \tag{11.5}$$

Hence,

$$\text{var}(\hat{\boldsymbol{\delta}}) \approx \frac{1}{n}\boldsymbol{C}\boldsymbol{\Sigma}\boldsymbol{C}^\top,$$

and the approximate standard errors of the $\hat{\delta}_j$s are the square roots of the diagonal elements of $n^{-1}\boldsymbol{C}\boldsymbol{\Sigma}\boldsymbol{C}^\top$.

A direct estimator of the conditional inter-quantile difference (11.1) can be obtained by modelling the two quantiles jointly while constraining their separation. Let

$$Q_{Y|\boldsymbol{X}}(p_r) = \alpha + \boldsymbol{x}^\top\gamma, \quad Q_{Y|\boldsymbol{X}}(p_s) = \alpha + \boldsymbol{x}^\top\gamma + \boldsymbol{x}^\top\boldsymbol{\delta}.$$

Then

$$Q_{Y|\boldsymbol{X}}(p_s) - Q_{Y|\boldsymbol{X}}(p_r) = \boldsymbol{x}^\top\boldsymbol{\delta},$$

so that $\boldsymbol{\delta}$ directly parameterises the conditional inter-quantile difference.

An estimator of $(\boldsymbol{\delta}, \alpha, \boldsymbol{\gamma})$ is obtained as

$$(\hat{\boldsymbol{\delta}}, \hat{\alpha}, \hat{\gamma}) = \underset{\boldsymbol{\delta}, \alpha, \gamma}{\text{argmin}} \sum_{i=1}^{n} \left\{\psi_{p_r}\left(y_i - \alpha - \boldsymbol{x}_i^\top\gamma\right) + \psi_{p_s}\left(y_i - \alpha - \boldsymbol{x}_i^\top\gamma - \boldsymbol{x}_i^\top\boldsymbol{\delta}\right)\right\}.$$

This can be viewed as a pair of quantile regressions sharing a common baseline component, with their separation governed by $\boldsymbol{x}^\top\boldsymbol{\delta}$. Unlike separate estimation at p_r and p_s, this approach imposes the structural restriction that the inter-quantile difference is linear in $\boldsymbol{x}$, and therefore the two approaches need not coincide. We do not pursue this formulation further, as it imposes structure on the quantiles that is unnecessary for our purposes.

An interesting question is whether Model (11.1) necessarily implies linear specifications for the individual quantiles $Q_{Y|\boldsymbol{X}}(p_r)$ and $Q_{Y|\boldsymbol{X}}(p_s)$. The answer is negative. The model only restricts the difference between the two conditional quantiles, not their levels. Indeed, let $g(\boldsymbol{x})$ be any function such that

$$Q_{Y|\boldsymbol{X}}(p_r) = g(\boldsymbol{x}).$$

Then

$$Q_{Y|\boldsymbol{X}}(p_s) = g(\boldsymbol{x}) + \boldsymbol{x}^\top\boldsymbol{\delta}$$

satisfies (11.1) for any choice of $g(\boldsymbol{x})$. Hence the model is compatible with arbitrarily non-linear specifications for the individual quantile functions, provided that their difference is linear in $\boldsymbol{x}$. Conversely, if both quantiles admit linear representations,

$$Q_{Y|\boldsymbol{X}}(p_r) = \boldsymbol{x}^\top\boldsymbol{\beta}(p_r), \qquad Q_{Y|\boldsymbol{X}}(p_s) = \boldsymbol{x}^\top\boldsymbol{\beta}(p_s),$$

then (11.1) follows with $\boldsymbol{\delta} = \boldsymbol{\beta}(p_s) - \boldsymbol{\beta}(p_r)$. Therefore, the usual linear quantile regression specification is a special case of Model (11.1), which is strictly weaker since it imposes

R Code 11.1: Inter-quantile regression for forced expiratory volume as a function of sex using the Childhood Respiratory Disease Study dataset.

```
p <- 0.1
fit <- rq(fev ~ sex, tau = c(p, 1-p), data = fev)
coef(fit)
>              tau= 0.1 tau= 0.9
> (Intercept)     1.552    3.236
> sexmale         0.113    1.048
print(delta <- apply(coef(fit), 1, diff))
> (Intercept)      sexmale
>       1.684        0.935
```

structure only on the inter-quantile difference and leaves the individual quantiles otherwise unrestricted.

In R Code 11.1, an example is illustrated using the FEV dataset. We set $p_r = 0.1$ and $p_s = 0.9$, hence the quantile difference is $IPR(0.1)$. The estimated intercept is $\hat{\delta}_0 = 1.68$ l/s, interpreted as the spread (at the level $p = 0.1$) of the FEV distribution among females. The estimated slope is $\hat{\delta}_1 = 0.93$ l/s, which is the effect of male sex on the spread of the conditional distribution. Boys have a wider distribution as measured by $IPR(0.1)$.

The asymptotic standard errors under NID assumptions are reported in R Code 11.2. The blocks of the matrix $\mathbf{\Sigma}$ in (11.5) are assembled from the output of `summary.rq(..., covariance = TRUE)`, which is a list with two sublists, one for each quantile. The diagonal blocks of $\mathbf{\Sigma}$ are readily obtained from `summary.rq(..., covariance = TRUE)[[j]]$cov`, while the cross-covariance block is assembled using `summary.rq(..., covariance = TRUE)[[j]]$Hinv` for the estimate of $\hat{\boldsymbol{H}}^{-1}$. R Code 11.2 also shows the bootstrap standard errors using the `boot` package with 1000 replicates. The two methods yield estimates of similar magnitude. A 95% normal-approximation confidence interval based on the bootstrap standard error for δ_1 is therefore obtained as $\hat{\delta}_1 \pm \Phi^{-1}(0.975) \cdot SE(\hat{\delta}_1) = 0.93 \pm 1.96 \cdot 0.10$, which yields an interval from 0.74 to 1.13 l/s.

11.1.2 Quantile ratios

Ratios are more delicate than differences, as they are nonlinear and can be highly skewed, especially when the denominator is small. In analogy with the quantile difference, we define the quantile ratio $Q_Y(p_s)/Q_Y(p_r)$, where $Q_Y(p_r) \neq 0$, and, without loss of generality, $0 < p_r < p_s < 1$. The conditional specification of quantile ratios proposed by Farcomeni and Geraci (2024) is

$$g\left(\frac{Q_{Y|\mathbf{X}}(p_s)}{Q_{Y|\mathbf{X}}(p_r)}\right) = \boldsymbol{x}^\top \boldsymbol{\rho}, \tag{11.6}$$

where g is a known monotone transformation. The $(k+1) \times 1$ vector $\boldsymbol{\rho} = (\rho_0, \rho_1, \ldots, \rho_k)^\top$ collects the regression coefficients associated with $\boldsymbol{x} = (1, x_1, \ldots, x_k)^\top$. The intercept ρ_0 is interpreted as the (transformed) quantile ratio conditional on $x_j = 0$, $j = 1, \ldots, k$, while ρ_j represents the *quantile ratio treatment effect* (on the scale of g) associated with x_j. Although $\boldsymbol{\rho}$ depends on p_r and p_s, we suppress this dependence.

If $g(u) = u$, Model (11.6) is compatible with linear quantile models only when the quantile ratio is constant. Modelling flexibility is gained by introducing nonlinear transformations g. For example, if $g(u) = \log(u)$, then $\log\{Q_Y(p_s)/Q_Y(p_r)\} = Q_{\log Y}(p_s) - Q_{\log Y}(p_r)$. In

R Code 11.2: Asymptotic and bootstrap standard errors for the inter-quantile regression on forced expiratory volume using the Childhood Respiratory Disease Study dataset.

```
# Asymptotic SE (nid)
xx <- crossprod(fit$x)
fit.s  <- summary(fit,  se = "nid", covariance = TRUE)

H1 <- fit.s[[1]]$Hinv
H2 <- fit.s[[2]]$Hinv

S12 <- (min(p,1-p) - p*(1-p)) * H1 %*% xx %*% t(H2)

Sigma <- rbind(
  cbind(fit.s[[1]]$cov,S12),
  cbind(t(S12),fit.s[[2]]$cov)
)

C <- cbind(-diag(2), diag(2))
V <- C %*% Sigma %*% t(C)
print(SEa <- sqrt(diag(V)))
> [1] 0.07105924 0.11665887
# Bootstrap SE
library(boot)
delta_fun <- function(data, indices, p) {
  d <- data[indices, ]
  b1 <- coef(rq(fev ~ sex, tau = p, data = d))
  b2 <- coef(rq(fev ~ sex, tau = 1 - p, data = d))
  return(b1 - b2)
}
set.seed(123)
B <- boot(data = fev, statistic = delta_fun, R = 1000, p = 0.1)
print(SEb <- apply(B$t, 2, sd))
> [1] 0.06061194 0.10152643
```

the special case $p_r = 1 - p_s = p < 0.5$, this coincides with the inter-quantile range $IPR(p)$ of $\log Y$. Another useful transformation in applied research is $g(u) = \log(u-1)$, which corresponds to an inverse-logit transformation of $1/u$. This transformation ensures that $Q_Y(p_s) > Q_Y(p_r)$, thereby preventing quantile crossing between p_r and p_s. Moreover, each element of the vector $\boldsymbol{\rho}$ can be roughly interpreted as an adjusted per-unit log-fold increase in the excess ratio $Q_Y(p_s)/Q_Y(p_r) - 1$.

Estimating treatment effects for quantile ratios is generally more involved than for quantile differences. Model (11.6) entails the following:

$$\begin{aligned} Q_{Y|\mathbf{X}}(p_s) &= Q_{Y|\mathbf{X}}(p_r) \cdot g^{-1}(\boldsymbol{x}^\top \boldsymbol{\rho}), \\ Q_{Y|\mathbf{X}}(p_r) &= Q_{Y|\mathbf{X}}(p_s) \cdot \{g^{-1}(\boldsymbol{x}^\top \boldsymbol{\rho})\}^{-1}. \end{aligned} \tag{11.7}$$

When g is nonlinear, as in the case $g(u) = \log(u-1)$, Farcomeni and Geraci (2024) propose alternating nonlinear quantile estimation in an iterative algorithm and provide the following heuristic argument. Define $\xi_j \equiv Q_Y(p_j \mid \boldsymbol{x})$, $j \in \{r, s\}$, as the true quantile of Y conditional on $\boldsymbol{x}$, and let $\hat{\xi}_j$ denote its prediction. Correspondingly, $\hat{\xi}_{i,j}$ denotes the predicted quantile for the ith observation. The parameter $\boldsymbol{\rho}$ in (11.7) can be estimated by either fitting $Q_{Z|\mathbf{X}}(p_s) = g^{-1}\{\boldsymbol{x}^\top\boldsymbol{\gamma}(p_s)\}$, where $Z = Y/\hat{\xi}_r$, or $Q_{Z|\mathbf{X}}(p_r) = \{g^{-1}(\boldsymbol{x}^\top\boldsymbol{\gamma}(p_r))\}^{-1}$, where $Z = Y/\hat{\xi}_s$. Say we use the former model, for given $\hat{\xi}_r$, and obtain the solution $\hat{\boldsymbol{\gamma}}(p_s)$ using nonlinear estimation (Section 9.1). Now we turn to the second model in (11.7) and demonstrate that $\hat{\boldsymbol{\gamma}}(p_s)$ is, in general, a stationary point for

$$\hat{\boldsymbol{\gamma}}_n(p_r) = \underset{\boldsymbol{\gamma}\in\mathbb{R}^{k+1}}{\operatorname{argmin}} \sum_{i=1}^{n} \psi_{p_r}(r_i(\boldsymbol{\gamma})), \tag{11.8}$$

where $r_i(\boldsymbol{\gamma}) = z_i - \{g^{-1}(\boldsymbol{x}_i^\top\boldsymbol{\gamma})\}^{-1}$ and $z_i = y_i/\hat{\xi}_{i,s}$. Now, by plugging the estimate $\hat{\boldsymbol{\gamma}}(p_s)$ obtained previously and the corresponding predictions $\hat{\xi}_{i,s}$ in the very first iteration of the algorithm to solve (11.8), we have (see Equation (9.3))

$$\begin{aligned}\sum_{i=1}^{n} \psi_{p_r}\left(r_i\left(\hat{\boldsymbol{\gamma}}(p_s) + \lambda\boldsymbol{\zeta}_\gamma\right)\right) &= \sum_{i=1}^{n} \psi_{p_r}\left(y_i/\hat{\xi}_{i,s} - 1/g^{-1}\{\boldsymbol{x}_i^\top\hat{\boldsymbol{\gamma}}(p_s) + \lambda\boldsymbol{x}_i^\top\boldsymbol{\zeta}_\gamma\}\right)\\ &= \sum_{i=1}^{n} \psi_{p_r}\left(\frac{y_i/\hat{\xi}_{i,r}}{g^{-1}\{\boldsymbol{x}_i^\top\hat{\boldsymbol{\gamma}}(p_s)\}} - \frac{1}{g^{-1}\{\boldsymbol{x}_i^\top\hat{\boldsymbol{\gamma}}(p_s) + \lambda\boldsymbol{x}_i^\top\boldsymbol{\zeta}_\gamma\}}\right).\end{aligned} \tag{11.9}$$

Since, heuristically, $Q_{Y/\hat{\xi}_r}(p_r \mid \boldsymbol{x}) \approx \xi_r/\hat{\xi}_r$, the empirical p_r-quantile of $Y/\hat{\xi}_r$ is approximately equal to 1. This suggests that (11.9) attains a stationary point at $\lambda = 0$ since

$$\sum_{i=1}^{n} \psi_{p_r}\left(\frac{y_i/\hat{\xi}_{i,r}}{g^{-1}\{\boldsymbol{x}_i^\top\hat{\boldsymbol{\gamma}}(p_s)\}} - \frac{1}{g^{-1}\{\boldsymbol{x}_i^\top\hat{\boldsymbol{\gamma}}(p_s)\}}\right) \propto \sum_{i=1}^{n} \psi_{p_r}\left(y_i/\hat{\xi}_{i,r} - 1\right),$$

in which case the algorithm stops immediately and returns $\hat{\boldsymbol{\gamma}}(p_r) = \hat{\boldsymbol{\gamma}}(p_s)$. Of course, if the initial estimate $\hat{\xi}_r$ does not yield $\hat{Q}_{Y/\hat{\xi}_r}(p_r) = 1$, the estimation will need to be refined iteratively as discussed in Farcomeni and Geraci (2024).

In practice, Farcomeni and Geraci (2024) recommend linearising the estimation problem in order to scale inference to large datasets. They also show that the resulting estimator is consistent and asymptotically normal (see Box 11.1 for technical details). When the estimation of $\boldsymbol{\rho}$ relies on linear programming algorithms, the calculation of the limiting variance-covariance $\boldsymbol{\Omega}$ in Theorem 8 must still account for the nonlinear transformation g. Alternatively, standard error estimation can be carried out by means of an ordinary xy-pair bootstrap, which, in simulation studies, proved to work well even at smaller sample sizes.

We end this section with an analysis of systolic blood pressure (SBP) as a function of age and sex, using the National Health and Nutrition Examination Survey dataset from the `NHANES` package (Pruim, 2015). The dataset includes 10,000 participants from the 2009-2010 and 2011-2012 survey cycles. Figure 11.1 shows the 5th, 50th and 95th percentiles of SBP[1] conditional on age, by sex, obtained from a transformation-based quantile regression as defined in Equations (9.6) and (9.7). Notably, the lowest regression quantile is approximately linear for both sexes. In contrast, higher quantiles exhibit increasing curvature: slightly convex for females and concave for males. This suggests that nonlinear effects become more pronounced at higher quantiles.

[1]The variable name `BPSysAve` refers to the value obtained according to the NHANES protocol. If only one blood pressure measurement was obtained, that measurement is used. Otherwise, the first measurement is excluded when calculating the average. Consequently, if exactly two measurements were obtained, the second measurement is used as the average. See the NHANES Blood Pressure Examination documentation (`https://wwwn.cdc.gov/Nchs/Data/Nhanes/Public/2009/DataFiles/BPX_F.htm`).

Box 11.1: Consistency and asymptotic normality of the conditional quantile ratio estimator

Let $\hat{\boldsymbol{\rho}}$ denote the quantile ratio regression estimator resulting from the procedure described in the present section.

Theorem 7. *Assume standard regularity conditions for quantile regression models. Let $\boldsymbol{X}$ denote the $n \times (k+1)$ design matrix and assume that $n^{-1}\boldsymbol{X}^\top \boldsymbol{X}$ is positive definite and converges to a positive definite matrix $\boldsymbol{C}$, and that the initial predictions $\hat{\xi}$ are consistent. Then, $\hat{\boldsymbol{\rho}}$ converges in probability to $\boldsymbol{\rho}$.*

Theorem 8. *Under the assumptions of Theorem 7, there exists a positive definite matrix $\boldsymbol{\Omega}$ such that $\sqrt{n}(\hat{\boldsymbol{\rho}} - \boldsymbol{\rho})$ converges in distribution to a multivariate Gaussian random variable with mean $\mathbf{0}$ and covariance matrix $\boldsymbol{\Omega}$.*

For a proof of Theorems 7 and 8, as well as for practical strategies for standard error estimation, see Farcomeni and Geraci (2024). The theoretical properties of quantile ratio regression are very similar to those of classical quantile regression, including the rate of convergence.

Consider the following model:

$$\log\left(\frac{Q_{Y|\boldsymbol{X}}(p_s)}{Q_{Y|\boldsymbol{X}}(p_r)} - 1\right) = \boldsymbol{x}^\top \boldsymbol{\rho},$$

with $p_r = 0.05$ and $p_s = 0.95$, and age, sex, and their interaction as covariates. R Code 11.3 illustrates the function `qrr` from the `Qtools` package, which, by default, estimates this model. An exponentiated coefficient $\exp(\rho_j)$ represents a multiplier of the excess quantile

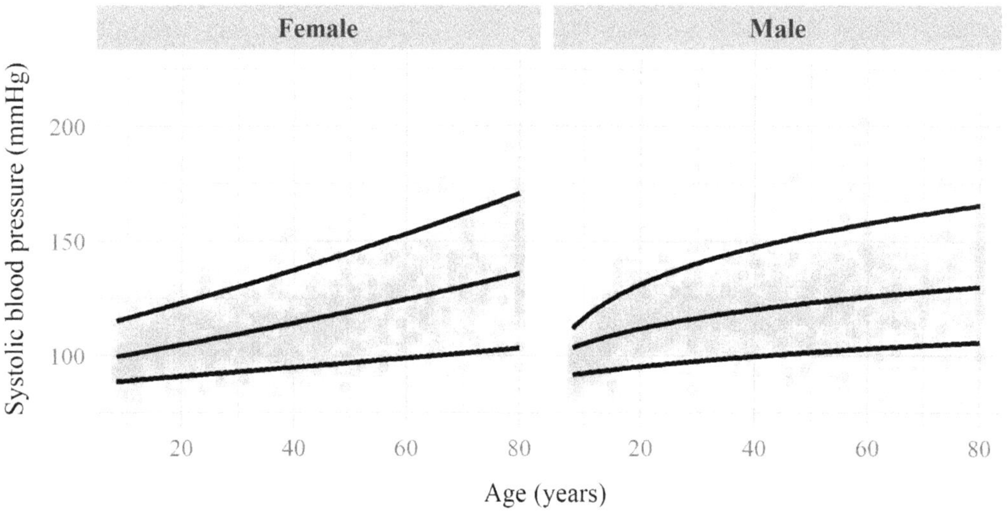

Figure 11.1
Predicted quantiles ($p \in \{0.05, 0.5, 0.95\}$) of systolic blood pressure conditional on age and sex in the `NHANES` dataset using transformation-based quantile regression.

R Code 11.3: Quantile ratio regression analysis of systolic blood pressure in the NHANES dataset.

```
data(NHANES, package = "NHANES")
dds <- NHANES[,c("Age", "Gender", "BPSysAve")]
dds <- subset(dds, complete.cases(dds))
range(dds$Age)
> [1]  8 80
dds$Age.c <- dds$Age - min(dds$Age)
p <- 0.05
fit <- qrr(BPSysAve ~ Age.c*Gender, data = dds, tau = c(p,1-p))
fit
>
> Quantile ratio regression 0.95:0.05
>
> Coefficients linear predictor:
>      (Intercept)            Age.c       Gendermale Age.c:Gendermale
>     -1.153044775      0.010568079      0.114952857     -0.002395144
>
> Degrees of freedom: 8551 total; 8547 residual
exp(coef(fit))
>      (Intercept)            Age.c       Gendermale Age.c:Gendermale
>        0.3156741        1.0106241        1.1218205        0.9976077
# age effect among males
exp(coef(fit)['Age.c'] + coef(fit)['Age.c:Gendermale'])
>    Age.c
> 1.008206
set.seed(123)
summary(fit, se = "boot", R = 200)
>
> Quantile ratio regression 0.95:0.05
>
> Coefficients linear predictor:
>                     Estimate    Std.Err  t value Pr(>|t|)
> (Intercept)      -1.1530448  0.0330686 -34.8682  < 2e-16 ***
> Age.c             0.0105681  0.0007226  14.6250  < 2e-16 ***
> Gendermale        0.1149529  0.0474741   2.4214  0.01548 *
> Age.c:Gendermale -0.0023951  0.0012018  -1.9930  0.04629 *
> ---
> Signif. codes:  0 '***' 0.001 '**' 0.01 '*' 0.05 '.' 0.1 ' ' 1
>
> Degrees of freedom: 8551 total; 8547 residual
```

ratio $Q_{Y|\boldsymbol{X}}(p_s)/Q_{Y|\boldsymbol{X}}(p_r) - 1$. Starting with 8-year-old females (the reference group), the 95th percentile of SBP is approximately $1 + \exp(\hat{\rho}_0) = 1.32$ times the 5th percentile. For women, each additional year of age increases the excess quantile ratio by about 1.1% indicating that the 0.95:0.05 SBP ratio widens with age. At age 8 years, males have about 12.2% larger excess quantile ratio than females, reflecting greater inequality in SBP at that age. However, this inequality grows more slowly with age among men: each additional year multiplies the excess quantile ratio by about 0.8%, slightly less than among women. The statistical evidence supporting differential age trajectories is modest, as indicated by the standard error and p-value of the interaction term shown in the final part of R Code 11.3. In this example, inference is based on 200 bootstrap replicates using `Qtools::summary.qrr(..., se = "boot")`. Alternatively, standard errors and p-values may be obtained by setting `se = "approximate"`, which computes the approximate (smoothed) asymptotic covariance matrix $\boldsymbol{\Omega}$ introduced in Theorem 8 using `numDeriv::jacobian`.

In summary, SBP variability increases with age, and males exhibit higher overall dispersion. The age-related increase appears slightly stronger among females, which is qualitatively in line with commonly reported sex differences in SBP trajectories.

11.2 Location, scale, and shape

In this section, we extend the LSS quantile-based measures that were introduced in Section 3.1.4. We consider them conditional on covariates. Estimation proceeds with a simple approach,[2] whereby we specify a model for $Q_{Y|\mathbf{X}}(p)$, for example, linear or transformation-based, and then predict $\hat{Q}_{Y|\mathbf{X}}(p)$ to obtain the conditional LSS measures at specific values of $\mathbf{x}$. Uncertainty is then assessed via bootstrap.

Estimation of conditional LSS can be carried out by using the function `qlss.formula`. The conditional model is specified in the argument `formula`, while the probability p is given in `probs`. (As seen in Equation (3.4), the other probabilities of interest to obtain the decomposition of the conditional quantiles are $1-p$, 0.25, 0.5, and 0.75.) The argument `type` specifies the required type of regression model, that is, `rq` for linear models and `rqt` for transformation-based models. The function `qlss` will take any additional argument to be passed to `quantreg::rq` or `tsrq` (e.g., `subset`, `weights`, etc.) (Geraci, 2016b).

Consider the `NHANES` dataset and assume that the transformation model as defined in Equations (9.6) and (9.7) holds for the quantiles $p \in \{0.05, 0.25, 0.5, 0.75, 0.95\}$. In R Code 11.4, the conditional LSS summary of the distribution of SBP conditional on age is calculated separately for females and males. The specification `probs = 0.05` entails the calculation of the conditional inter-quantile range $Q_{Y|\mathbf{X}}(0.95) - Q_{Y|\mathbf{X}}(0.05)$, the skewness and tail indices

$$\frac{Q_{Y|\mathbf{X}}(0.95) + Q_{Y|\mathbf{X}}(0.05) - 2Q_{Y|\mathbf{X}}(0.5)}{Q_{Y|\mathbf{X}}(0.95) - Q_{Y|\mathbf{X}}(0.05)}$$

and

$$\frac{Q_{Y|\mathbf{X}}(0.95) - Q_{Y|\mathbf{X}}(0.05)}{Q_{Y|\mathbf{X}}(0.75) - Q_{Y|\mathbf{X}}(0.25)},$$

respectively. The conditional median $Q_{Y|\mathbf{X}}(0.5)$ and conditional inter-quartile range $Q_{Y|\mathbf{X}}(0.75) - Q_{Y|\mathbf{X}}(0.25)$ are calculated by default.

The output, which is of class `"qlss"`, is a named list with the same LSS measures seen in the case of unconditional quantiles. However, these are now conditional on the covariates specified in the formula. By default, the predictions are the fitted values, which are averaged over observations for printing purposes. An optional data frame for predictions can be given via the argument `newdata` in `predict.qlss`. If `interval = TRUE`, the latter computes confidence intervals at the specified `level` using `R` bootstrap replications. The

[2]We do not pursue the more ambitious goal of estimating 'treatment effects' for the more complex shape statistics, since these involve differences and ratios of differences.

R Code 11.4: Location scale and shape of systolic blood pressure conditional on age and sex in the `NHANES` dataset.

```
fit1 <- qlss(formula = BPSysAve ~ Age, data = NHANES,
subset = NHANES$Gender == "female", tsf = "bc", probs = 0.05)
fit2 <- qlss(formula = BPSysAve ~ Age, data = NHANES,
subset = NHANES$Gender == "male", tsf = "bc", probs = 0.05)
fit1
> call:
> qlss.formula(formula = BPSysAve ~ Age, probs = 0.05, data = NHANES,
>     subset = NHANES$Gender == "female", tsf = "bc")
>
> Conditional Quantile-Based Location, Scale, and Shape
> -- Values are averaged over observations --
>
> ** Location **
> Median
> [1] 115.5207
> ** Scale **
> Inter-quartile range (IQR)
> [1] 16.92974
> Inter-quantile range (IPR)
>     0.05
> 44.48753
> **Shape**
> Skewness index
>       0.05
> 0.09547958
> Shape index
>     0.05
> 2.612101
fit2
> call:
> qlss.formula(formula = BPSysAve ~ Age, probs = 0.05, data = NHANES,
>     subset = NHANES$Gender == "male", tsf = "bc")
>
> Conditional Quantile-Based Location, Scale, and Shape
> -- Values are averaged over observations --
>
> ** Location **
> Median
> [1] 118.8695
> ** Scale **
> Inter-quartile range (IQR)
> [1] 17.15927
> Inter-quantile range (IPR)
>     0.05
> 46.20076
> **Shape**
> Skewness index
>      0.05
> 0.1244986
> Shape index
>     0.05
> 2.678242
```

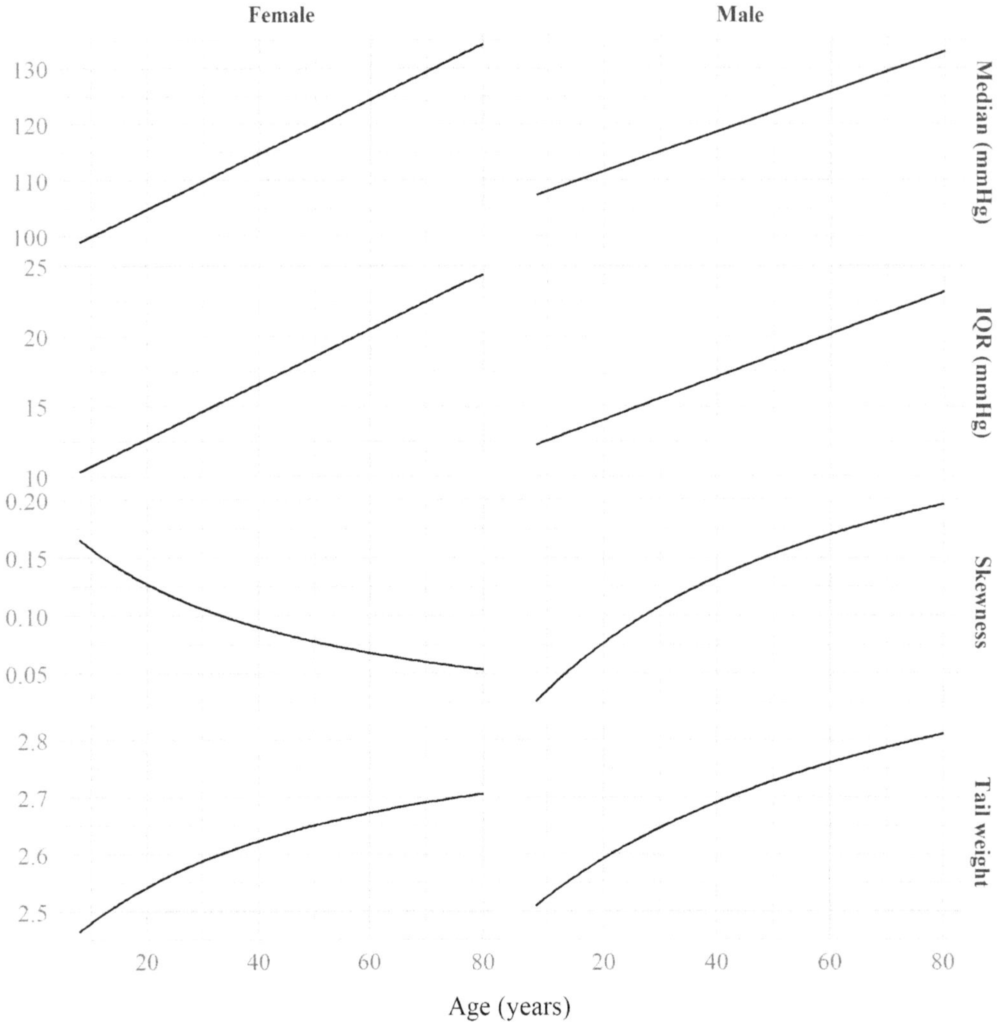

Figure 11.2
Location, scale and shape of systolic blood pressure conditional on age in the NHANES dataset.

conditional LSS measures can be conveniently plotted using the `plot.qlss` function as shown in the code below. The argument `z` is required and specifies the covariate used for plotting. Finally, the argument `whichp` specifies one probability among those given in `probs` that should be used for plotting (e.g., $p = 0.05$ in the code fragment below).

```
set.seed(123)
x <- seq(8, 80, by = 0.5)
fit1p <- predict(fit1, newdata = data.frame(Age = x),
  interval = TRUE, level = 0.90, R = 500)
plot(fit1p, z = x, whichp = 0.05, interval = TRUE)
```

Figure 11.2 shows that both the median and the IQR of SBP increase approximately linearly with age in both sexes. However, among males these statistics start at higher values and end at lower values, indicating a smaller slope. The distribution of SBP is right-skewed in both sexes, but the magnitude of the skewness decreases with age among females and increases among males in a slightly nonlinear fashion. The conditional tail weight index is, on average, equal to 2.6 for both males and females, and it increases monotonically from about 2.5 to about 2.7 in females and 2.8 in males, remaining well above the tail-weight threshold of a normal distribution (1.9) throughout.

11.3 Conditional comparison function

Another way of comparing distributions is by mapping one distribution onto the probability scale of another. Section 2.2.3 introduced the comparison CDF

$$v(p; F, G) = G(F^{-1}(p)), \qquad 0 < p < 1,$$

which maps a quantile p of the distribution F to its corresponding cumulative probability in distribution G, interpreted as the *relative rank* of the pth quantile of Y on the reference scale G. This transformation captures relative differences between the two distributions in terms of probability content, and reduces to the identity $v(p) = p$ when $F = G$. Moreover, its derivative corresponds to the density ratio $d(p; F, G) = g\left(F^{-1}(p)\right) / f\left(F^{-1}(p)\right)$ (see Equations (2.4) and (2.5)) evaluated at matched quantiles, providing a local measure of discrepancy between F and G.

In the more general regression setting

$$v(p; \boldsymbol{x}_0, \boldsymbol{x}_1) = F_{Y|\boldsymbol{X}=\boldsymbol{x}_1}(F^{-1}_{Y|\boldsymbol{X}=\boldsymbol{x}_0}(p)) \qquad (11.10)$$

the contrast is between two subpopulations as defined by $\boldsymbol{x}$. For simplicity, consider a scalar covariate X, specifically $X \in \{0, 1\}$, and let F_0 and F_1 denote $F_{Y|X=0}$ and $F_{Y|X=1}$, respectively. Using the same ideas presented in Section 3.2.2, let $\Delta(p) = F_1^{-1}(p) - F_0^{-1}(p)$ be the QTE at level p. By monotonicity of F_1, it follows that

$$\begin{aligned}
\Delta(p) < 0 &\iff v(p) = F_1(F_0^{-1}(p)) > p \\
\Delta(p) = 0 &\iff v(p) = p \\
\Delta(p) > 0 &\iff v(p) < p.
\end{aligned}$$

The sign of the quantile treatment effect is reflected in whether $v(p)$ lies above or below the identity line. In this sense, the conditional comparison function provides a unifying framework connecting the measures developed in this chapter to the theoretical concept in Section 2.2.3. The following proposition considers the case of a scalar continuous covariate X and provides a formal link between the comparison function and the sign and magnitude of the quantile treatment effect.

Proposition 2. *Let*

$$v(p; x_0, x_1) = F_{Y|X=x_1}\left\{F^{-1}_{Y|X=x_0}(p)\right\}, \qquad 0 < p < 1,$$

and define the conditional quantile function

$$\xi(p, x) = F^{-1}_{Y|X=x}(p).$$

Assume that $F_{Y|X=x}(y)$ is differentiable with respect to x in a neighbourhood of $y = \xi(p,x)$, that $\xi(p,x)$ is differentiable with respect to x for fixed p, and that the conditional density

$$f_{Y|X=x}(y) = \frac{\partial}{\partial y} F_{Y|X=x}(y)$$

exists and is continuous in a neighbourhood of $y = \xi(p,x)$. Then

$$\begin{aligned}\lim_{\delta \to 0} \frac{\upsilon(p;x,x+\delta) - \upsilon(p;x,x)}{\delta} &= \left. \frac{\partial}{\partial x_1} \upsilon(p;x,x_1) \right|_{x_1=x} \\ &= -\, f_{Y|X=x}\{\xi(p,x)\} \, \frac{\partial}{\partial x} \xi(p,x). \end{aligned} \tag{11.11}$$

Proof. We have

$$\begin{aligned}\lim_{\delta \to 0} \frac{\upsilon(p;x,x+\delta) - \upsilon(p;x,x)}{\delta} &= \lim_{\delta \to 0} \frac{F_{Y|X=x+\delta}\{\xi(p,x)\} - F_{Y|X=x}\{\xi(p,x)\}}{\delta} \\ &= \left. \frac{\partial}{\partial x} F_{Y|X=x}(y) \right|_{y=\xi(p,x)}, \end{aligned}$$

because $\upsilon(p;x,x+\delta) = F_{Y|X=x+\delta}\{\xi(p,x)\}$ and $\upsilon(p;x,x) = p = F_{Y|X=x}\{\xi(p,x)\}$.

Let

$$H(x,y) = F_{Y|X=x}(y).$$

Since

$$H\{x, \xi(p,x)\} = p,$$

differentiating both sides with respect to x gives

$$\frac{\partial H}{\partial x}\{x,\xi(p,x)\} + \frac{\partial H}{\partial y}\{x,\xi(p,x)\} \frac{\partial}{\partial x} \xi(p,x) = 0.$$

Because

$$\frac{\partial H}{\partial y}\{x,\xi(p,x)\} = f_{Y|X=x}\{\xi(p,x)\},$$

it follows that

$$\frac{\partial H}{\partial x}\{x,\xi(p,x)\} = -\, f_{Y|X=x}\{\xi(p,x)\} \frac{\partial}{\partial x} \xi(p,x).$$

□

The proposition implies that the local change of the comparison distribution, when moving from x to a nearby value $x + \delta$, is governed by the slope of the conditional quantile function at level p (i.e., the quantile treatment effect), rescaled by the conditional density at that quantile. Thus, the derivative of $\upsilon(p;x,x_1)$ at $x_1 = x$ provides a density-weighted local quantile effect. In particular, if $\xi(p,x)$ increases with x, then $\upsilon(p;x,x+\delta)$ falls below the identity line for small positive δ, and conversely.

Remark 9. *A first-order expansion of $\upsilon(p;x_0,x_1)$ around $x_1 = x$ yields*

$$\left. \frac{\partial}{\partial x_1} \upsilon(p;x,x_1) \right|_{x_1=x} = -\, f_{Y|X=x}\{\xi(p,x)\} \, \frac{\partial}{\partial x} \xi(p,x),$$

where $\xi(p,x) = F^{-1}_{Y|X=x}(p)$.

Under the linear conditional quantile model

$$\xi(p, x) = \beta_0(p) + \beta_1(p)x,$$

this simplifies to

$$\frac{\partial}{\partial x_1} \upsilon(p; x, x_1)\bigg|_{x_1 = x} = -\, f_{Y|X=x}\{\xi(p, x)\}\, \beta_1(p) = -\frac{\beta_1(p)}{q(p, x)},$$

where

$$q(p, x) = \frac{\partial}{\partial p}\xi(p, x) = \frac{1}{f_{Y|X=x}\{\xi(p, x)\}}$$

is the conditional sparsity function. Thus, the local departure of the comparison distribution from the identity line reflects the quantile regression slope at level p, rescaled by the local density (or, equivalently, by the inverse sparsity).

To illustrate the comparison distribution in a practical setting, we consider SBP from the `NHANES` study. Let $F_{Y|\mathrm{sex},a}$ denote the conditional distribution of SBP at age a for females ($\mathrm{sex} = 0$) or males ($\mathrm{sex} = 1$). The comparison distribution as defined in (11.10) between females and males at a given age a is given by

$$\upsilon(p; \boldsymbol{x}_0, \boldsymbol{x}_1) = F_{Y|1,a}\left\{Q_{Y|0,a}(p)\right\},$$

where $\boldsymbol{x}_0 = (0, a)^\top$ and $\boldsymbol{x}_1 = (1, a)^\top$. This quantity maps the pth SBP quantile among females of age a to its percentile rank in the male conditional distribution at the same age. Figure 11.3 illustrates this idea for ages 20, 40, and 60. The top panel shows the conditional quantile functions for females and males, the bottom left panel shows the quantile difference

$$\Delta(p) = Q_{Y|1,a}(p) - Q_{Y|0,a}(p),$$

and the bottom right panel displays the comparison distribution $\upsilon(p)$, all estimated from sex-specific linear models of the type $Q_Y(p \mid \mathrm{age}) = \beta_0(p) + \beta_1(p)\mathrm{age}$.

The conditional quantile functions show higher SBP among males at all quantile levels ($\Delta(p) > 0$), particularly at younger ages. Quantile differences indicate that this gap is larger in the upper part of the distribution at age 20 and becomes more uniform across quantiles with increasing age. The comparison distributions lie below the identity line for all ages, indicating that female quantiles correspond to lower percentile ranks in the male distribution ($\upsilon(p) < p$). The departure from the identity line decreases with age, suggesting a reduction in sex differences in SBP at older ages.

Figure 11.4 displays the quantity $\upsilon(p) - p$ as a function of the quantile level p and age a. Values close to zero indicate similar conditional SBP distributions between females and males. Positive values indicate that female quantiles fall at lower percentiles in the male distribution, implying higher SBP among males. Systematic deviations from zero reveal how the sex contrast varies across the distribution and changes with age.

Finally, we obtain a model-based estimate of the local derivative (11.11) for the comparison distribution at age $a = 40$ using $-\hat{q}^{-1}(p)\hat{\beta}_1(p)$, where $\hat{q}$ is the sex-specific estimate of the sparsity q at age $a = 40$ obtained as the difference quotient of the conditional quantile function (8.13) (see also `Qtools::sparsity`) and $\hat{\beta}_1(p)$ is the sex-specific age coefficient obtained from the same linear model. This is exemplified using the females subset in R Code 11.5.

In Figure 11.5, the resulting estimates are smoothed via LOESS for visual aid. The derivative is negative across all quantiles for both sexes, indicating that around age 40 a small increase in age locally shifts the conditional SBP distribution upward. The magnitude

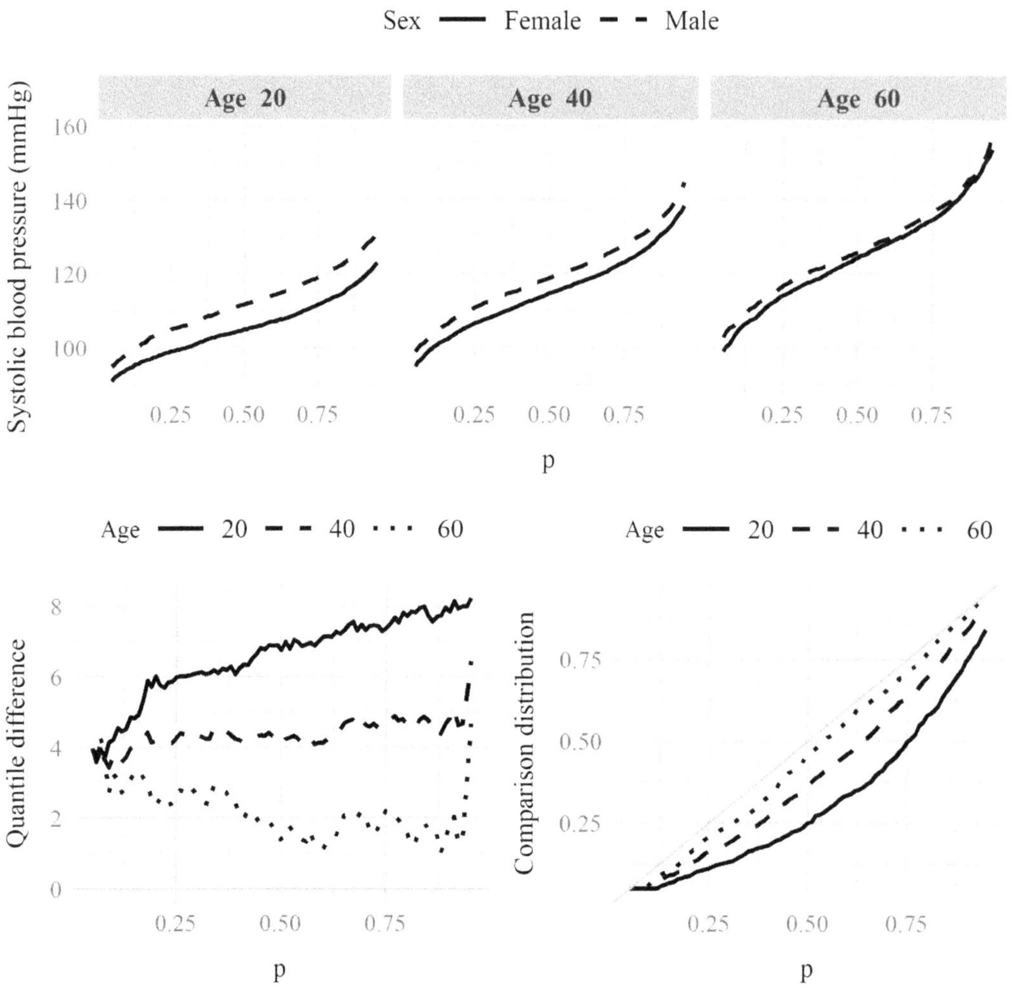

Figure 11.3
Systolic blood pressure by sex at ages 20, 40, and 60 in the `NHANES` dataset. Top: conditional quantile functions for females and males. Bottom left: quantile difference of males vs females. Bottom right: comparison distribution for males vs females with the identity line as reference.

of the derivative is larger for intermediate quantiles, consistent with higher conditional density in the centre of the distribution. The effect is more pronounced among females, particularly in the central part of the distribution, whereas it is comparatively attenuated among males. Differences between females and males are smaller and less systematic in the tails.

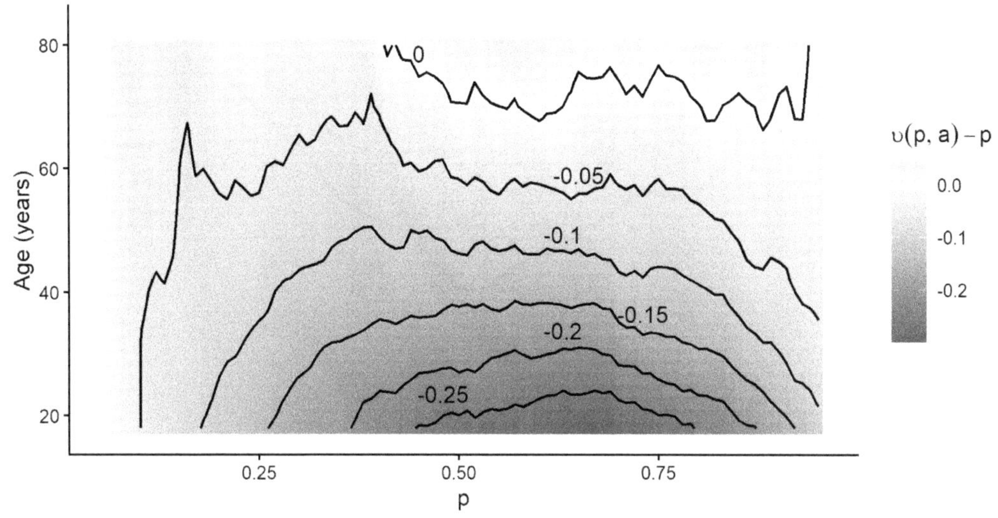

Figure 11.4
Heatmap of $\upsilon(p) - p$ comparing systolic blood pressure among females and males as a function of the quantile level p and age in the NHANES dataset.

R Code 11.5: Estimation of the derivative of the comparison function from conditional quantile regression using sparsity estimates obtained by finite differences.

```
p_grid <- seq(.05, .95, .01)
dds <- subset(NHANES, Gender == "female")
fit <- rq(BPSysAve ~ Age, tau = p_grid, data = dds)
# sparsity via difference quotient of the conditional quantile function
n <- length(fit$y)
h <- bandwidth.rq(p_grid, n, hs = TRUE)
if(any(p_grid + h > 1) | any(p_grid - h < 0)) warning("h is too large")
bhi <- mapply(function(p, x, y) rq.fit.fnb(x, y, p)$coef, p = p_grid + h,
  MoreArgs = list(x = fit$x, y = fit$y))
blo <- mapply(function(p, x, y) rq.fit.fnb(x, y, p)$coef, p = p_grid - h,
  MoreArgs = list(x = fit$x, y = fit$y))
x0 <- 40 # predict at age 40
dQ <- as.numeric(matrix(c(1,x0), nrow = 1) %*% (bhi - blo))
q <- pmax((dQ - 1e-11)/(2 * h), 1e-6)
# derivative of comparison function
deriv <- -coef(fit)[2,]/q
```

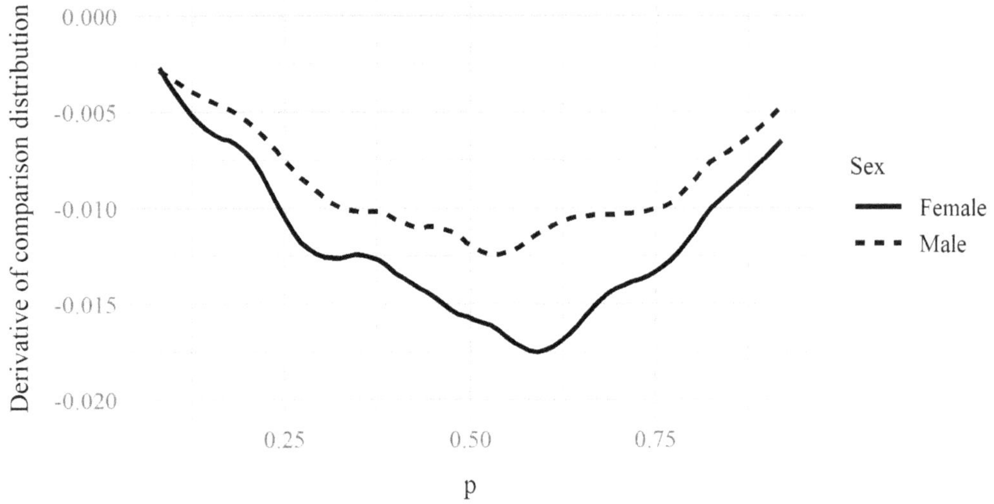

Figure 11.5
Local derivative of the comparison distribution at age 40 for systolic blood pressure in the NHANES dataset, shown separately for females and males as a function of the quantile level p.

11.4 L-statistics

The preceding sections considered quantile differences and ratios as simple operations that compare selected parts of the conditional distribution. This framework can be extended by considering linear combinations of quantiles, commonly known as L-statistics. L-statistics extend quantile contrasts by allowing linear combinations of multiple quantiles, providing flexible summaries of distributional features. By aggregating information across broader portions of the distribution, they yield smooth and stable functionals. Because they are based on quantiles rather than moments, they are typically more robust to outliers. Some, such as the median and inter-quantile ranges, also have high breakdown points.

Let $w(\cdot)$ denote a fixed weight function such that $\int_0^1 w(p)\,\mathrm{d}p = 1$. Also, let $0 < p_1 < \cdots < p_m < 1$ denote a set of quantile levels and let $w_1, \ldots, w_m$ be fixed weights.[3] In the conditional setting, L-statistics are obtained by replacing population quantiles with conditional quantiles, typically estimated via quantile regression. Specifically, an L-functional of

[3] In this section the weights are treated as fixed and chosen a priori to target interpretable summaries. In more advanced settings, however, they may be selected to improve efficiency under specific distributional assumptions and need to be estimated. Such extensions require additional assumptions on the conditional density and its estimation (Koenker, 2005, Chapter 5), and are beyond the present scope. An alternative to averaging quantiles as in (11.12) or (11.13) is to minimise a weighted sum of quantile objective functions across several quantile levels

$$\sum_{j=1}^{m} \sum_{i=1}^{n} w_j \psi_{p_j}(y_i - \boldsymbol{x}_i^\top \boldsymbol{\beta})$$

This yields a related class of estimators with similar efficiency potential but different optimal weights (e.g., see Koenker, 2005).

the conditional distribution of Y given $\boldsymbol{X} = \boldsymbol{x}$ is defined as

$$L(\boldsymbol{x}) = \int_0^1 w(p) Q_{Y|\boldsymbol{X}=\boldsymbol{x}}(p)\,\mathrm{d}p, \tag{11.12}$$

which is analogous to (2.11). The conditional equivalent of (2.12) is given by replacing the integral in (11.12) with a finite weighted sum

$$L(\boldsymbol{x}) = \sum_{j=1}^{m} w_j\, Q_{Y|\boldsymbol{X}=\boldsymbol{x}}(p_j). \tag{11.13}$$

Many familiar distributional summaries can be expressed in this form. For example, the α-trimmed conditional mean corresponds to

$$\mu_\alpha(\boldsymbol{x}) = \frac{1}{1-2\alpha} \int_\alpha^{1-\alpha} Q_{Y|\boldsymbol{X}=\boldsymbol{x}}(p)\,\mathrm{d}p,$$

with $w(p) = (1-2\alpha)^{-1} I(\alpha \le p \le 1-\alpha)$.

Because L-statistics are linear functions of conditional quantiles, their large-sample properties follow directly from the joint asymptotic normality of quantile regression estimators. Consider the discrete form in (11.13). If we assume a linear relationship $Q_{Y|\boldsymbol{X}}(p) = \boldsymbol{x}^\top \boldsymbol{\beta}(p)$, then (11.13) becomes

$$L(\boldsymbol{x}) = \sum_{j=1}^{m} w_j\, \boldsymbol{x}^\top \boldsymbol{\beta}(p_j). \tag{11.14}$$

Define the stacked parameter vector

$$\boldsymbol{\theta} = \left(\boldsymbol{\beta}(p_1)^\top, \ldots, \boldsymbol{\beta}(p_m)^\top\right)^\top.$$

and let

$$\hat{\boldsymbol{\theta}} = \left(\hat{\boldsymbol{\beta}}(p_1)^\top, \ldots, \hat{\boldsymbol{\beta}}(p_m)^\top\right)^\top$$

denote the stacked vector of regression quantiles for $p_1, \ldots, p_m$. Then the L-statistic estimator

$$\hat{L}(\boldsymbol{x}) = \sum_{j=1}^{m} w_j\, \boldsymbol{x}^\top \hat{\boldsymbol{\beta}}(p_j) \tag{11.15}$$

satisfies, for large n, the approximate distribution

$$\sqrt{n}\{\hat{L}(\boldsymbol{x}) - L(\boldsymbol{x})\} \,\dot{\sim}\, \mathcal{N}\left(0, \sum_{r=1}^{m}\sum_{s=1}^{m} w_r w_s \boldsymbol{x}^\top \boldsymbol{\Sigma}_{rs} \boldsymbol{x}\right), \tag{11.16}$$

under standard regularity conditions and for fixed $\boldsymbol{x}$, where $\boldsymbol{\Sigma}_{rs}$ is the asymptotic covariance between $\hat{\boldsymbol{\beta}}(p_r)$ and $\hat{\boldsymbol{\beta}}(p_s)$.[4]

[4] Under the smooth form (11.12) and suitable regularity conditions,

$$\sqrt{n}\{\hat{L}(\boldsymbol{x}) - L(\boldsymbol{x})\} \,\dot{\sim}\, \mathcal{N}\left(0, \int_0^1 \int_0^1 w(r)w(s)\boldsymbol{x}^\top \boldsymbol{\Sigma}(r,s)\boldsymbol{x}\,\mathrm{d}r\,\mathrm{d}s\right),$$

where $\boldsymbol{\Sigma}(r, s)$ denotes the asymptotic covariance kernel of the quantile regression process. Thus, the discrete theory developed below extends immediately to smooth L-statistics by replacing finite sums with integrals. However, although this is elegant at the theoretical level, in practice it is harder to estimate the full covariance kernel than the finite block covariance matrix. See Koenker (2005, Chapter 5) for related formulations based on smooth L-statistics.

R Code 11.6: Conditional lower-tail and upper-tail means of income, and their 90:10 ratio, by sex using the IPUMS income data, United States, 2024.

```
pL <- seq(0.01, 0.10, by = 0.01)
pU <- seq(0.90, 0.99, by = 0.01)
dpL <- diff(c(0, pL))
dpU <- diff(c(pU, 1))

## Linear quantile model by sex
fit1 <- rq(incwage ~ sex, tau = c(0.1,0.9), data = ipums, method = "fn")
fitL <- rq(incwage ~ sex, tau = pL, data = ipums, method = "fn")
fitU <- rq(incwage ~ sex, tau = pU, data = ipums, method = "fn")

## Predicted conditional quantiles on original scale
res1 <- data.frame(sex = factor(c("Female","Male"),
levels = levels(ipums$sex)))
Qhat <- predict(fit1, newdata = res1)
res1$Q10 <- Qhat[,1]
res1$Q90 <- Qhat[,2]
res1$Quantile_ratio <- res1$Q90/res1$Q10
res1$SE_ratio <- sqrt(qr_var(fit1, res1, se = "nid"))

## Tail means using a Riemann approximation
Qhat <- predict(fitL, newdata = res1)
res1$lower_mean <- rowSums(Qhat*dpL)/sum(dpL)
Qhat <- predict(fitU, newdata = res1)
res1$upper_mean <- rowSums(Qhat*dpU)/sum(dpU)
res1$Tail_ratio <- res1$upper_mean/res1$lower_mean
res1$SE_Tratio <- sqrt(qtr_var(fitL, fitU, newdata = res1,
se = "iid", wL = dpL, wU = dpU))
```

One specification of the formulation in (11.12) that is useful in inequality studies defines tail-average functionals with $w(p) = \alpha^{-1}I(p \leq \alpha)$ and $w(p) = \alpha^{-1}I(p \geq 1 - \alpha)$, $\alpha < 0.5$. Say Y is income and $\alpha = 0.1$, then

$$\mu_L(\boldsymbol{x}) = 0.1^{-1} \int_0^{0.1} Q_{Y|\boldsymbol{X}=\boldsymbol{x}}(p)\, \mathrm{d}p, \qquad \mu_U(\boldsymbol{x}) = 0.1^{-1} \int_{0.9}^{1} Q_{Y|\boldsymbol{X}=\boldsymbol{x}}(p)\, \mathrm{d}p.$$

These functionals summarise, respectively, the average income in the bottom and top deciles of the conditional distribution. Their ratio provides an interpretable measure of inequality. Large values of the 90:10 *tail-mean ratio* $\mu_U(\boldsymbol{x})/\mu_L(\boldsymbol{x})$ indicate that the upper tail of the conditional income distribution is far from the lower tail relative to the overall scale of the outcome. In parallel, one could also take the 90:10 *quantile ratio* $Q_{Y|\boldsymbol{X}}(0.9)/Q_{Y|\boldsymbol{X}}(0.1)$, which serves the analogous purpose of measuring inequality, though it gives a different, yet complementary, information.

An illustration is given in R Codes 11.6 and 11.7, using pre-tax wage and salary income from the Integrated Public Use Microdata Series (IPUMS USA) 2024 data. The sample

R Code 11.7: Conditional lower-tail and upper-tail means of income, and their 90:10 ratio, by age and sex using the IPUMS income data, United States, 2024.

```
## Linear quantile model by age, sex and their interaction
fit2 <- rq(incwage ~ age*sex, tau = c(0.1,0.9), data = ipums, method = "fn")
fitL <- rq(incwage ~ age*sex, tau = pL, data = ipums, method = "fn")
fitU <- rq(incwage ~ age*sex, tau = pU, data = ipums, method = "fn")

## Predicted conditional quantiles on original scale
res2 <- expand.grid(
  age = c(21, 30, 40, 50, 65),
  sex = factor(c("Female","Male"), levels = levels(ipums$sex))
)
Qhat <- predict(fit2, newdata = res2)
res2$Q10 <- Qhat[,1]
res2$Q90 <- Qhat[,2]
res2$Quantile_ratio <- res2$Q90/res2$Q10
res2$SE_ratio <- sqrt(qr_var(fit2, res2, se = "nid"))

## Tail means using a Riemann-sum approximation
Qhat <- predict(fitL, newdata = res2)
res2$lower_mean <- rowSums(Qhat*dpL)/sum(dpL)
Qhat <- predict(fitU, newdata = res2)
res2$upper_mean <- rowSums(Qhat*dpU)/sum(dpU)
res2$Tail_ratio <- res2$upper_mean/res2$lower_mean
res2$SE_Tratio <- sqrt(qtr_var(fitL, fitU, newdata = res2,
se = "iid", wL = dpL, wU = dpU))
```

(n = 609,601) was restricted to people aged 21 to 65 years with at least 3 years of college education. In these calculations, the integrals are approximated on a fine grid of quantile levels $p \in \{0.01, 0.02, \dots, 0.1, 0.90, 0.91, \dots, 0.99\}$, hence $\mu_L(\boldsymbol{x}) \approx 0.1^{-1} \sum_{j=1}^{10} Q_{Y|\boldsymbol{X}=\boldsymbol{x}}(p_j) \cdot (p_j - p_{j-1})$ and $\mu_U(\boldsymbol{x}) \approx 0.1^{-1} \sum_{j=11}^{20} Q_{Y|\boldsymbol{X}=\boldsymbol{x}}(p_j) \cdot (p_j - p_{j-1})$, with $p_0 = 0$ and $p_{20} = 1$. Point estimates and standard errors are obtained using (11.15) and (11.16), respectively.

In the first analysis (R Code 11.6), quantiles are estimated conditional on sex for all ages using a linear model. The two inequality measures lead to very similar conclusions regarding sex differences (Table 11.1). Both the 90:10 quantile ratio and the ratio of tail means indicate slightly higher inequality among females than males, although the differences between sexes are modest. This suggests that, in these data, the choice between pointwise and tail-aggregated summaries does not materially affect the comparison across sexes. This result is also confirmed by the Gini coefficient, which can be expressed as a functional of the quantile function,

$$G = \frac{1}{\mu} \int_0^1 (2p - 1) Q_Y(p)\, \mathrm{d}p, \qquad \mu = \int_0^1 Q_Y(p)\, \mathrm{d}p.$$

calculated by sex in the code fragment below with the `R` package `reldist`.

Table 11.1
Tail quantiles, 90:10 quantile ratio, lower- and upper-tail means (μ_L and μ_U, respectively), and 90:10 tail ratio of income by sex and by sex and age, with standard error (SE) of the ratios, using the IPUMS income data, United States, 2024.

Sex	Age	$Q(0.1)$	$Q(0.9)$	Quantile ratio	SE	μ_L	μ_U	Tail ratio	SE
Female	21-65	15,000	150,000	10.00	0.05	8,050	228,100	28.34	0.64
Male	21-65	27,200	240,000	8.82	0.08	14,830	399,100	26.91	0.36
Female	21	11,900	103,333	8.68	0.14	6,605	127,880	19.36	0.47
	30	13,541	124,333	9.18	0.09	7,292	168,759	23.14	0.34
	40	15,365	147,667	9.61	0.06	8,055	214,182	26.59	0.25
	50	17,188	171,000	9.95	0.08	8,818	259,604	29.44	0.29
	65	19,924	206,000	10.34	0.13	9,962	327,737	32.90	0.51
Male	21	17,067	105,217	6.17	0.10	9,038	173,073	19.15	0.37
	30	21,921	160,000	7.30	0.06	11,879	258,539	21.76	0.22
	40	27,315	220,870	8.09	0.04	15,036	353,502	23.51	0.13
	50	32,709	281,739	8.61	0.05	18,192	448,464	24.65	0.12
	65	40,800	373,043	9.14	0.08	22,927	590,908	25.77	0.19

```
library(reldist)
tapply(ipums$incwage, ipums$sex, gini)
>      Female      Male
> 0.4244849 0.4457584
```

In the second analysis (R Code 11.7), a linear quantile regression model is fitted with sex, age, and their interaction as covariates. The results (Table 11.1) illustrate the difference between pointwise and tail-aggregated measures of inequality.

The 90:10 quantile ratio varies moderately with age and suggests broadly similar patterns across sexes. In contrast, the ratio of upper-to-lower tail mean increases markedly with age, particularly among females, indicating that changes in inequality are driven by the behaviour of the full tails rather than by shifts at specific quantile levels. While the quantile ratio captures the distance between two points of the distribution, the tail-mean ratio reflects the aggregate mass of the upper and lower deciles, and is therefore more sensitive to changes in the shape and spread of the tails. In these data, this distinction reveals a stronger age gradient in inequality than would be inferred from quantile ratios alone.

12

Regression models for distribution functions

12.1 Binomial regression

The binomial distribution is intimately connected to quantile inference as it naturally arises from the sample CDF, which is instrumental in the derivation of the asymptotic distribution of sample quantiles. This distribution is also tied to quantiles from a modelling perspective as an alternative (or even complement) to quantile regression.

In applied research, variables are often subjected to threshold mechanisms to conveniently analyse and summarise observations into proportions above or below a given cutoff. The cutoff itself may be grounded in field-specific (e.g., biological, clinical, or policy-related) rationales. As an example, consider the birth weight data introduced in Section 1.2. Epidemiological and clinical research, as well as public health interventions, are frequently aimed at low birth weight (LBW) infants as these experience higher risk of mortality and morbidity compared to normal weight infants. This motivates the introduction of a binary indicator capturing whether the response lies below the cutoff. The international cutoff for LBW is 2500 grams. This cutoff was set by the World Health Organization in 1950 (Expert Group on Prematurity, 1950) following the study of the Finnish pediatrician Dr. Arvo Ylppö in 1919 (Ylppö, 1919), who observed a high mortality rate among these 'premature' infants. The same cutoff has been since applied to babies of any sex, race, and country of origin, despite the fact that these factors (along with others) influence every single aspect of the birth weight distribution (Rooth, 1980).

Let Y denote the birth weight and let $F_Y(y) = \Pr(Y \leq y)$ be its distribution function. Denote the cutoff of interest by c, with $F_Y(c) = \pi$, and define

$$D = \begin{cases} 1 & \text{if } Y \leq c \\ 0 & \text{otherwise.} \end{cases}$$

It follows that D is Bernoulli with parameter $\pi \equiv \mathrm{E}(D) = \Pr(D = 1)$. Replacing the continuous response Y with the indicator $D = I(Y \leq c)$ fundamentally changes the object of inference. The analysis no longer concerns the distribution of Y itself, but only the probability that the response falls below the threshold c. In probabilistic terms, binomial regression models the single quantity $F_Y(c) = \Pr(Y \leq c)$, that is, one point of the cumulative distribution function of Y. The remainder of the distribution, including its shape, spread, and behaviour away from the cutoff, is no longer directly modelled.

In the presence of covariates, binomial regression is the typical model of choice. To simplify the exposition, assume that X is binary and let $\pi_x = \Pr(D = 1 \mid X = x)$, $x \in \{0, 1\}$. In Figure 12.1, the two plots show birth weight observations for males (left) and females (right) among 2001 singletons born at term in the US.[1] The cutoff at 2500 g separates the LBW infants from the others, resulting in two clearly different LBW proportions: 2.07%

[1]To avoid overcrowding, the plot displays a random sample of 10,000 observations drawn from approximately 3 million records.

DOI: 10.1201/9781003326410-12

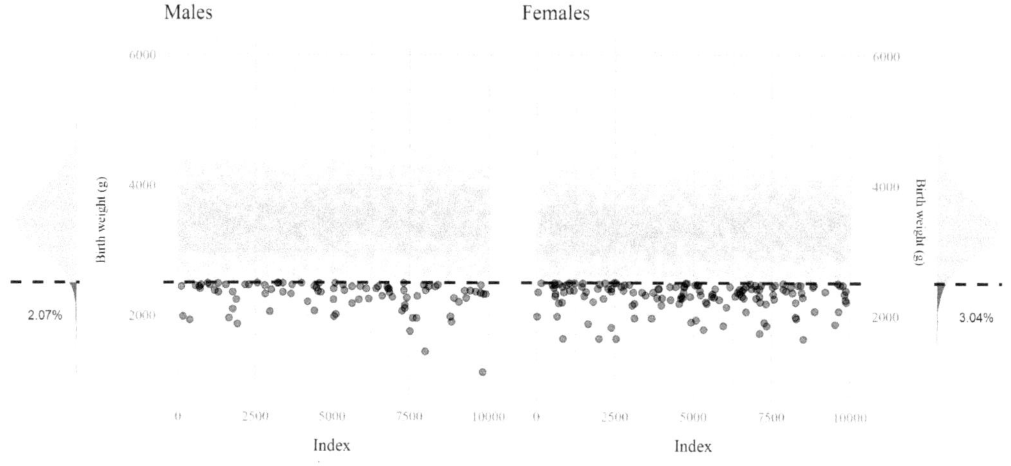

Figure 12.1
Low birth weight among born-at-term singletons in the United States, 2001.

among boys and 3.04% among girls. As expected, these proportions are equal to the estimated probabilities from the logistic regression of the LBW indicator on sex (R Code 12.1). The same cutoff therefore corresponds to two different conditional quantile levels, since $F_{Y|X}(2500 \mid x)$ differs by group. In other words, the same clinical threshold corresponds to different degrees of rarity in the underlying distributions, meaning that a fixed LBW cutoff does not represent the same level of risk across groups.

The association between X and D can be measured by the risk difference $\pi_1 - \pi_0$, or the risk ratio π_1/π_0, or, very commonly, by the odds ratio ω_1/ω_0, where $\omega_x = \pi_x/(1 - \pi_x)$. If X is not associated with Y, then we should expect any of

$$\pi_1 - \pi_0 = 0 \tag{12.1}$$
$$\log(\pi_1) - \log(\pi_0) = 0 \tag{12.2}$$
$$\log(\omega_1) - \log(\omega_0) = 0. \tag{12.3}$$

We can consider the relationship between quantile regression and binomial regression by looking at the definition of QTE as given in Section 8.2. The QTE is the quantity $\Delta(p)$ such that $F_{Y|X}(y \mid x = 1) = F_{Y|X}\{y - \Delta(p) \mid x = 0\}$. If we let $y = c$ and $p = F_{Y|X}(c \mid x = 1)$, then a null QTE leads to $F_{Y|X}(c \mid x = 1) = F_{Y|X}(c \mid x = 0)$, which implies a null effect as in (12.1)–(12.3) (note that $F_{Y|X}(c \mid x) = \pi_x$ by definition).

However, if we do not find a meaningful or statistically significant association between X and D, does this apply also to X and Y? Not necessarily, since the dichotomisation may obscure the association with Y and gravely affect the statistical power of statistics derived from the binomial model. Suppose

$$Y = \theta_1 + \theta_2 x + (\theta_3 + \theta_4 x)\varepsilon, \tag{12.4}$$

where $\varepsilon \sim \mathcal{N}(0, 1)$ and x is the treatment indicator. The QTE associated with model (12.4) is therefore $\Delta(p) = \theta_2 + \theta_4\Phi^{-1}(p)$. Also, define the cutoff $c(p) = Q_{Y|X}(p \mid X = 1) = (\theta_1 + \theta_2) + (\theta_3 + \theta_4)\Phi^{-1}(p)$, as a function of p, with $p \in (0, 1)$. The conditional probabilities $\Pr\{Y \leq c(p) \mid X\}$ are

$$\pi_0(p) = \Phi(\{c(p) - \theta_1\}/\theta_3) = \Phi(\{\theta_2 + (\theta_3 + \theta_4)\Phi^{-1}(p)\}/\theta_3)$$

R Code 12.1: Probabilities and quantiles of the birth weight distribution conditional on sex among singletons born in the United States, 2001.

```
# singletons born at term only
dd <- subset(USbirths2001, USbirths2001$ga >= 37 &
   USbirths2001$multiple == 0)
nrow(dd) # about 3 million
> [1] 3003342
# fit logistic regression on LBW indicator
# conditional on sex (x = 1 if female)
y <- dd$bw
d <- 1*(y < 2500)
x <- dd$female
fit <- glm(d ~ x, family = binomial)

# conditional probabilities of birthweight less than 2500
xx <- cbind(1, c(0,1))
eta <- xx %*% coef(fit)
phat <- as.numeric(1/(1 + exp(-eta)))
names(phat) <- c("males", "females")
phat
>       males    females
> 0.02070983 0.03036235
# verify that the estimated probabilities correspond to
# the same empirical quantile (i.e., 2500)
round(quantile(y[x == 0], probs = phat[1], names = FALSE))
> [1] 2500
round(quantile(y[x == 1], probs = phat[2], names = FALSE))
> [1] 2500
```

and

$$\pi_1(p) = \Phi(\{c(p) - (\theta_1 + \theta_2)\}/\{\theta_3 + \theta_4\}) = \Phi(\Phi^{-1}(p)) = p.$$

Without loss of generality, we set $\theta_1 = 0$, $\theta_2 = 0$, and $\theta_4 = 1$ (heteroskedastic errors). We consider two alternative scenarios, corresponding to either unit baseline variance ($\theta_3 = 1$) or inflated baseline variance ($\theta_3 = 5$). In both scenarios, $\Delta(p) = \Phi^{-1}(p)$ while the cutoff $c(p)$ is an increasing function of p. The plots in Figure 12.2 show the QTE and the log relative risk for $p = i/100$, $i = 1, \ldots, 99$ that result from these settings. Both effects are null at $p = 0.5$, a point that marks a sign change.

First, we see that, for $p < 0.5$, the treatment shifts the distribution to the left, implying higher risk among the treated group ($x = 1$). For $p > 0.5$, the QTE becomes positive, reversing the association between X and D. Although the magnitude of the QTE increases monotonically, the magnitude of the log relative risk not only increases minimally but also trends back towards the null effect as p approaches the upper tail of the distribution. As baseline variance increases, the entire relative risk curve flattens towards the null value, while the QTE remains unchanged. In short, an unsuitably high cutoff or elevated baseline noise could dilute the relative risk, masking a meaningful association.

Models based on a single cutoff therefore capture only one point of the conditional distribution function and discard all information about its behaviour elsewhere.

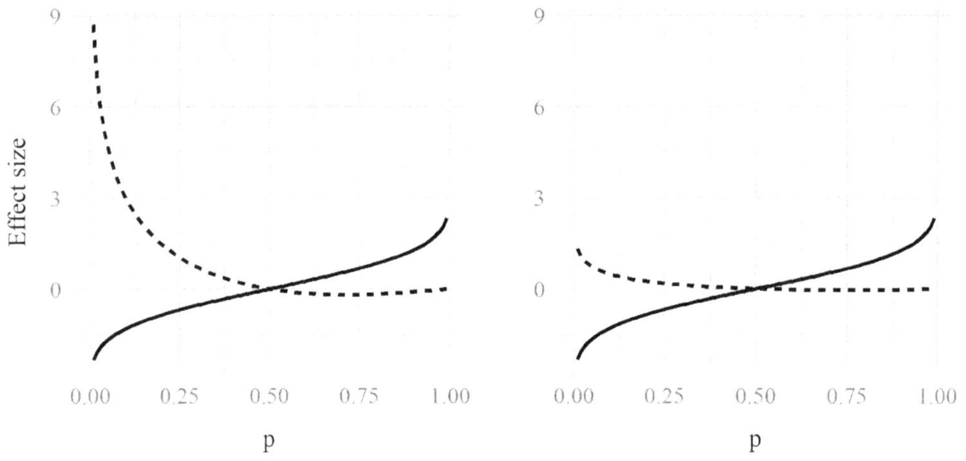

Figure 12.2
Quantile treatment effect (solid line) and log relative risk (dashed line) as a function of p for unit baseline variance (left) and inflated baseline variance (right).

12.2 Distributional regression

Suppose we have a sample (x_i, y_i), $i = 1, \ldots, n$. The goal is to estimate the conditional distribution $F_{Y|X}(y) = \Pr(Y \leq y \mid X = x)$ with a sequence of parametric, semiparametric and nonparametric estimators (e.g., an example of nonparametric estimator is given in Equation (10.9)). Distributional regression, as presented in Section 10.3, consists in fitting binomial regressions on the data (x_i, d_{ij}), $i = 1, \ldots, n$, separately for each $j = 1, \ldots, s$, where $d_{ij} = I(y_i \leq z_j)$. The value z_j, $j = 1, \ldots, s$, represents the jth distinct observation of the response variable Y that occurs in the sample, with $z_j < z_{j+1}$ for all $j = 1, \ldots, s - 1$. If the response is continuous and rounding is kept to a minimum, then usually $s = n$.

For example, consider the LS specification of model 12.4 with a single, continuous covariates x, $\theta_1 = 1$, $\theta_2 = 2$, $\theta_3 = 1$, and $\theta_4 = 0$. Then $F_{Y|X}(y \mid x) = \Phi(y - 1 - 2x)$. The estimated CDFs using quantile and distributional regression as in R Code 12.2 are shown in Figure 12.3. Overall, both approaches provide an accurate approximation of the true conditional distribution function (as evinced by the similar mean absolute error), but they do so through different parameterisations: distributional regression models the distribution function directly, whereas quantile regression estimates its inverse. However, this implementation of distributional regression, based on repeated calls to `glm` across thresholds, is substantially more computationally demanding than estimation based on quantile regression, although more efficient implementations are possible.

Besides the CDFs, it may be instructive to compare the regression coefficients. Since $\partial F_{Y|X}(y)/\partial x = -\theta_2 \phi(\{y - \theta_1 - \theta_2 x\}/\theta_3)$ and $\partial F_{Y|X}(y)/\partial y = \phi(\{y - \theta_1 - \theta_2 x\}/\theta_3)$, we have

$$-\frac{\partial F_{Y|X}(y)/\partial x}{\partial F_{Y|X}(y)/\partial y} = \theta_2 = \frac{\partial F^{-1}_{Y|X}(p)}{\partial x}. \tag{12.5}$$

Equation (12.5) highlights an important unifying insight. The slope parameter θ_2 has a consistent interpretation across different perspectives on the conditional distribution. On the one hand, it governs the sensitivity of the CDF to changes in the covariate x, scaled by

R Code 12.2: Distributional and quantile regression.

```
set.seed(123)
n <- 500
x <- sort(runif(n))
y <- (m <- 1 + 2*x) + (e <- rnorm(n))
length(unique(y)) == n
> [1] TRUE
z <- sort(y)
range(pnorm(e))
> [1] 0.00247881 0.99644571
p <- seq(0.01, 0.99, by = 0.005)
F <- outer(m, z, function(mm, zz) pnorm(zz - mm))

# rq
t0 <- proc.time()['elapsed']
fit.rq <- rq(y ~ x, tau = p)
Fhat_fun.rq <- predict(fit.rq, type = "Fhat", stepfun = TRUE)
Fhat.rq <- t(sapply(Fhat_fun.rq, function(F, z) F(z), z = z))
t1 <- proc.time()['elapsed']
print(t1 - t0)
> elapsed
>    0.22
# glm
t0 <- proc.time()['elapsed']
fit.glm <- lapply(z, function(zz, y, x)
  glm(1*(y <= zz) ~ x, family = binomial(link = "probit")),
  y = y, x = x)
Fhat.dr <- sapply(fit.glm, function(obj) obj$fitted)
Fhat_fun.dr <- apply(Fhat.dr, 1, function(u, z) stepfun(z, c(0,u)), z = z)
t1 <- proc.time()['elapsed']
print(t1 - t0)
> elapsed
>    3.14
mean(abs(Fhat.rq - F))
> [1] 0.01897596
mean(abs(Fhat.dr - F))
> [1] 0.01637087
```

the local density at y. On the other hand, it coincides with the derivative of the conditional quantile function $F_{Y|X}^{-1}(p)$ with respect to x under the model, meaning that θ_2 directly describes how any quantile of Y shifts as x changes.

The models discussed so far describe the conditional distribution of the response variable given the covariates. In many applications, however, the primary interest lies not in conditional quantities but in how the overall distribution of the response changes when the distribution of covariates changes. This distinction between conditional and marginal interpretation is fundamental in regression modelling and motivates approaches such as unconditional quantile regression.

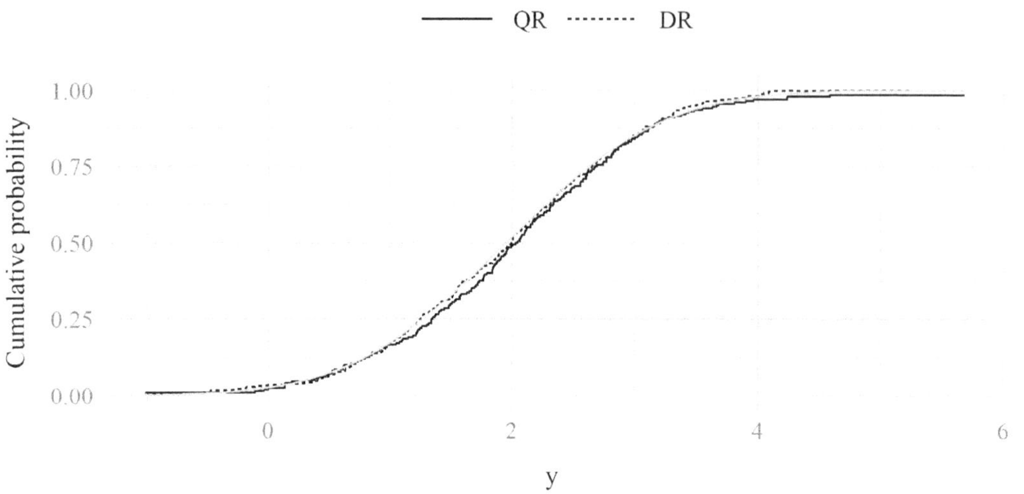

Figure 12.3
Estimated cumulative distribution functions, conditional on a selected value of x, obtained using distributional (dashed black line) and quantile (solid black line) regression. The true function is shown as a grey line.

12.3 Unconditional quantile regression

So far we have focused on conditional quantities, which describe how the outcome varies within strata of observed covariates and that are typically summarised by regression parameters. Yet many questions of substantive interest concern the distribution of an outcome in the population as a whole rather than within covariate-defined subgroups. A prominent example is the distribution of wealth or wages in society. Changes in the sociodemographic composition of the population can alter this marginal distribution. Because such marginal shifts shape economic inequality and carry substantial political and social implications, understanding how population composition translates into changes in the overall distribution of outcomes is often as important as understanding the conditional mechanisms themselves.

Let $\pi \equiv \mathrm{E}(D) = \Pr(D = 1)$ and $\pi_x = \Pr(D = 1 \mid X = x)$, $x \in \{0, 1\}$. Suppose that X is binary with $\Pr(X = 1) = u$. By the law of total expectation, we can write the marginal (i.e., not conditional on x) probability π as a function of u, namely

$$\begin{aligned} \pi &= \Pr(D = 1 \mid X = 0)\Pr(X = 0) + \Pr(D = 1 \mid X = 1)\Pr(X = 1) \\ &= \pi_0(1 - u) + \pi_1 u = \pi_0 + (\pi_1 - \pi_0)u \equiv \pi(u). \end{aligned} \tag{12.6}$$

If the population composition with respect to X were to change by $\mathrm{d}u$, the prevalence π would change by $\mathrm{d}\pi(u) = (\pi_1 - \pi_0)\mathrm{d}u$. Thus, the difference $\pi_1 - \pi_0$ represents the change in the population prevalence of $D = 1$ associated with a marginal change in the proportion u of individuals with $X = 1$ in the population. More generally, this type of argument can be used to study how changes in the distribution of covariates affect marginal features of the response distribution, including its quantiles.

Consider a linear probability model on the LBW indicator

$$\pi_x \equiv \mathrm{E}(D \mid X) = \beta_0 + \beta_1 x,$$

R Code 12.3: Linear probability regression for low birth weight prevalence as a function of multiple-birth status among all infants born in the United States, 2001.

```
nrow(USbirths2001) # about 3.5 million
> [1] 3477802
# define variables
y <- USbirths2001$bw
d <- 1*(y < 2500)
x <- 1*(USbirths2001$multiple == 1)
# multiple birth prevalence in the population
print(px <- mean(x))
> [1] 0.03240064
# LBW prevalence in the population (unconditional)
print(LBW <- mean(d))
> [1] 0.07888919
# LBW prevalence conditional on multiple status
print(pi0 <- mean(d[x == 0])) # compare with lm intercept
> [1] 0.06218056
print(pi1 <- mean(d[x == 1]))
> [1] 0.5778689
# LBW prevalence difference between groups
print(dp <- pi1 - pi0) # compare with lm slope
> [1] 0.5156883
# probability linear model
fit <- lm(d ~ x)
fit
>
> Call:
> lm(formula = d ~ x)
>
> Coefficients:
> (Intercept)            x
>     0.06218      0.51569
# expected percentage points change of LBW in the population
# for a change du in the singleton/multiple composition
du <- 0.1
(dp*du)*100
> [1] 5.156883
```

where $x = 1$ indicates a multiple birth. R Code 12.3 illustrates this model applied to the 2001 US birth weight data. The results, obtained with `lm`, shows $\hat{\beta}_0 = \hat{\pi}_0 = 0.0622$ and $\hat{\beta}_1 = \hat{\pi}_1 - \hat{\pi}_0 = 0.5157$. The conditional interpretation of $\hat{\beta}_1$ is that the LBW risk for an infant from a multiple birth is 51.6 percentage points higher than for a singleton. But it can also be interpreted as the expected change in the LBW prevalence among the population should the singleton-multiple composition change. For example, it would increase by 5.2 percentage points going from 7.9% to 13.1%, should multiple births increase by 10.0 percentage points.

This simple calculation illustrates how regression parameters link changes in population composition to marginal outcomes. The same idea can be extended beyond the mean to other features of the marginal distribution. In particular, one may ask how the quantiles of the outcome distribution in the population would change if the distribution of covariates were to shift. This leads naturally to the study of unconditional quantiles, which summarise the outcome distribution in the population as a whole rather than within covariate-defined subgroups.

The work by Firpo et al. (2009) elaborates this idea by introducing what they call *unconditional* quantile regression. Building on the influence function of Hampel (1974), they define the recentred influence function (RIF)

$$\mathrm{RIF}\left(Y;\xi_p,F_Y\right)=\xi_p+\frac{p-I\left(Y\leq\xi_p\right)}{f_Y\left(\xi_p\right)} \tag{12.7}$$

Intuitively, the influence function measures the effect of an infinitesimal contamination at point y on the statistic. Now imagine slightly perturbing the population by increasing the weight of an observation Y. The influence function measures how much the quantile would move because of that perturbation. If $Y\leq\xi_p$, then $I\left(Y\leq\xi_p\right)=1$, so the influence is $(p-1)/f_Y\left(\xi_p\right)$, which is negative. Adding more weight to an observation below the quantile pushes the quantile down. If $Y>\xi_p$, then $I\left(Y\leq\xi_p\right)=0$, so the influence is $p/f_Y\left(\xi_p\right)$, which is positive. Adding more weight to an observation above the quantile pushes the quantile up. The denominator $f_Y\left(\xi_p\right)$ plays the same role we have already seen previously: when the density is high, the quantile is stable and moves little; when it is low, the quantile moves more.

Given that $\mathrm{E}(\mathrm{RIF}\left(Y;\xi_p,F_Y\right))=\xi_p$, regressing (12.7) on X approximates how covariates shift the unconditional quantile ξ_p. This decomposition separates the variability due to the second-stage regression from the variability induced by estimating the first-stage components ξ_p and $f_Y(\xi_p)$. Its implementation is illustrated in R Code 12.4 to compare singletons and multiple births, building on the example in R Code 12.3; see Firpo et al. (2009) for additional methodological details on the OLS and related estimators. The contrast between

R Code 12.4: Unconditional and conditional quantile regression for low birth weight quantiles as a function of multiple-birth status among all infants born in the United States, 2001.

```
# use LBW prevalences (conditional and unconditional) as quantile levels
p <- round(c(pi0, LBW, pi1), 3)
# fit unconditional and conditional QR
res <- matrix(NA, length(p), 3)
res[,1] <- p
colnames(res) <- c("p", "UQR", "QR")
for(j in 1:length(p)){
xi <- quantile(y, p[j])
fhat <- density(y)
w <- xi + (p[j] - 1*(y <= xi))/approxfun(fhat$x, fhat$y)(xi)
res[j,2] <- round(lm(w ~ x)$coef[2])
res[j,3] <- round(rq(y ~ x, tau = p[j], method = "fn")$coef[2])
}
res
>          p  UQR    QR
> [1,] 0.062 -3586 -1425
> [2,] 0.079 -3021 -1389
> [3,] 0.578  -514  -959
```

conditional and unconditional quantile regression is particularly striking around the LBW region. At the population quantile corresponding to the 2500 g threshold ($p = 0.079$), the conditional quantile regression suggests that infants in multiple births weigh about 1.4 kg less than singletons at the same within-group rank. In contrast, the unconditional quantile regression effect of 3.0 kg suggests that the corresponding population quantile would decrease as the proportion of multiple births increases, reflecting a downward shift of the overall birth-weight distribution. Numerically, the estimated slope implies a decrease of about 302 g for each 10 percentage-point increase in the prevalence of multiple births. This phenomenon grows stronger the deeper we go into the lower tail of the distribution. For example, at $p = 0.062$, corresponding to the LBW prevalence among singletons, the unconditional effect reaches about 3.6 kg. The larger unconditional effects reflect the fact that multiples are heavily concentrated in the lower tail of the birthweight distribution (58% are LBW), so changes in their prevalence substantially reshape the population distribution itself. These results have clear policy implications because the prevalence of multiple births is partly shaped by medical practice, especially assisted reproductive technologies.

The three approaches discussed in this chapter provide complementary perspectives on birthweight disparities. Binomial regression focuses on the probability of crossing a clinically relevant threshold such as low birthweight, offering a simple and interpretable measure of risk. Distributional regression and conditional quantile regression extend this perspective by showing how differences between groups vary across the conditional distribution. Unconditional quantile regression shifts the focus further to the population level, showing how changes in the composition of births can reshape the lower tail of the overall distribution. Together, these approaches highlight how different modelling frameworks answer different but related questions about health inequality, from individual risk to the distributional consequences of population change.

13

Quantile regression with correlated observations

13.1 Clustered data

Dependent data arise naturally in clustered designs, where observations within the same unit are correlated. Ignoring within-cluster dependence has different consequences for estimation and inference. Under suitable conditions, point estimates of fixed effects may remain consistent if the conditional quantile function is correctly specified, as the objective function still targets the same population quantities. However, efficiency is generally reduced, and, more importantly, standard errors are typically underestimated because the effective sample size is smaller than the nominal number of observations. Consequently, confidence intervals are too narrow and hypothesis tests are anti-conservative. Bias in the point estimates may arise if the unobserved cluster-specific effects are correlated with the covariates, in which case the omission of random effects leads to a misspecified conditional quantile function.

Several approaches have been proposed to extend quantile regression to dependent data. A useful distinction is between distribution-free and likelihood-based methods. Distribution-free approaches include fixed-effects models (Koenker, 2004; Lamarche, 2010; Galvao and Montes-Rojas, 2010; Galvao, 2011), which eliminate unobserved heterogeneity through unit-specific intercepts, and estimating equation methods, which target population-level quantiles using a working correlation structure; weighted formulations have also been considered to account for clustering (Lipsitz et al., 1997; Karlsson, 2008; Fu and Wang, 2012). Likelihood-based approaches, on the other hand, specify an auxiliary distribution, typically the asymmetric Laplace (Section 6.3.1), and introduce random effects to model dependence, yielding conditional, cluster-specific quantiles.[1]

Fixed- and mixed-effects modelling is arguably the most common approach and is widely implemented in standard software. In mixed-effects models, a cluster is a grouping of observations associated with a unit-specific latent effect. Random-effects formulations treat these effects as draws from a common distribution, implying that clusters are exchangeable and that inference extends beyond the observed units. This interpretation is natural when clusters can be viewed as a sample from a larger population. When, instead, the groups correspond to a fixed and exhaustive set of categories, such as administrative areas, this assumption is less compelling. In such settings, fixed-effects models, which treat group-specific effects as parameters rather than random draws, provide a more appropriate representation. A further distinction is conceptual. Fixed-effects models assign a separate parameter to each cluster, effectively treating clusters as entities of intrinsic interest. This leads to a parameter dimension that increases with the number of clusters, which may be undesirable when the latter is large or grows with the sample size. Random-effects models, instead,

[1] In this section, the term conditional is used in two distinct senses: it may refer to conditioning on the covariates, as opposed to unconditional quantities considered in previous chapters, or to conditioning on the random effects, yielding cluster-specific quantities as opposed to marginal or population-level ones; the intended meaning will be clear from the context.

DOI: 10.1201/9781003326410-13

treat cluster-specific effects as realisations from a common distribution, thereby capturing between-cluster variability through a finite number of parameters.

A linear quantile mixed model (LQMM) (Geraci and Bottai, 2014) combines random effects with quantile regression. Consider clustered data in the form $(y_{ij}, \boldsymbol{x}_{ij}, \boldsymbol{z}_{ij})$, $j = 1, \ldots, n_i$, $i = 1, \ldots, M$, where $N = \sum_{i=1}^{M} n_i$ is the total sample size, and $\boldsymbol{x}_{ij}$ and $\boldsymbol{z}_{ij}$ denote, respectively, the $q \times 1$ and $r \times 1$ vectors corresponding to the jth rows of the design matrices $\boldsymbol{X}_i$ and $\boldsymbol{Z}_i$. The LQMM specifies the conditional quantile function of Y_{ij} given the $r \times 1$ vector of cluster-specific random effects $\boldsymbol{u}_i = (u_{i1}, \ldots, u_{ir})^{\top}$ as

$$Q_{Y_{ij}|\boldsymbol{u}_i}(p) = \boldsymbol{x}_{ij}^{\top}\boldsymbol{\beta}_p + \boldsymbol{z}_{ij}^{\top}\boldsymbol{u}_i. \tag{13.1}$$

(For simplicity, we omit including the vectors $\boldsymbol{x}_{ij}$ and $\boldsymbol{z}_{ij}$ in the conditioning set as they are treated as fixed covariates.) The random effects $\boldsymbol{u}_i$ are assumed to follow, independently for $i = 1, \ldots, M$, a distribution centred at $\boldsymbol{0}$ with variance-covariance matrix $\boldsymbol{\Psi}_p$, and to be independent of the within-cluster errors. Conceptually, random effects represent unobserved heterogeneity across clusters, and the induced correlation arises from the fact that observations within the same cluster share the same latent component. Thus, correlation is a by-product of the model, whereas heterogeneity is its substantive interpretation. The $q \times 1$ vector of fixed coefficients $\boldsymbol{\beta}_p$ collects the regression effects common to all clusters. It also determines the conditional quantile for the modal cluster, that is, the cluster with random effects equal to zero: $Q_{Y_{ij}|\boldsymbol{0}}(p) = \boldsymbol{x}_{ij}^{\top}\boldsymbol{\beta}_p$.

Estimation is based on the link between quantile regression and the asymmetric Laplace distribution. In particular, it is assumed that, conditionally on $\boldsymbol{u}_i$,

$$Y_{ij} = \mu_{p,ij}(\boldsymbol{u}_i) + \epsilon_{p,ij}, \tag{13.2}$$

where $\epsilon_{p,ij} \sim \mathcal{AL}(0, \sigma_p, p)$, $\mu_{p,ij}(\boldsymbol{u}_i) = \boldsymbol{x}_{ij}^{\top}\boldsymbol{\beta}_p + \boldsymbol{z}_{ij}^{\top}\boldsymbol{u}_i$, $\sigma_p > 0$, and $p \in (0, 1)$ is fixed. This assumption is used as a convenient device for estimation and does not imply that the true distribution is asymmetric Laplace. The idea of estimating sample quantiles using the AL device has been discussed in Section 6.3.1.

There are similarities and dissimilarities between model (13.1) and the traditional linear mixed-effects (LME) model $Y_{ij} = \boldsymbol{x}_{ij}^{\top}\boldsymbol{\beta} + \boldsymbol{z}_{ij}^{\top}\boldsymbol{u}_i + \epsilon_{ij}$, with $\boldsymbol{u}_i \sim \mathcal{N}(\boldsymbol{0}, \boldsymbol{\Omega})$ and $\epsilon_{ij} \sim \mathcal{N}(0, \omega_{\epsilon}^2)$. First of all, conditionally on the random effects, both models impose a restriction on the error term. However, the LME model requires $\mathrm{E}(\epsilon_{ij}|\boldsymbol{x}_{ij}, \boldsymbol{u}_i) = 0$, while the AL-based specification of the error given in (13.2) leads to $Q_{\epsilon_{p,ij}|\boldsymbol{x}_{ij},\boldsymbol{u}_i}(p) = 0$ or, equivalently, $\Pr(\epsilon_{p,ij} < 0|\boldsymbol{x}_{ij}, \boldsymbol{u}_i) = p$. Moreover, the parameter $\boldsymbol{\beta}_p$ is allowed to vary with the quantile level p, while $\boldsymbol{\beta}$ in the LME model is not (except for a location shift). Finally, in both approaches the variance-covariance matrix of the random effects gives a measure of the variability of $\boldsymbol{u}_i$ around $\mathrm{E}_{\boldsymbol{u}_i}\{\mu_{p,ij}(\boldsymbol{u}_i)\}$ but, again, estimates are allowed to differ by p only for the quantile mixed-effects model.

Interpretation

Before turning to inferential details, it is helpful to discuss the interpretation of the parameters in model (13.1). To fix ideas, consider the standard random intercept model

$$Y_{ij} = \mu + u_i + \epsilon_{ij}, \quad u_i \perp\!\!\!\perp \epsilon_{ij}, \tag{13.3}$$

where $u_i \sim \mathcal{N}(0, \omega_u^2)$ and $\epsilon_{ij} \sim \mathcal{N}(0, \omega_{\epsilon}^2)$, $j = 1, \ldots, n_i$, $i = 1, \ldots, M$. The conditional distribution of $Y_{ij} \mid u_i$ is $\mathcal{N}(\mu + u_i, \omega_{\epsilon}^2)$, while the marginal (with respect to u) distribution is $\mathcal{N}(\mu, \omega_u^2 + \omega_{\epsilon}^2)$. The parameters of model (13.3) have a well-established interpretation: μ is the population mean,[2] and ω_u^2 measures the dispersion of the cluster-specific random ef-

[2]The mean can be viewed as arising from either $\mathrm{E}_Y(Y) = \mu$ or $\mathrm{E}_u\{\mathrm{E}(Y \mid u)\} = \mu$.

fects (heterogeneity) around μ. Together with ω_ϵ^2, it determines the intra-cluster correlation, $\text{ICC} = \omega_u^2/(\omega_u^2 + \omega_\epsilon^2)$. The error variance ω_ϵ^2 is typically treated as a nuisance parameter.

We now turn to the quantiles implied by model (13.3), considering both conditional (on u) and marginal perspectives. The pth conditional quantile in cluster i is $\xi_i^{(u)} \equiv Q_{Y_{ij}|u_i}(p) = \mu + u_i + \omega_\epsilon \Phi^{-1}(p)$, for all $j = 1, \ldots, n_i$. Because the Y_{ij}s are independent given u_i, $\xi_i^{(u)}$ can be consistently (for large n_i) estimated by the sample quantile based on $Y_{i1}, \ldots, Y_{in_i}$. Averaging M independent such estimators yields, under $u \sim \mathcal{N}\left(0, \omega_u^2\right)$, an estimate of $\text{E}_u\{Q_{Y_{ij}|u_i}(p)\} = \mu + \omega_\epsilon \Phi^{-1}(p)$, which corresponds to the common quantile p targeted by (13.1). This differs from the population-level (marginal) quantile, $Q_Y(p) = \mu + \sqrt{\omega_u^2 + \omega_\epsilon^2}\,\Phi^{-1}(p)$.

In R Code 13.1, we give a numerical example. The data are generated as in (13.3) with $\mu = 0$, $\omega_u^2 = 1$, and $\omega_\epsilon^2 = 1$ (hence $\text{ICC} = 0.5$). The sample sizes for both independent clusters and within-cluster observations are fixed to large values to reduce sampling noise. As expected, the marginal 0.75th sample quantile approximates the corresponding population-level true value. This is consistent with Oberhofer and Haupt (2005), who showed that the sample quantile remains consistent and asymptotically normal for dependent observations under suitable conditions. R Code 13.1 also shows that the average of the cluster-specific quantiles is close to the theoretical value, as is the estimate from the intercept-only LQMM model. The LQMM estimate aligns with the average of cluster-specific quantiles, not with the marginal quantile.

Thus, it is important to distinguish carefully between marginal quantiles and averaged cluster-specific quantiles. Even in the simple Gaussian random intercept model, the two target fundamentally different features of the distribution: the former describes the population-level variability, while the latter reflects the typical within-cluster behaviour after averaging out heterogeneity. In other words, the quantile of a mixture is not, in general, the mixture of quantiles. Estimation based on the asymmetric Laplace likelihood (13.2) introduces a further source of discrepancy, which is purely technical and stems from the likelihood approximation rather than from this conditional-marginal distinction, as discussed in the next section.

Approximation to the asymmetric Laplace bias

In LQMMs, the asymmetric Laplace distribution is used as a working likelihood to facilitate estimation; however, after integrating over the random effects, the resulting objective function is no longer equivalent to the quantile loss. Consequently, under model misspecification, such as Gaussian conditional errors, the estimator converges to a pseudo-true value determined by the assumed likelihood rather than to the target conditional quantile. This may lead to systematic bias relative to the average cluster-specific quantile, even in simple settings. The resulting bias does not generally admit a closed-form expression and depends on the degree of misspecification, but can be assessed empirically through simulation or by a first-order approximation.

Consider the random-intercept model in (13.3). Let $\beta_p = \mu + \omega_\epsilon \Phi^{-1}(p)$ denote the target quantity $E_u\{Q_{Y_{ij}|u_i}(p)\}$, and let β_p^* be the pseudo-true value defined implicitly by the population score equation

$$E\{\dot{\ell}_i(\beta_p^*; \omega_u^2)\} = 0. \tag{13.4}$$

The expectation is taken under the true data-generating process, where the random effects have variance ω_u^2, which in general differs from the parameter $\boldsymbol{\Psi}_p$ of the working model. A first-order Taylor expansion of the population score around β_p and $\omega_u^2 = 0$ leads to the

R Code 13.1: Marginal and cluster-specific quantiles.

```
library(lqmm)
set.seed(123)

M <- 4000
n <- 20
mu <- 0
omega_u <- 1
omega_e <- 1
p <- 0.75
id <- rep(1:M, each = n)

u <- rnorm(M, 0, omega_u)
e <- rnorm(M*n, 0, omega_e)
y <- mu + rep(u, each = n) + e

# true population-level quantile
mu + sqrt(omega_e^2 + omega_u^2)*qnorm(p)
> [1] 0.9538726
# marginal sample quantile
quantile(y, probs = p, names = FALSE)
> [1] 0.9630072
# true average of cluster-specific quantiles
# under symmetric random effects distribution
mu + omega_e*qnorm(p)
> [1] 0.6744898
# average cluster-specific sample quantiles
print(xibar <- mean(tapply(y, id, quantile, probs = p)))
> [1] 0.6403186
# lqmm estimate
fit <- lqmm(y ~ 1, random = ~1, group = id, tau = p,
data = data.frame(y = y, id = id), nK = 11)
coef(fit)
> (Intercept)
>   0.6482559
```

first-order bias approximation[3]

$$\beta_p^* - \beta_p = -\omega_u^2 \frac{\left.\dfrac{\partial}{\partial \omega_u^2} E\{\dot{\ell}_i(\beta_p; \omega_u^2)\}\right|_{\omega_u^2=0}}{E\{\ddot{\ell}_i(\beta_p; 0)\}} + o(\omega_u^2), \tag{13.5}$$

provided $E\{\ddot{\ell}_i(\beta_p; 0)\} \neq 0$, hence locally of order ω_u^2.

[3]Although $\omega_u^2 = 0$ is a boundary point, a formal expansion around the degenerate case of vanishing between-cluster variability, obtained by writing $u_i = \omega_u z_i$ with $z_i \sim \mathcal{N}(0, 1)$, suggests that the leading bias term is of order ω_u^2.

Two limiting cases provide useful approximations.

Case $n_i = 1$.

When each cluster contains a single observation, the random intercept and the idiosyncratic error cannot be separated from the data. A useful benchmark is therefore obtained by comparing the marginal Gaussian quantile with the target cluster-specific quantile. Since

$$Y_{ij} \sim \mathcal{N}\left(\mu, \omega_u^2 + \omega_\epsilon^2\right),$$

the marginal p-th quantile is

$$\mu + \sqrt{\omega_u^2 + \omega_\epsilon^2}\Phi^{-1}(p),$$

whereas the average cluster-specific target is

$$\beta_p = \mu + \omega_\epsilon \Phi^{-1}(p).$$

Thus, in this limiting case, the bias is approximated by

$$\beta_p^* - \beta_p \approx \left(\sqrt{\omega_u^2 + \omega_\epsilon^2} - \omega_\epsilon\right)\Phi^{-1}(p). \tag{13.6}$$

The difference is non-zero whenever $\omega_u^2 > 0$ and $p \neq 0.5$. Its magnitude increases with the between-cluster variability and with the distance of p from 0.5. For small ω_u^2, a first-order expansion yields

$$\beta_p^* - \beta_p \approx \frac{\omega_u^2}{2\omega_\epsilon}\Phi^{-1}(p). \tag{13.7}$$

Thus the leading bias is of order ω_u^2 and has the sign of $\Phi^{-1}(p)$. It is positive above the median, negative below the median, and vanishes at $p = 0.5$. The factor $\omega_u^2/(2\omega_\epsilon)$ (reminiscent of the shrinkage factor in linear mixed models) shows that the magnitude of the approximation increases with the between-cluster variability relative to the within-cluster scale.

Case $n_i \to \infty$.

When the number of observations per cluster increases, the random intercept u_i can be estimated with increasing accuracy, and the model approaches a conditional formulation. In this limit,

$$\beta_p^* \to \beta_p.$$

Intermediate regime.

For finite n_i, the pseudo-true value β_p^* generally lies between the marginal benchmark obtained when $n_i = 1$ and the conditional target obtained as $n_i \to \infty$, although this monotonic behaviour is best regarded as a heuristic rather than a closed-form result. Its behaviour reflects the amount of information available within each cluster to disentangle the random effects from the idiosyncratic errors. As n_i increases, the contribution of within-cluster information becomes more prominent, and the estimator is expected to move towards the conditional target β_p. In particular, the approximation suggests positive bias for $p > 0.5$ and negative bias for $p < 0.5$, with the bias vanishing at the median and tending to disappear as within-cluster information increases.

More generally, the magnitude of the bias is driven by the relative importance of between-cluster variability and within-cluster noise, as well as by the distance of p from 0.5. In the absence of a closed-form expression, its extent is best assessed through first-order approximations, such as (13.7), or via simulation under the assumed data-generating mechanism.

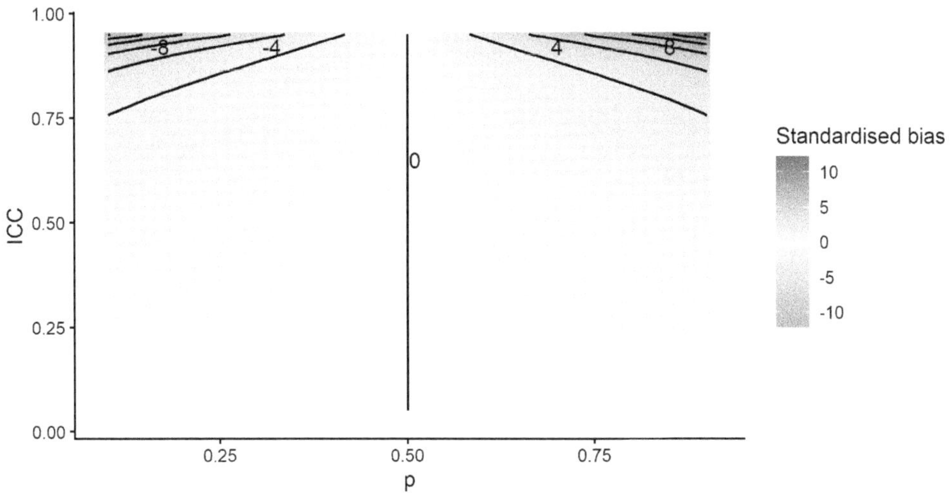

Figure 13.1
Heatmap of the standardised bias in linear quantile mixed models when $n_i = 1$.

If we standardise (13.7) by ω_ϵ and express the right-hand side as a function of the ICC, we have

$$\frac{\beta_p^* - \beta_p}{\omega_\epsilon} \approx \frac{1}{2}\frac{ICC}{1 - ICC}\Phi^{-1}(p).$$

Figure 13.1 displays the standardised bias as a function of the quantile level p and the ICC in the limiting case $n_i = 1$. The magnitude becomes substantial for moderate-to-high ICC values and in the tails. This is not surprising, as estimation in the tails under strong within-cluster dependence, and in the absence of within-cluster replication, is inherently less precise.

Estimation

Since the $\boldsymbol{u}_i$s in (13.2) are unobserved, the goal is to evaluate the integral

$$f(\boldsymbol{y}_i; \boldsymbol{\beta}_p, \boldsymbol{\Psi}_p, \sigma_p) = \left\{\frac{p(1-p)}{\sigma_p}\right\}^{n_i} \int_{\mathbb{R}^r} \exp\left\{-\frac{1}{\sigma_p}\sum_{j=1}^{n_i} \psi_p\left\{y_{ij} - \mu_{p,ij}(\boldsymbol{u}_i)\right\}\right\} f(\boldsymbol{u}_i; \boldsymbol{\Psi}_p)\,\mathrm{d}\boldsymbol{u}_i, \tag{13.8}$$

where $\boldsymbol{y}_i = (y_{i1}, \dots, y_{in_i})^\top$, for $i = 1, \dots, M$, and then maximise

$$\ell(\boldsymbol{\beta}_p, \boldsymbol{\Psi}_p, \sigma_p) = \sum_{i=1}^{M} \ell_i(\boldsymbol{\beta}_p, \boldsymbol{\Psi}_p, \sigma_p), \tag{13.9}$$

where $\ell_i(\boldsymbol{\beta}_p, \boldsymbol{\Psi}_p, \sigma_p) = \log f(y_i; \boldsymbol{\beta}_p, \boldsymbol{\Psi}_p)$ is the ith cluster contribution to the log-likelihood.

The integral in (13.8) does not admit, in general, a closed-form expression, although simple cases can be handled analytically. One possible strategy is to approximate the integral numerically. Moreover, there is a direct link between the assumed distribution of $\boldsymbol{u}$ and the resulting quadrature rule: normal random effects lead to Gauss–Hermite quadrature, whereas adopting a double-exponential distribution yields a Gauss–Laguerre rule. Suppose we assume normal random effects. Upon change of variable $\boldsymbol{v} = \boldsymbol{\Psi}_p^{1/2}\boldsymbol{u}$, the r-dimensional

Gaussian quadrature for (13.8), obtained by successive application of one-dimensional rules, yields the approximation (up to a proportionality constant)

$$\sum_{k_1=1}^{K} \cdots \sum_{k_r=1}^{K} \exp\left\{-\frac{1}{\sigma_p}\sum_{j=1}^{n_i} \psi_p\left\{y_{ij} - \mu_{p,ij}(\boldsymbol{v})\right\}\right\} \prod_{l=1}^{r} w_{k_l}, \tag{13.10}$$

where K is the number of quadrature nodes, $\boldsymbol{v} = (v_{k_1}, \ldots, v_{k_r})^\top$ the abscissas, and w_{k_l} the weights, with $k_l = 1, \ldots, K$ and $l = 1, \ldots, r$. The function in (13.10) is nonsmooth (but Lipschitz), hence optimisation relies on Clarke derivatives. Details of the iterative algorithm implemented in the `lqmm` package are given in Geraci (2014) and Geraci and Bottai (2014). At convergence, the random effects can be obtained via best linear prediction, namely

$$\boldsymbol{Z}_i \hat{\boldsymbol{u}}_i = \mathrm{E}\left(\boldsymbol{Z}_i \boldsymbol{u}_i\right) + \left\{\mathrm{cov}\left(\boldsymbol{Z}_i \boldsymbol{u}_i, \boldsymbol{y}_i\right)\right\} \mathrm{var}(\boldsymbol{y}_i)^{-1} \left\{\boldsymbol{y}_i - \mathrm{E}\left(\boldsymbol{y}_i\right)\right\}.$$

The product rule suffers from the well-known curse of dimensionality, as the number of function evaluations grows exponentially with r. In practice, however, numerical experiments suggest that satisfactory accuracy is often achieved with fewer than 20 nodes. Gaussian quadrature is not the only estimation strategy. A widely used alternative is the expectation-maximisation (EM) algorithm and its variants, including the Monte Carlo EM (Geraci and Bottai, 2007). The expectations computed in the E-step depend on the current parameter values, leading to an iterative estimation scheme. As discussed in Geraci (2019b), claims that EM-based approaches provide superior estimation for LQMMs are not supported by empirical evidence. Direct comparisons show that LQMM estimation based on numerical integration and nonsmooth optimisation attains comparable or smaller bias and root mean squared error, substantially lower computational cost, and greater numerical stability. By contrast, EM-type algorithms may exhibit slow convergence, higher computational burden, and nonnegligible rates of convergence failure. Overall, these results indicate that LQMM methods offer a more efficient and reliable approach to estimation in practice.

Another estimation strategy, often pursued in random-effects modelling, consists in introducing a Taylor expansion of the integrand that may lead to an analytical solution of the integral. An added difficulty of quantile regression is the presence of a (strictly) non-differentiable loss function. A smooth approximation of ψ_p as discussed in Section 5.4 can be an effective fix, particularly if the conditional quantile function is nonlinear or nonparametric (Geraci, 2019a,c). Assuming normality for $\boldsymbol{u}$, consider the function

$$h(\boldsymbol{u}_i) = 2\sum_{j=1}^{n_i} \psi_p\left\{y_{ij} - \mu_{p,ij}(\boldsymbol{u}_i)\right\} + \boldsymbol{u}_i^\top \tilde{\boldsymbol{\Psi}}_p^{-1} \boldsymbol{u}_i, \tag{13.11}$$

where $\tilde{\boldsymbol{\Psi}}_p = \sigma_p^{-1}\boldsymbol{\Psi}_p$, and define $\boldsymbol{r}_i = (r_{i1}, \ldots, r_{in_i})^\top$ as the vector of residuals $r_{ij} = y_{ij} - \mu_{p,ij}(\boldsymbol{u}_i)$, $j = 1, \ldots, n_i$, $i = 1, \ldots, M$. If the function ψ_p in (13.11) is replaced with the smooth approximation $\kappa_{c,p}$ (for given tuning parameter c) in (5.19), then the function h becomes

$$h\left(\boldsymbol{u}_i\right) = \boldsymbol{r}_i^\top \boldsymbol{A}_i \boldsymbol{r}_i + \boldsymbol{b}_i^\top \boldsymbol{r}_i + \boldsymbol{e}_i^\top \boldsymbol{1}_{n_i} + \boldsymbol{u}_i^\top \tilde{\boldsymbol{\Psi}}_p^{-1} \boldsymbol{u}_i, \tag{13.12}$$

where $\boldsymbol{A}_i$ is an $n_i \times n_i$ diagonal matrix with diagonal elements $\{\boldsymbol{A}_i\}_{jj} = (1 - s_{ij}^2)/c$, $\boldsymbol{b}_i$ and $\boldsymbol{e}_i$ are two $n_i \times 1$ vectors with elements

$$b_{ij} = s_{ij}((2p-1)s_{ij} + 1)$$

and

$$e_{ij} = \frac{1}{2}\left\{(1-2p)cs_{ij} - (1-2p+2p^2)cs_{ij}^2\right\},$$

respectively, and $\boldsymbol{s}_i = (s_{i1}, \ldots, s_{in_i})^\top$ with

$$s_{ij} = \begin{cases} -1 & \text{if } r_{ij} \leq -(1-p)c, \\ 0 & \text{if } -(1-p)c < r_{ij} < pc, \\ 1 & \text{if } r_{ij} \geq pc. \end{cases}$$

Moreover, let

$$\widehat{\boldsymbol{u}}_i = \underset{\boldsymbol{u}_i}{\operatorname{argmin}}\, h\left(\boldsymbol{u}_i\right) \tag{13.13}$$

be the mode of $\boldsymbol{u}_i$, conditional on the model's parameters. For a given value of c, this can be obtained by means of penalised least-squares. A second-order approximation of (13.12) around $\widehat{\boldsymbol{u}}_i$ is given by

$$h\left(\boldsymbol{u}_i\right) \simeq h_i + \dot{\boldsymbol{h}}_i^\top \left(\boldsymbol{u}_i - \widehat{\boldsymbol{u}}_i\right) + \left(\boldsymbol{u}_i - \widehat{\boldsymbol{u}}_i\right)^\top \ddot{\boldsymbol{H}}_i \left(\boldsymbol{u}_i - \widehat{\boldsymbol{u}}_i\right),$$

where $h_i = h\left(\widehat{\boldsymbol{u}}_i\right)$, $\dot{\boldsymbol{h}}_i = h'\left(\widehat{\boldsymbol{u}}_i\right)$, and $\ddot{\boldsymbol{H}}_i = h''\left(\widehat{\boldsymbol{u}}_i\right)/2$. Note that $\dot{\boldsymbol{h}}_i$ is zero at $\boldsymbol{u}_i = \widehat{\boldsymbol{u}}_i$.

The smoothed log-likelihood

$$\ell(\boldsymbol{\beta}_p, \boldsymbol{\Psi}_p, \sigma_p) = \sum_{i=1}^{M} \log \left\{ \frac{p(1-p)}{\sigma_p} \right\}^{n_i} \frac{1}{(2\pi)^{r/2} |\boldsymbol{\Psi}_p|^{1/2}} \int_{\mathbb{R}^r} \exp\left\{ -\frac{1}{2\sigma_p} h(\boldsymbol{u}_i) \right\} \mathrm{d}\boldsymbol{u}$$

has the following Laplacian approximation:

$$\begin{aligned} \ell_{\mathrm{LA}}\left(\boldsymbol{\theta}_p; \boldsymbol{y}\right) &= N \log\left\{ \frac{p(1-p)}{\sigma_p} \right\} - \frac{M}{2} \log|\boldsymbol{\Psi}_p| - \frac{1}{2\sigma_p} \sum_{i=1}^{M} h_i + \sum_{i=1}^{M} \log \int_{\mathbb{R}^r} (2\pi\sigma_p)^{-r/2} \\ &\quad \times \exp\left\{ -\frac{1}{2\sigma_p} \left(\boldsymbol{u}_i - \widehat{\boldsymbol{u}}_i\right)^\top \ddot{\boldsymbol{H}}_i \left(\boldsymbol{u}_i - \widehat{\boldsymbol{u}}_i\right) \right\} \mathrm{d}\boldsymbol{u}_i \\ &= N \log\left\{ \frac{p(1-p)}{\sigma_p} \right\} - \frac{1}{2} \left\{ \sum_{i=1}^{M} \log|\boldsymbol{\Psi}_p \ddot{\boldsymbol{H}}_i| + \sigma_p^{-1} \sum_{i=1}^{M} h_i \right\}. \end{aligned} \tag{13.14}$$

The maximisation of (13.14) is carried out iteratively by means of general-purpose optimisers like `optim`. As a byproduct, predicted random effects are available at convergence.

13.1.1 Clustering induced by group-specific heterogeneity

In many applications, observations can be naturally partitioned into groups such as individuals within families, patients within hospitals, or measurements within biological species. Units belonging to the same group tend to be more similar than units from different groups, owing to shared traits, unobserved factors, or common environments. This induces within-group dependence, which, if ignored, can lead to misleading inference and inefficient estimation.

In this section, we consider an application using observations on depth distribution of staghorn corals, which make up a large proportion of coral species richness, across different latitudes. Among the factors corals need for growth, solar radiation and temperature are those more strongly linked to latitude. Some studies offer evidence for poleward range expansions of staghorn corals associated with increasing sea temperatures. However, Muir et al. (2015) argue that of the main environmental factors influencing coral growth, only solar radiation is not expected to vary markedly under global warming. Incoming radiation at the sea surface declines with latitude, largely because of shorter winter days and lower solar angles, and it is further reduced as it penetrates the water column. As a result, light availability imposes a natural constraint on the maximum depth of phototrophic corals. Warming enables coral expansion, but light constrains it. Therefore, Muir et al. (2015) investigate the latitudinal gradient in the 'maximum' (97.5th percentile) depth limits of staghorn corals.

The original dataset (Muir et al., 2015) comprises more than 14,000 records of coral species occurrence at different depths and latitudes. For illustrative purposes, we selected a subset of the most frequent 20 species that covers about half of the original data ($N = 7119$). R Code 13.2 shows the summary of the dataset, which contains the species, latitude (decimal degrees), and depth (meters) of the corals. While maximum depth at different latitudes is the quantity of interest, its estimation relies on extreme quantiles, which, as we have seen in Section 6.3.3, are subject to limitations of their own. However, in this case, we must account for the correlation within clusters (i.e., species), which effectively reduces the amount of independent information. As a rough indication, we borrow the concept of design effect from standard cluster sampling theory. The effective sample size can then be approximated as $N_e = N \cdot \{1 + (\bar{n} - 1) \cdot \mathrm{ICC}\}^{-1}$, where $\bar{n} = M^{-1} \sum_{i=1}^{M} n_i$ is the average cluster size. In our subset, we obtain $\bar{n} = 356$ and ICC $= 0.31$, which yields $N_e = 64$. With such a small effective sample size, pushing the estimation to $p = 0.975$ would rely on very limited independent tail information, unlike in Muir et al. (2015), who used a much larger dataset. We therefore adopt a more conservative choice, $p = 0.9$. This argument is heuristic. Dependence in the tails may differ from dependence near the centre; indeed, if we extend this concept to the LQMM estimates, then the ICC is given by $\hat{\psi}_u^2/(\hat{\psi}_u^2 + \hat{\psi}_\epsilon^2) = 48.02/(48.02 + 156.25) = 0.24$. The residual variance $\hat{\psi}_\epsilon^2 = \hat{\sigma}_{0.9}^2 \cdot (1 - 2p + 2p^2)/\{p^2 \cdot (1 - p)^2\}$ is a model-based estimate from the AL with scale parameter $\hat{\sigma}_p = 1.24$, where the latter estimate is available via the output `fit.lqmm$scale`.

In R Code 13.3, we adopt a quadratic specification for latitude, in line with Muir et al. (2015). The orthogonal basis generated by `poly` is used to enhance numerical stability. A random intercept is introduced to account for the correlation of observations within `Species`. The first two arguments, `fixed` and `random`, are formula objects that define, respectively, the fixed and the random parts of the linear predictor, while the grouping factor is defined in the argument `group`. In formulas, we have

$$Q_{Y_{ij}|u_i}(p) = \beta_{p,0} + \beta_{p,1} x_{ij,1} + \beta_{p,2} x_{ij,2} + u_i,$$

where Y_{ij} is the coral depth, $x_{ij,1}$ and $x_{ij,2}$ are the basis of the polynomial for latitude, and u_i is a random intercept for the ith species.

The joint effect of the regression coefficients $\hat{\boldsymbol{\beta}}_{0.9} = (18.07, 34.24, -86.55)^\top$ describes a curved relationship for the average (or modal) cluster-specific quantile. Because the polynomial basis is orthogonal, the individual coefficients are not directly interpretable on the original scale; instead, their joint effect implies a maximum near the equator and progressively shallower limits as the magnitude of latitude increases.

As a crude diagnostic, we compare the predicted LQMM random intercepts with the deviations of cluster-specific intercepts obtained by fitting separate quantile regressions for each species (R Code 13.4). Specifically, for each species we estimate an intercept and subtract the overall mean intercept, yielding a set of unpooled deviations. The correlation between the two sets of quantities is 0.98, indicating that they capture essentially the same between-species variation. However, the deviations from the identity line (Figure 13.2) increase in magnitude with the size of the effect, with positive and negative values tending to lie above and below the identity line, respectively. This reflects the shrinkage induced by the mixed model, whereby information is borrowed across clusters and more extreme species-specific effects are pulled towards the overall trend.

R Code 13.2: Summary statistics and effective sample size for the Staghorn Corals dataset.

```
library(nlme)
library(lqmm)

# summary stats for staghorn corals dataset
summary(corals)
>              Species          Latitude            Depth
>  A. muricata     : 549   Min.   :-31.833   Min.   : 0.000
>  A. nasuta       : 542   1st Qu.:-18.233   1st Qu.: 3.000
>  A. valida       : 521   Median : -8.333   Median : 6.000
>  A. cerealis     : 438   Mean   : -7.080   Mean   : 7.643
>  A. hyacinthus   : 421   3rd Qu.:  1.750   3rd Qu.:10.000
>  A. solitaryensis: 382   Max.   : 33.693   Max.   :65.000
>  (Other)         :4266
# calculate ICC and effective sample size with lme
fit.lme <- lme(Depth ~ 1, random = ~ 1|Species, data = corals)
# get random intercept variance, residual variance
omega <- as.numeric(VarCorr(fit.lme)[,1])
# ICC = sigma_u^2 / (sigma_u^2 + sigma_e^2)
print(ICC <- omega[1]/sum(omega))
> [1] 0.3124307
# average cluster size
print(nbar <- mean(table(corals$Species)))
> [1] 355.95
# total sample size
print(N <- nrow(corals))
> [1] 7119
# design effect
DEFF <- 1 + (nbar - 1)*ICC
# effective sample size
print(Ne <- N/DEFF)
> [1] 63.62086
# calculate ICC with lqmm at p = 0.9
ctrl <- lqmmControl(method = "df", LP_max_iter = 1000, LP_tol_ll = 1e-5)
fit.lqmm <- lqmm(fixed  = Depth ~ 1, random = ~ 1, group  = Species,
tau = 0.9, nK = 7, type = "normal", data = corals, control = ctrl)
print(psi_u <- VarCorr(fit.lqmm))
> (Intercept)
>    48.02465
print(psi_e <- varAL(fit.lqmm$scale, tau = 0.9))
> [1] 156.2533
psi_u/(psi_u + psi_e)
> (Intercept)
>   0.2350946
```

In Figure 13.3, we show the predicted depth at $p = 0.9$ for the modal ($u = 0$) species, together with the 95% bootstrap percentile pointwise bands obtained via `boot.lqmm` using $B = 1000$ replications. These bands reflect the estimation uncertainty around the modal

R Code 13.3: Linear quantile mixed model for depth as a function of latitude using the Staghorn Corals dataset.

```
library(lqmm)

# set control parameters for estimation
ctrl <- lqmmControl(method = "df", LP_max_iter = 1000, LP_tol_ll = 1e-5)
# 2nd degree orthogonal polynomial on latitude
P <- poly(corals$Latitude, degree = 2)
corals$lat1 <- P[,1]; corals$lat2 <- P[,2]
# fit LQMM
fit <- lqmm(fixed  = Depth ~ lat1 + lat2, random = ~ 1,
group  = Species, tau = 0.9, nK = 7, type = "normal", data = corals,
control = ctrl)
fit
> Call: lqmm(fixed = Depth ~ lat1 + lat2, random = ~1, group = Species,
>     tau = 0.9, nK = 7, type = "normal", data = corals, control = ctrl)
>
> Quantile 0.9
>
> Fixed effects:
> (Intercept)          lat1          lat2
>       18.07         34.24        -86.55
>
> Covariance matrix of the random effects:
> (Intercept)
>       33.32
>
> Residual scale parameter: 1.201 (standard deviation 12.08)
> Log-likelihood: -25595
>
> Number of observations: 7119
> Number of groups: 20
```

quantile curve. By contrast, Figure 13.4 displays both the modal and the cluster-specific quantiles, providing a visual representation of the between-cluster heterogeneity around the 90th percentile.

13.1.2 Clustering induced by repeated measurements

Longitudinal and cohort studies follow the same units over time, recording repeated measurements on each individual. This design provides information not only on the distribution of the outcome at a given time, but also on its evolution within units, allowing the study of trajectories, temporal patterns, and individual-specific responses. By observing how outcomes change within the same unit, longitudinal data enable the separation of within-unit variability from between-unit heterogeneity, offering a richer understanding of the underlying processes than cross-sectional data. Such a structure naturally induces dependence among repeated measurements from the same unit, which can be conveniently modelled through random effects that capture unobserved, unit-specific characteristics.

R Code 13.4: Random effects prediction using the Staghorn Corals dataset.

```
print(u <- ranef(fit))
>                    (Intercept)
> A. aculeus          5.4706052
> A. austera          0.5196915
> A. cerealis         1.5426991
> A. digitifera      -4.1424742
> A. divaricata       0.5915945
> A. florida          1.0916571
> A. granulosa       13.3351439
> A. humilis         -2.6658815
> A. hyacinthus      -1.3852027
> A. intermedia      -2.1652899
> A. latistella       1.3639099
> A. loripes          2.6733837
> A. lutkeni          0.5809411
> A. millepora       -3.5191645
> A. muricata        -1.3789267
> A. nasuta          -2.3700092
> A. secale          -0.7314098
> A. solitaryensis    2.6578757
> A. tenuis          -0.8914950
> A. valida          -0.7739285
# fit a QR model for each species
xx <- by(corals, corals$Species, function(x)
  coef(rq(Depth ~ poly(Latitude, 2), data = x, tau = 0.9)))
xx <- do.call("rbind", xx)
# calculate deviation of each species intercept from the mean intercept
delta <- xx[,1] - mean(xx[,1])
# correlation between RQ deviations and LQMM predicted random effects
cor(delta, u[[1]])
> [1] 0.9800138
```

Having introduced the linear formulation in the previous section, we now turn, for didactic purposes, to its nonlinear extension called nonlinear quantile mixed model (NLQMM). This is defined as (Geraci, 2019c)

$$Q_{Y_{ij}|\boldsymbol{u}_i}(p) = f\left(\boldsymbol{\phi}_{p,ij}, \boldsymbol{x}_{ij}\right), \qquad j = 1, \ldots, n_i, \quad i = 1, \ldots, M, \tag{13.15}$$

where f is a nonlinear, smooth function of the $s \times 1$ random parameter $\boldsymbol{\phi}_{p,ij} = \boldsymbol{F}_{ij}\boldsymbol{\beta}_p + \boldsymbol{G}_{ij}\boldsymbol{u}_i$, $\boldsymbol{F}_{ij}$ and $\boldsymbol{G}_{ij}$ are two given design matrices of dimensions $s \times q$ and $s \times r$, respectively, which in general contain elements of the covariates $\boldsymbol{x}_{ij}$.

To stress the functional dependence of the quantiles on the $q \times 1$ fixed parameter $\boldsymbol{\beta}_p$ and on the $r \times 1$ random parameter $\boldsymbol{u}_i$, we write $f\left(\boldsymbol{\phi}_{p,ij}, \boldsymbol{x}_{ij}\right) \equiv f_{ij}\left(\boldsymbol{\beta}_p, \boldsymbol{u}_i\right)$. For estimation purposes, model (13.15) can be equivalently written as

$$Y_{ij} = f_{ij}\left(\boldsymbol{\beta}_p, \boldsymbol{u}_i\right) + \epsilon_{p,ij}, \tag{13.16}$$

conditionally on $\boldsymbol{u}_i$, where the usual asymmetric Laplace device for $\epsilon_{p,ij}$ is used. Moreover, we assume $\boldsymbol{u}_i \sim \mathcal{N}\left(\boldsymbol{0}_r, \Psi_p\right)$, independent from $\epsilon_{p,ij}$. Estimation is carried out by means of an iterative algorithm to maximise the Laplacian approximation (13.14).

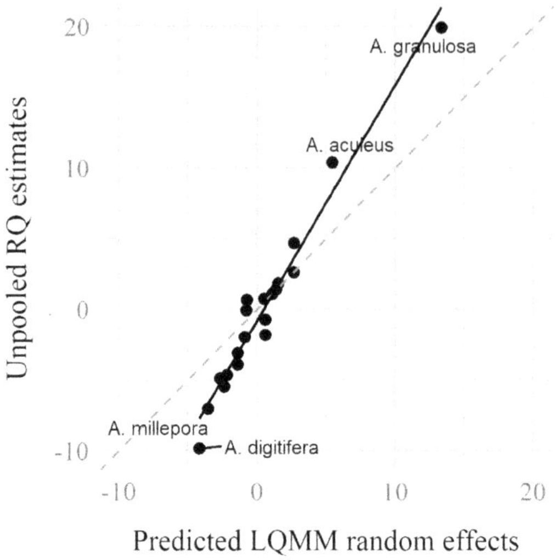

Figure 13.2
Random intercepts predicted via linear quantile mixed model (LQMM) versus deviations of species-specific intercepts estimated via quantile regression (QR) at $p = 0.9$. The dashed line denotes equality; departures reflect shrinkage towards zero.

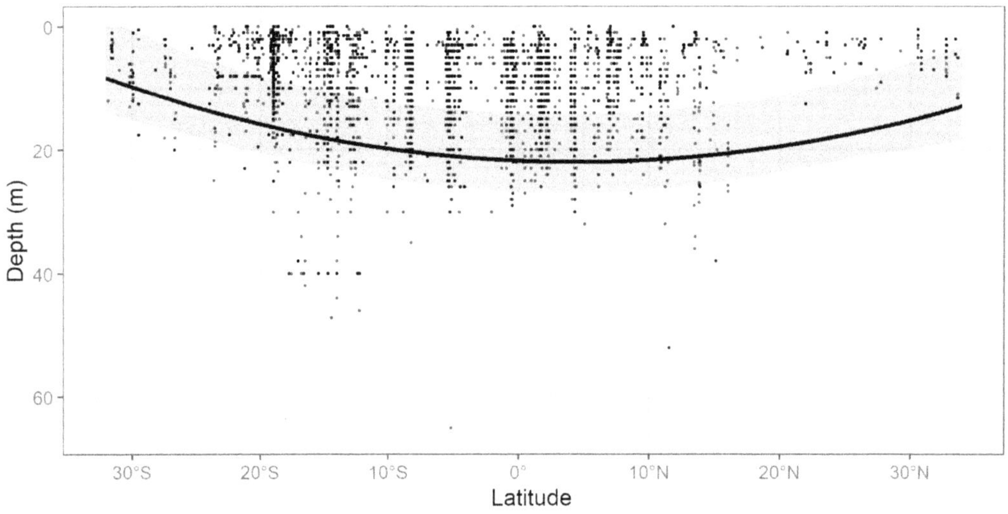

Figure 13.3
Predicted modal quantile at $p = 0.9$ for coral depth limits conditional on latitude using the Staghorn Corals dataset.

Some caveats about interpretation, similar to those discussed in the previous section, are in order, particularly with respect to the traditional nonlinear mixed-effects (NLME) model $Y_{ij} = f_{ij}(\boldsymbol{\beta}, \boldsymbol{u}_i) + \epsilon_{ij}$, with $\boldsymbol{u}_i \sim \mathcal{N}(\mathbf{0}_q, \boldsymbol{\Omega})$ and $\epsilon_{ij} \sim \mathcal{N}\left(0, \omega_\epsilon^2\right)$. The fixed effects can be interpreted, in general, as the average value of the cluster-specific parameters, that is, $\mathrm{E}_{\boldsymbol{u}_i}(\boldsymbol{\phi}_{ij})$, or as the regression parameters of the 'modal' cluster, that is, a cluster with a

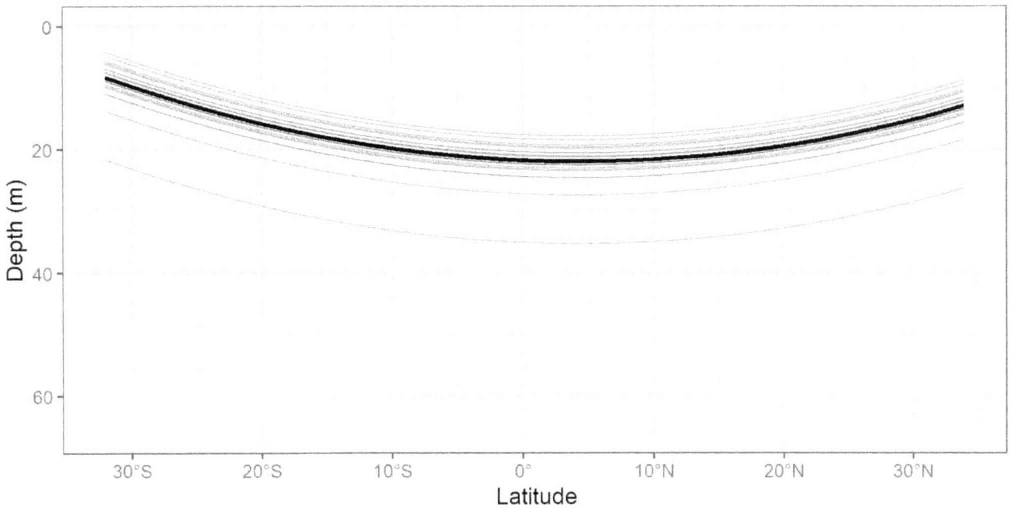

Figure 13.4
Predicted modal (black) and cluster-specific (grey) quantiles at $p = 0.9$ for coral depth limits conditional on latitude using the Staghorn Corals dataset.

zero random-effect vector. In general, neither model (13.15) nor the NLME model provide fixed parameters that can be interpreted as, respectively, regression quantiles or regression means for the population. This is because random effects are allowed to enter nonlinearly in the model. (Similarly, several generalised linear mixed models with nonlinear link functions lack marginal interpretability (Ritz and Spiegelman, 2004; Gory et al., 2021).) In contrast, as we have seen in Section 13.1, the fixed effects of a linear (normal) mixed model remain the same after the random effects are integrated out, whereas, in general, this is not true for the fixed effects of the linear quantile mixed models in Geraci and Bottai (2014).

We illustrate an application using data from the World Bank consisting of internet adoption rates (% of population) in $M = 213$ economies (countries, territories, administrative regions) between 1990 and 2024. The total sample size $N = 6078$. The dataset was stratified by income: low- and middle-income countries (LMICs) ($N = 3737$) and high-income countries ($N = 2341$). A summary of the dataset is given in R Code 13.5, while temporal trajectories in adoptions rates are shown in Figure 13.5.

Diffusion of a new technology in a market notoriously follows a logistic (sigmoid) function. Adoption starts slowly, accelerates as network effects kick in, and eventually saturates. Therefore, we consider a logistic model

$$Q_{Y_{ij}|\boldsymbol{u}_i}(p) = \frac{\theta_{p,1} + u_{1i}}{1 + \exp\{(\theta_{p,2} + u_{2i} - x_{ij})/\theta_{p,3}\}}, \tag{13.17}$$

where x_{ij} is the year of observation and

$$\begin{aligned}
\theta_{p,1} &= \alpha_{p,1} + \beta_{p,1} z_i, \\
\theta_{p,2} &= \alpha_{p,2} + \beta_{p,2} z_i, \\
\theta_{p,3} &= \alpha_{p,3} + \beta_{p,3} z_i,
\end{aligned}$$

with z_i defined as a dummy for high-income country. The interpretation of the parameters, conditionally on $\boldsymbol{u}_i = \boldsymbol{0}$, is as follows: $\theta_{p,1}$ (the asymptote) is the market potential; $\theta_{p,2}$ (the

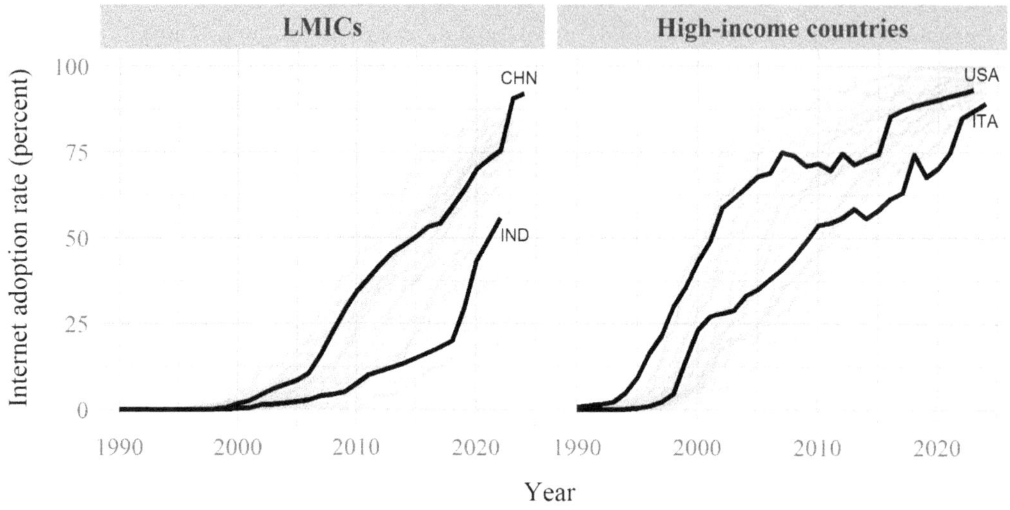

Figure 13.5
Trends in internet adoption rates by low- and middle-income countries (LMICs) and high-income countries over time using the World Bank Internet Adoption dataset. Grey lines represent all countries, while selected countries are highlighted.

R Code 13.5: Summary statistics for the World Bank Internet Adoption dataset.

```
summary(internet)
>     country            iso3            percentage              year
>  Austria:  35    AUT      :  35    Min.     :  0.000    Min.    :1990
>  Belgium:  35    BEL      :  35    1st Qu.:   1.645    1st Qu.:2000
>  Brazil :  35    BRA      :  35    Median :  18.069    Median :2008
>  Denmark:  35    DEU      :  35    Mean    :  31.209    Mean    :2008
>  France :  35    DNK      :  35    3rd Qu.:  59.892    3rd Qu.:2016
>  Germany:  35    ESP      :  35    Max.    :100.000    Max.    :2024
>  (Other):5868    (Other):5868
>                      incomegr
>  LMICs                   :3737
>  High-income countries:2341
>
>
>
>
>
```

midpoint or point of inflection) is where the market starts to slow down; and the reciprocal of $\theta_{p,3}$ (the scale) is the adoption speed. The parameters $\beta_{j,p}$, $j = 1, 2, 3$, measure the contrast between the two income groups, with $\alpha_{j,p}$, $j = 1, 2, 3$, as the baseline (LMICs). Conditional on calendar year, low quantile economies represent laggards while high quantile economies represent leaders. Quantiles index cross-economy heterogeneity in adoption levels,

Table 13.1
Estimated fixed effects and standard errors (SE) from three nonlinear quantile mixed models as in Equation (13.17) ($p \in \{0.1, 0.5, 0.9\}$) and the analogous nonlinear mixed-effects model for the mean (NLME), fitted to the World Bank Internet Adoption dataset.

	1st decile		Median		9th decile		NLME	
	Est.	SE	Est.	SE	Est.	SE	Est.	SE
α_1	31.3	4.7	77.1	2.8	89.8	12.9	74.2	2.1
β_1	48.8	9.3	9.5	2.8	4.0	12.9	12.7	3.2
α_2	2016.8	1.3	2016.7	0.5	2013.1	1.5	2016.6	0.4
β_2	−8.2	2.3	−11.1	0.6	−10.6	1.6	−11.1	0.6
α_3	2.7	0.5	3.5	0.1	4.3	0.6	4.0	0.0
β_3	0.5	0.8	−0.3	0.2	−1.0	0.6	−0.5	0.1

not stages of diffusion. The random effects for the asymptote and the midpoint capture the heterogeneity between countries relative to market potential and point of inflection.[4]

As with other nonlinear regression models, the NLQMM in (13.17) requires starting values. Some nonlinear models in `R` are facilitated by self-starting functions (see Section 9.1), and the logistic model is one of them. However, since the fixed parameters in (13.17) are functions of a covariate, `SSlogis` cannot provide starting values as it does in the unconditional case, but must itself be supplied with initial values. In R Code 13.6, we show two similar approaches, the first of which exploits the presence of a binary covariate. By applying `nls` separately to the two categories of the dummy variable z defined above, the starting values for the intercepts α are obtained from the estimates for $z = 0$, while those for the slopes β are obtained by differencing the estimates for $z = 1$ and $z = 0$. The second approach is simply to make an educated guess of such starting values. In both cases, the preliminary use of `nls` serves only as a fine-tuning step and is not strictly necessary; in principle, the `SSlogis` model can be supplied directly to `nlqmm`.

R Code 13.7 shows an instance of the use of the `nlqmm` function from the `nlqmm` package (Geraci, 2020). The code fits the logistic specification in (13.17) at the median, with fixed effects for the asymptote (`Asym`), midpoint (`xmid`), and scale (`scal`) parameterised by income group, and random effects on the asymptote and midpoint to capture country-specific deviations. The output reports the estimated fixed effects, which quantify differences between high-income countries and LMICs, and the covariance matrix of the random effects, which summarises between-country heterogeneity in the trajectories. In this example, the random-effects variance-covariance matrix $\boldsymbol{\Psi}_{0.5}$ is specified as a general positive definite matrix, allowing for correlation between the random effects; simpler specifications, such as diagonal matrices or multiples of the identity, can be used to impose independence. The reported tuning parameter corresponds to the constant c in the smooth approximation (5.19) to the quantile loss function, with smaller values indicating a closer approximation to the original loss.

The predicted quantiles from model (13.17) for $p \in \{0.1, 0.5, 0.9\}$ are shown in Figure 13.6, while estimates and bootstrap standard errors are reported in Table 13.1. The quantile trajectories for the modal country indicate that LMICs started later, so that the midpoint for high-income economies occurs earlier (negative β_2). The effect is remarkably

[4]The model that includes a third random effect for scale yields a very small estimate of the corresponding variance and similar estimates of the parameters from the smaller model (13.17).

R Code 13.6: Two approaches for obtaining starting values for the logistic model with covariate-dependent parameters in (13.17).

```
# define z
internet$z <- 1*(internet$incomegr == "High-income countries")

## first approach

# z = 0
dd0 <- subset(internet, internet$z == 0)
fit0 <- nls(percentage ~ SSlogis(year, Asym, xmid, scal), data = dd0)

# z = 1
dd1 <- subset(internet, internet$z == 1)
fit1 <- nls(percentage ~ SSlogis(year, Asym, xmid, scal), data = dd1)

ss0 <- as.list(coef(fit0))
ss1 <- as.list(coef(fit1))
ss <- rep(NA, 6)
ss[c(1,3,5)] <- as.numeric(ss0)
ss[c(2,4,6)] <- as.numeric(ss1) - as.numeric(ss0)
names(ss) <- paste0("b", 1:6)
round(ss, 3)
>       b1       b2       b3       b4       b5       b6
>   89.523   -0.036 2018.296  -12.706    5.300   -0.957
## second approach

fit <- nls(percentage ~ SSlogis(year,
  Asym0 + Asym1*z,
  xmid0 + xmid1*z,
  scal0 + scal1*z), data = internet,
  start = list(Asym0 = 80, Asym1 = 10, xmid0 = 2005, xmid1 = 2,
  scal0 = 1, scal1 = 1))

ss <- coef(fit)
names(ss) <- paste0("b", 1:6)
round(ss, 3)
>       b1       b2       b3       b4       b5       b6
>   89.523   -0.036 2018.296  -12.706    5.300   -0.957
```

stable across quantiles, suggesting a structural lag of about 10 years rather than a tail-specific phenomenon. High-income economies also exhibit higher adoption ceilings (positive β_1), although the magnitude of this difference decreases substantially from the lower to the upper quantiles. In particular, the much stronger effect at $p = 0.1$ than at $p = 0.9$ indicates that income status plays a major role among persistently lagging economies, while it becomes much less relevant among frontrunners.

The diffusion parameter shows a more nuanced pattern. At the lower quantile, the estimate of β_3 is positive but imprecise, providing little evidence of differential diffusion speed between groups. At the median and upper quantiles, however, β_3 becomes negative,

R Code 13.7: Nonlinear quantile mixed model for adoption percentage as a function of year and income status in the World Bank Internet Adoption dataset using the logistic model (13.17).

```
#pak::pak("marco-geraci/nlqmm")
library(nlqmm)

# fit nlqmm for the median
nlqmm(model = percentage ~ Asym/(1 + exp((xmid - year)/scal)),
  fixed = Asym + xmid + scal ~ z, random = Asym + xmid ~ 1,
  group = iso3, covariance = "pdSymm", tau = 0.5, data = internet,
  start = as.list(ss))

> Quantile 0.5
>
> Fixed effects:
> Asym1.(Intercept)             Asym2.z
>           77.0507              9.4779
> xmid1.(Intercept)             xmid2.z
>         2016.7074            -11.0836
> scal1.(Intercept)             scal2.z
>            3.5470             -0.2728
>
> Covariance matrix of the random effects:
>                  Asym.(Intercept) xmid.(Intercept)
> Asym.(Intercept)  54.282          -13.817
> xmid.(Intercept) -13.817            4.051
>
> Residual scale parameter:  1.872
> Log-likelihood: -20028
> Tuning parameter: 0.0008575
>
> Number of observations: 6078
> Number of groups: 213
```

suggesting that high-income economies tend to experience faster diffusion once adoption is more advanced. Overall, this points to a convergence in dynamics at the lower tail, with clearer differences emerging as countries move towards higher levels of adoption.

Figure 13.7 illustrates country-specific trajectories for the four countries highlighted in Figure 13.5. Italy and the US, for example, appear to be mature markets that have largely reached their plateaus. However, they exhibit markedly different distributional patterns. For Italy, the relatively large separation between the ninth decile and the median indicates a pronounced right-skewness in the model-implied distribution of adoption rates. In contrast, the US displays an opposite pattern of asymmetry, with a more compressed upper tail. Interpreted probabilistically, this suggests that, under the fitted model, high levels of adoption are more systematically represented across the distribution for the US, whereas in Italy they correspond to more extreme, less typical outcomes. China appears to follow a pattern similar to that of the US.

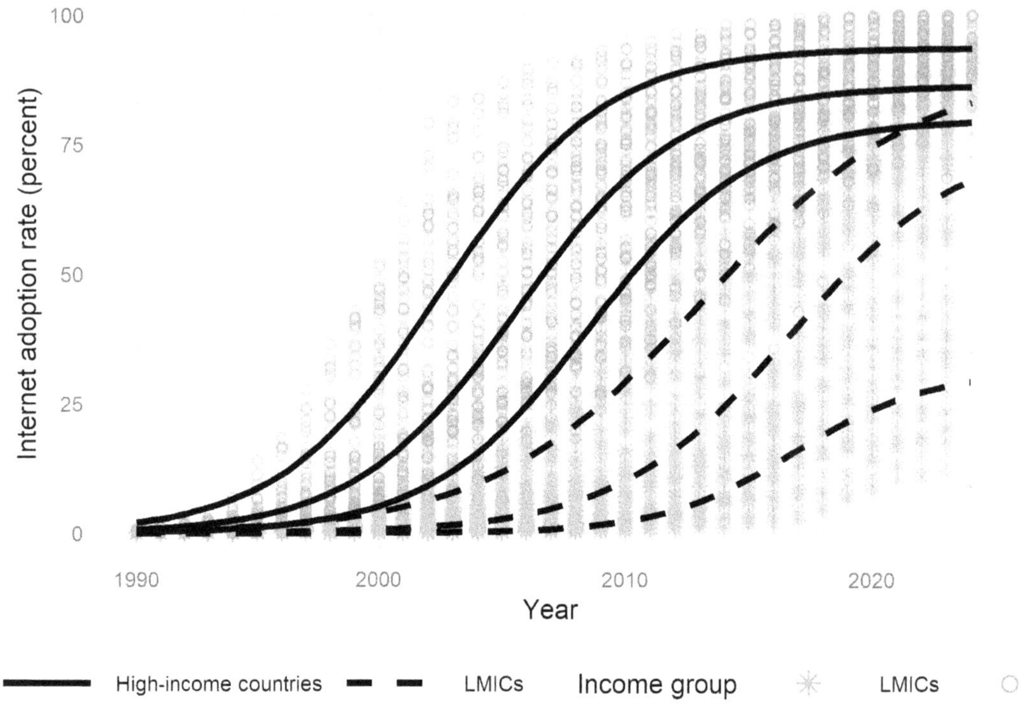

Figure 13.6
Predicted quantiles ($p \in \{0.1, 0.5, 0.9\}$) of internet adoption rates for the modal country, conditional on year and income group, based on the nonlinear quantile mixed model in Equation (13.17) fitted to the World Bank Internet Adoption dataset.

13.2 Time series

In many applications, observations are not independent but exhibit dependence across an index set. A prominent example is time series, where observations are indexed by time. The notion of a quantile remains well defined at each index, but estimation and inference must account for the underlying dependence structure.

Let $\{Y_t\}_{t \in \mathbb{Z}}$ be a stochastic process. For each t, denote by F_t the marginal distribution function of Y_t, and by $Q_t(p)$ its pth quantile. When the process is strictly stationary, $F_t \equiv F$ and $Q_t(p) \equiv Q(p)$ do not depend on t, although the observations remain dependent.

A natural starting point is the empirical quantile based on $\{Y_1, \ldots, Y_n\}$. Under suitable weak dependence conditions (e.g., mixing), consistency and asymptotic normality still hold, but the limiting variance reflects serial dependence. In particular,

$$\sqrt{n}\left\{\hat{Q}(p) - Q(p)\right\} \xrightarrow{d} \mathcal{N}\left(0, \sigma_p^2\right), \tag{13.18}$$

where σ_p^2 depends on the long-run variance of the indicator process $\{\mathbb{I}(Y_t \leq Q(p))\}$.

While marginal quantiles are informative, most applications require modelling the dynamics of the conditional distribution. Let $\mathcal{F}_{t-1}$ denote the σ-field generated by past

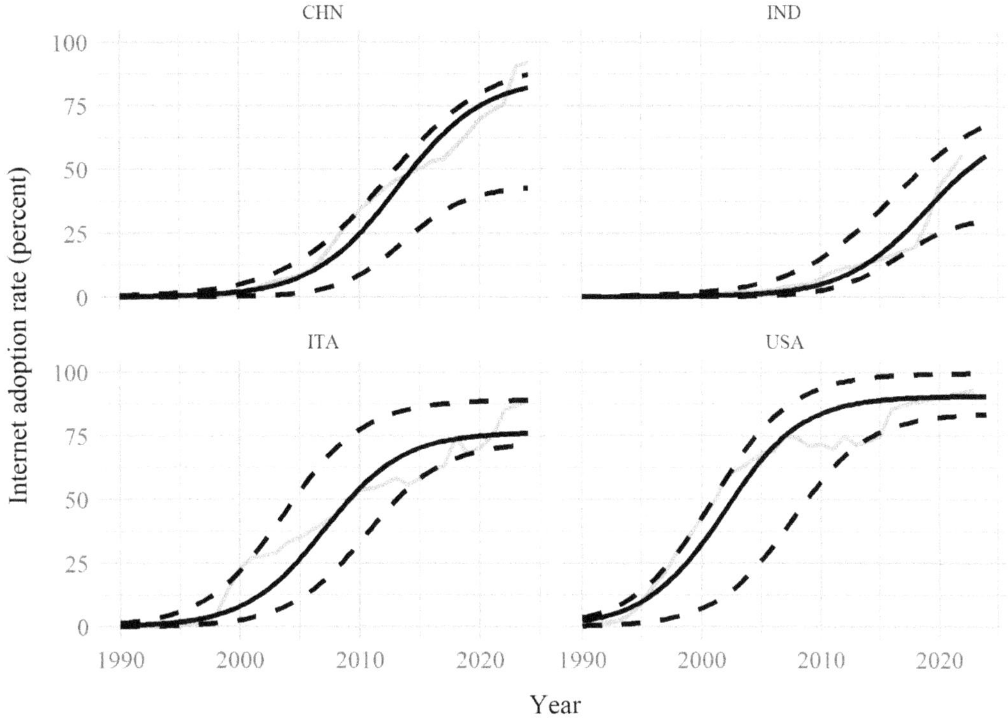

Figure 13.7
Observed internet adoption rates and fitted quantile curves for selected countries in the World Bank Internet Adoption dataset. The grey line shows the observed adoption rates, while the solid and dashed lines show the fitted median and the fitted 0.1 and 0.9 quantiles, respectively, from the nonlinear quantile mixed model in Equation (13.17).

observations. The conditional quantile is defined as

$$Q(p \mid \mathcal{F}_{t-1}) = \inf\{y : \Pr(Y_t \leq y \mid \mathcal{F}_{t-1}) \geq p\}. \tag{13.19}$$

Modelling $Q(p \mid \mathcal{F}_{t-1})$ provides a direct way to describe how different parts of the distribution evolve over time. A simple and widely used specification is the *quantile autoregressive* (QAR) model. In its linear form,

$$Q_{Y_t}(p \mid \mathcal{F}_{t-1}) = \alpha_0(p) + \alpha_1(p)Y_{t-1} + \cdots + \alpha_k(p)Y_{t-k}, \tag{13.20}$$

or, equivalently,

$$Q_{Y_t}(p \mid \mathcal{F}_{t-1}) = \boldsymbol{x}_t^\top \boldsymbol{\alpha}(p), \tag{13.21}$$

where $\boldsymbol{x}_t = (1, Y_{t-1}, \ldots, Y_{t-k})^\top$.

Estimation proceeds as in standard quantile regression by solving

$$\hat{\boldsymbol{\alpha}}(p) = \underset{\boldsymbol{\alpha} \in \mathbb{R}^{k+1}}{\operatorname{argmin}} \sum_{t=1}^{n} \psi_p \left(Y_t - \boldsymbol{x}_t^\top \boldsymbol{\alpha} \right),$$

yielding the *autoregression quantiles*.

R Code 13.8: Fitting a quantile autoregressive model to simulated AR(1) data.

```
set.seed(123)
n <- 500
u <- runif(n)
y <- rep(0, n)

a0 <- function(u) qnorm(u)
a1 <- function(u) 0.3 + 0.6*u

for(i in 2:n){
  y[i] <- a0(u[i]) + a1(u[i]) * y[i-1]
}

dat <- data.frame(y = y[-1], ylag = y[-n])

p <- seq(0.1, 0.9, by = 0.1)
fit <- rq(y ~ ylag, tau = p, data = dat)
```

The key feature of the QAR model is that the coefficients depend on p. This allows past values to affect not only the location but also the scale and shape of the conditional distribution. For example, if $\alpha_1(p)$ increases with p, large values of Y_{t-1} have a stronger effect on upper conditional quantiles than on lower ones. This leads to asymmetric persistence, often visible as increasing dispersion in the scatterplot of (Y_{t-1}, Y_t).

An equivalent representation highlights the underlying mechanism. Since $Q_{Y_t}(p \mid \mathcal{F}_{t-1})$ is a proper quantile function, one may write

$$Y_t = \alpha_0(U_t) + \alpha_1(U_t)Y_{t-1} + \cdots + \alpha_k(U_t)Y_{t-k},$$

where $U_t \overset{\text{IID}}{\sim} \mathcal{U}(0,1)$ and independent of the past. This expresses the process as a random-coefficient autoregression, with coefficients driven by a latent variable. The dependence of $\boldsymbol{\alpha}(\cdot)$ on p allows for different persistence behaviour across the distribution.

Under standard regularity conditions, the autoregressive quantile estimators inherit the large-sample behaviour of quantile regression with dependent data. In particular,

$$\sqrt{n}\,\{\hat{\boldsymbol{\alpha}}(p) - \boldsymbol{\alpha}(p)\} \xrightarrow{d} \mathcal{N}\left(0, \boldsymbol{\Sigma}(p)\right), \tag{13.22}$$

where the asymptotic covariance matrix involves the long-run covariance of the process

$$\boldsymbol{x}_t\big(p - \mathbb{I}(Y_t \le \boldsymbol{x}_t^\top \boldsymbol{\alpha}(p))\big).$$

Thus, as in the independent case, the rate of convergence is $\sqrt{n}$, but the variance reflects both the density at the quantile and the temporal dependence in the data. Estimation also becomes less precise in the tails, reflecting both the smaller effective information at extreme quantiles and the influence of the conditional density at the target quantile.

The simulation in R Code 13.8 illustrates a simple QAR(1) process

$$Q_{Y_t}(p \mid \mathcal{F}_{t-1}) = \alpha_0(p) + \alpha_1(p)Y_{t-1},$$

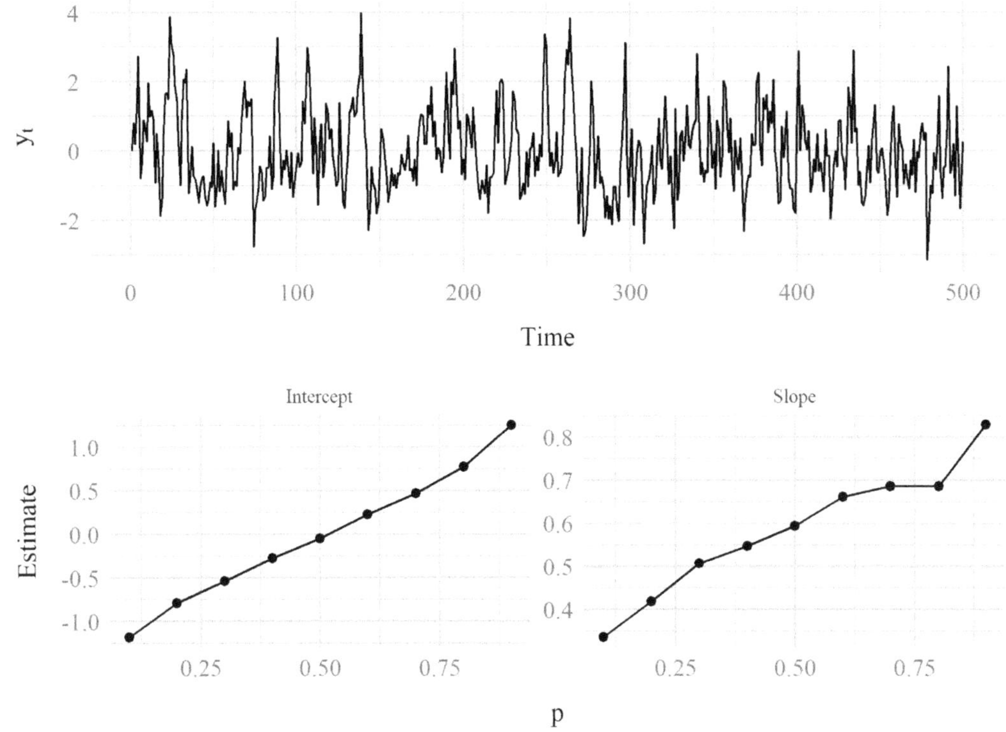

Figure 13.8
Graph depicting the time series data generated with R Code 13.8 (top) together with the estimated intercepts (bottom left) and slopes (bottom right).

in which the autoregressive coefficient increases with p. Consequently, the upper tail is more persistent than the centre or lower tail. Figure 13.8 shows the simulated series together with the estimated QAR coefficients. The slope estimates increase with p, reflecting stronger persistence in the upper tail. The increasing slope function confirms that the effect of Y_{t-1} is not constant across the conditional distribution of Y_t. In this example, large positive values tend to persist more strongly than smaller values, a feature that would be missed by a classical AR model with a single autoregressive coefficient. For an in-depth overview and discussion of quantile autoregressive models see Koenker (2005) and Koenker et al. (2018, Chapter 17).

In R Code 13.9, we illustrate quantile autoregression using the `UKgas` dataset, which records quarterly gas consumption in millions of therms (one million therms correspond to approximately 105 terajoules of energy) in the United Kingdom, 1960–1986. The series is transformed into a lagged representation, pairing each observation Y_t with its first lag Y_{t-1}. We then fit a QAR(1) model at three quantile levels, $p = 0.1, 0.5$, and 0.9, by estimating separate quantile regressions of Y_t on Y_{t-1}.[5]

The estimated coefficients reveal substantial heterogeneity across the distribution. The slope is approximately constant between the lower and median quantiles, but increases markedly at the upper quantile, indicating stronger persistence for large values of the series. In other words, high consumption levels tend to propagate more strongly over time

[5]Higher-order quantile autoregressive models, such as a QAR(r) model with predictors $Y_{t-1}, \ldots, Y_{t-r}$, are fitted by replacing `ylag` with the collection of lagged variables `ylag1`,. . . ,`ylagr` in the call to `rq()`.

R Code 13.9: Fitting a quantile autoregressive model using the UK Quarterly Gas Consumption dataset.

```
library(quantreg)
library(ggplot2)

data(UKgas)

y <- as.numeric(UKgas)
dat <- data.frame(
  y = y[-1],
  ylag = y[-length(y)]
)

# fit QAR(1)
fit <- rq(y ~ ylag, tau = c(0.1, 0.5, 0.9), data = dat)

# obtain standard errors
summary(fit, se = "nid")
>
> Call: rq(formula = y ~ ylag, tau = c(0.1, 0.5, 0.9), data = dat)
>
> tau: [1] 0.1
>
> Coefficients:
>             Value    Std. Error t value  Pr(>|t|)
> (Intercept) 28.07696  3.75448    7.47826  0.00000
> ylag         0.44661  0.02788   16.02046  0.00000
>
> Call: rq(formula = y ~ ylag, tau = c(0.1, 0.5, 0.9), data = dat)
>
> tau: [1] 0.5
>
> Coefficients:
>             Value    Std. Error t value  Pr(>|t|)
> (Intercept) 90.57330 33.18303    2.72951  0.00744
> ylag         0.44894  0.18073    2.48407  0.01457
>
> Call: rq(formula = y ~ ylag, tau = c(0.1, 0.5, 0.9), data = dat)
>
> tau: [1] 0.9
>
> Coefficients:
>             Value     Std. Error t value   Pr(>|t|)
> (Intercept) 291.63080  84.46376    3.45273   0.00080
> ylag          1.08955   0.11117    9.80110   0.00000
```

than typical or low values. In R Code 13.9, standard errors are based on local asymptotic approximations. More refined inference for time series would require methods that explicitly account for serial dependence. Also, this simple illustration ignores the strong trend and seasonal structure of the series; a substantive analysis would typically model these components explicitly.

A

Appendix A

A.1 Parametric distributions

In parametric statistics, the cumulative distribution function (CDF) $F_Y(y)$ and, correspondingly, the quantile function (QF) $Q_Y(p)$ of a random variable Y have a particular form and may be indexed by a quantity called *parameter*. There are many parametric families of discrete and continuous distributions. Here, we consider those that recur throughout the present book, some of which occupy a special place in quantile modelling. For each family, we define the corresponding probability mass function (PMF) or probability density function (PDF), as appropriate. Some distributions do not have, in general, a CDF or a QF in closed form, but some useful representations and approximations of these functions can be given. Also, the interested reader may refer to Steinbrecher and Shaw (2008) for a discussion on using differential equations to characterise quantile functions that lack a closed form.

We first introduce two special functions, the beta and gamma functions, which appear in several distribution functions.

Definition 15 (Beta function). *The incomplete beta function is defined as* $B(a,b,u) = \int_0^u t^{a-1}(1-t)^{b-1}\,\mathrm{d}t$.

For $u = 1$, the incomplete beta is called (complete) beta function, that is, $B(a,b,1) \equiv B(a,b)$. If a and b are positive integers, then $B(a,b) = \dfrac{(a-1)!\,(b-1)!}{(a+b-1)!}$. The *regularised beta function* is defined as the ratio $B_r(a,b,u) = \dfrac{B(a,b,u)}{B(a,b)}$.

Definition 16 (Gamma function). *The upper incomplete gamma function is defined as* $\Gamma(a,u) = \int_u^\infty t^{a-1}e^{-t}\,\mathrm{d}t$. *The lower incomplete gamma function is defined as* $\gamma(a,u) = \int_0^u t^{a-1}e^{-t}\,\mathrm{d}t$.

For $u = 0$, the upper incomplete gamma function reduces to the complete gamma function, that is, $\Gamma(a,0) = \Gamma(a)$. If a is a positive integer, then $\Gamma(a) = (a-1)!$. The *regularised gamma functions* are defined by the ratios $\Gamma_r(a,u) = \dfrac{\Gamma(a,u)}{\Gamma(a)}$ and $\gamma_r(a,u) = \dfrac{\gamma(a,u)}{\Gamma(a)}$. Also, $\Gamma_r(a,u) + \gamma_r(a,u) = 1$.

Definition 17 (Relationship between beta and gamma functions). *The beta and gamma functions are related by the following expression:*

$$\mathrm{B}(a,b) = \frac{\Gamma(a)\,\Gamma(b)}{\Gamma(a+b)}.$$

DOI: 10.1201/9781003326410-A

A.2 Models for discrete variables

Uniform

A *discrete uniform distribution* with parameter n belongs to the family of PMFs defined as

$$m_Y(y; n) = \begin{cases} \dfrac{1}{n} & \text{for } y \in \{1, 2, \ldots, n\} \\ 0 & \text{otherwise,} \end{cases}$$

where n is a positive integer. The corresponding CDF and QF are, respectively,

$$F_Y(y; n) = \begin{cases} 0 & \text{for } y < 1 \\ \dfrac{\lfloor y \rfloor}{n} & \text{for } 1 \leq y \leq n \\ 1 & \text{for } y > n \end{cases}$$

and

$$Q_Y(p; n) = \lceil np \rceil$$

for $p \in (0, 1]$.

Binomial

A *binomial distribution* with parameters n and π belongs to the family of PMFs defined as

$$m_Y(y; n, \pi) = \begin{cases} \binom{n}{y} \pi^y (1 - \pi)^{n-y} & \text{for } y \in \{0, 1, \ldots, n\} \\ 0 & \text{otherwise,} \end{cases}$$

where n is an integer, $n \geq 1$, and $0 \leq \pi \leq 1$. When $n = 1$, the binomial distribution reduces to the Bernoulli distribution and $\Pr(Y = 1) = \pi$. The CDF is given by $F_Y(y; n, \pi) = \sum_{j=0}^{\lfloor y \rfloor} m_Y(j; n, \pi)$ or equivalently as

$$\begin{aligned} F_Y(y; n, \pi) &= 1 - B_r(\lfloor y \rfloor + 1, n - \lfloor y \rfloor, \pi) \\ &= B_r(n - \lfloor y \rfloor, \lfloor y \rfloor + 1, 1 - \pi). \end{aligned}$$

Normal approximations, including continuity-corrected and Cornish–Fisher approximations, are available for the calculation of the binomial CDF and QF.

Poisson

A *Poisson distribution* with parameter λ belongs to the family of PMFs defined as

$$m_Y(y; \lambda) = \begin{cases} \dfrac{\exp(-\lambda)\lambda^y}{y!} & \text{for } y \in \{0, 1, 2, \ldots\} \\ 0 & \text{otherwise,} \end{cases}$$

where $\lambda > 0$. This distribution is denoted $\text{Pois}(\lambda)$.

The corresponding CDF is obtained, as usual, in terms of the PMF, $F_Y(y; \lambda) = \sum_{j=0}^{\lfloor y \rfloor} m_Y(j; \lambda)$, or in terms of the regularised gamma function

$$F_Y(y; \lambda) = \Gamma_r(\lfloor y + 1 \rfloor, \lambda).$$

Normal approximations and exponential inequalities are available for the calculation of the Poisson CDF and QF.

Negative binomial

A *negative binomial distribution* with parameters k and π belongs to the family of PMFs defined as

$$m_Y(y; k, \pi) = \begin{cases} \binom{y+k-1}{y} \pi^k (1-\pi)^y & \text{for } y \in \{0, 1, 2, \ldots\} \\ 0 & \text{otherwise,} \end{cases}$$

where $k = 1, 2, 3, \ldots$ and $0 < \pi \leq 1$. When $k = 1$, the negative binomial distribution reduces to the geometric distribution.

The corresponding CDF is obtained, as usual, in terms of the PMF or in terms of the regularised incomplete beta function

$$F_Y(y; k, \pi) = B_r(k, \lfloor y \rfloor + 1, \pi).$$

Numerical algorithms are available for the calculation of the negative binomial CDF and QF.

A.3 Models for continuous variables

Uniform

A random variable Y is said to follow a *uniform distribution* with parameters a and b, $Y \sim \mathcal{U}(a, b)$, if its PDF is given by

$$f_Y(y; a, b) = \begin{cases} \dfrac{1}{b-a} & \text{for } a \leq y \leq b \\ 0 & \text{otherwise,} \end{cases}$$

where $a, b \in \mathbb{R}$. The distribution $\mathcal{U}(0, 1)$ is called standard uniform.

The corresponding CDF and QF are, respectively,

$$F_Y(y; a, b) = \begin{cases} 0 & \text{for } y < a \\ \dfrac{y-a}{b-a} & \text{for } a \leq y \leq b \\ 1 & \text{for } y > b \end{cases}$$

and

$$Q_Y(p; a, b) = a + p(b-a)$$

for $p \in [0, 1]$.

Beta

A random variable Y is said to follow a *beta distribution* with parameters a and b,

$Y \sim \mathcal{B}(a, b)$, if its PDF is given by

$$f_Y(y; a, b) = \begin{cases} \dfrac{y^{a-1}(1-y)^{b-1}}{B(a,b)} & \text{for } 0 < y < 1 \\ 0 & \text{otherwise,} \end{cases}$$

where $a > 0$ and $b > 0$. Note that $\mathcal{B}(1, 1)$ is a standard uniform.

The corresponding CDF is given by the regularised beta function

$$F_Y(y; a, b) = B_r(a, b, y).$$

Normal

A random variable Y is said to follow a *normal distribution* with parameters μ and σ^2, $Y \sim \mathcal{N}(\mu, \sigma^2)$, if its PDF is given by

$$f_Y(y; \mu, \sigma) = \frac{1}{\sigma\sqrt{2\pi}} \exp\left\{ -\frac{1}{2}\left(\frac{y-\mu}{\sigma}\right)^2 \right\}$$

for $y \in \mathbb{R}$, where $\mu \in \mathbb{R}$ and $\sigma > 0$. The distribution $\mathcal{N}(0, 1)$ is called standard normal.

The corresponding CDF and QF are, respectively,

$$F_Y(y; \mu, \sigma) = \frac{1}{2} + \frac{1}{2} \operatorname{erf}\left(\frac{y-\mu}{\sigma\sqrt{2}}\right),$$

where $\operatorname{erf}(u) = \frac{2}{\sqrt{\pi}} \int_0^u e^{-t^2}\, \mathrm{d}t$, and

$$Q_Y(p; \mu, \sigma) = \mu + \sigma\sqrt{2}\, \operatorname{erf}^{-1}(2p-1)$$

for $p \in (0, 1)$. Note that $Q_Y(p; 0, 1)$ is known as *probit* function.

The inverse error function cannot be expressed in terms of elementary functions. However, rational approximations are available. For example, the QF of a standard normal is approximated by

$$Q_Y(p; 0, 1) \approx t - \frac{c_0 + c_1 t}{1 + c_2 t + c_3 t^2}$$

for $p \in (0.5, 1)$, where $t = \sqrt{-2\log(1-p)}$ and c_k, $k = 0, \dots, 3$, are known constants (see Equation 26.2.22 in Abramowitz and Stegun, 1972, p. 933).

Logistic

A random variable Y is said to follow a *logistic distribution* with parameters μ and σ, $Y \sim \text{Logistic}(\mu, \sigma)$, if its PDF is given by

$$f_Y(y; \mu, \sigma) = \frac{\exp\left(-\dfrac{y-\mu}{\sigma}\right)}{\sigma\left\{1 + \exp\left(-\dfrac{y-\mu}{\sigma}\right)\right\}^2}$$

for $y \in \mathbb{R}$, where $\mu \in \mathbb{R}$ and $\sigma > 0$. The distribution Logistic$(0, 1)$ is called standard logistic.

The corresponding CDF and QF are, respectively,

$$F_Y(y;\mu,\sigma) = \frac{1}{1+\exp\left(-\dfrac{y-\mu}{\sigma}\right)} = \frac{1}{2} + \frac{1}{2}\tanh\left(\frac{y-\mu}{2\sigma}\right)$$

and

$$Q_Y(p;\mu,\sigma) = \mu + \sigma \log\left(\frac{p}{1-p}\right)$$

for $p \in (0,1)$. Note that $Q_Y(p;0,1)$ is known as *logit* function.

Laplace

A random variable Y is said to follow a *Laplace distribution* with parameters μ and σ, $Y \sim \mathcal{L}(\mu,\sigma)$, if its PDF is given by

$$f_Y(y;\mu,\sigma) = \frac{1}{4\sigma}\exp\left\{-\frac{1}{2}\left|\frac{y-\mu}{\sigma}\right|\right\}$$

for $y \in \mathbb{R}$, where $\mu \in \mathbb{R}$ and $\sigma > 0$. This distribution is also known as the *double exponential distribution* (see the gamma and exponential distributions further below).

The corresponding CDF and QF are, respectively,

$$F_Y(y;\mu,\sigma) = \begin{cases} \dfrac{1}{2}\exp\left(\dfrac{y-\mu}{2\sigma}\right) & \text{for } y < \mu \\ 1 - \dfrac{1}{2}\exp\left(-\dfrac{y-\mu}{2\sigma}\right) & \text{for } y \geq \mu \end{cases}$$

and

$$Q_Y(p;\mu,\sigma) = \begin{cases} \mu + 2\sigma\log(2p) & \text{for } 0 < p < 0.5 \\ \mu - 2\sigma\log(2-2p) & \text{for } 0.5 \leq p < 1. \end{cases}$$

Asymmetric Laplace

A random variable Y is said to follow an *asymmetric Laplace distribution* with parameters μ, σ, and τ, $Y \sim \mathcal{AL}(\mu,\sigma,\tau)$, if its PDF is given by

$$f_Y(y;\mu,\sigma,\tau) = \frac{\tau(1-\tau)}{\sigma}\exp\left\{-h_\tau\left(\frac{y-\mu}{\sigma}\right)\right\}$$

for $y \in \mathbb{R}$, where $\mu \in \mathbb{R}$, $\sigma > 0$, $0 < \tau < 1$, and $h_\tau(u) = \tau\max(u,0) + (1-\tau)\max(-u,0)$. Note that $\mathcal{AL}(\mu,\sigma,0.5)$ is a (symmetric) Laplace distribution.

The corresponding CDF and QF are, respectively,

$$F_Y(y;\mu,\sigma,\tau) = \begin{cases} \tau\exp\left\{\dfrac{1-\tau}{\sigma}(y-\mu)\right\} & \text{for } y < \mu \\ 1-(1-\tau)\exp\left\{-\dfrac{\tau}{\sigma}(y-\mu)\right\} & \text{for } y \geq \mu \end{cases}$$

and

$$Q_Y(p;\mu,\sigma,\tau) = \begin{cases} \mu + \dfrac{\sigma}{1-\tau}\log\left(\dfrac{p}{\tau}\right) & \text{for } 0 < p < \tau \\ \mu - \dfrac{\sigma}{\tau}\log\left(\dfrac{1-p}{1-\tau}\right) & \text{for } \tau \leq p < 1. \end{cases}$$

Student's t

A random variable Y is said to follow a *Student's t-distribution* with parameter ν, $Y \sim t_\nu$, if its PDF is given by

$$f_Y(y;\nu) = \frac{1}{\sqrt{\nu}\,B\left(\frac{\nu}{2},\frac{1}{2}\right)}\left(1+\frac{y^2}{\nu}\right)^{-\frac{\nu+1}{2}}$$

for $y \in \mathbb{R}$, where $\nu > 0$.

The corresponding CDF and QF are, respectively,

$$F_Y(y;\nu) = \frac{1}{2} + \frac{1}{2}\left\{1 - B_r\left(\frac{\nu}{2},\frac{1}{2},g(y)\right)\right\}\operatorname{sgn}(y),$$

where $g(y) = \dfrac{\nu}{\nu+y^2}$, and (Shaw, 2006)

$$Q_Y(p;\nu) = \operatorname{sgn}\left(p-\frac{1}{2}\right)\sqrt{\nu\left\{\frac{1}{B_r^{-1}\left(\frac{\nu}{2},\frac{1}{2},h(p)\right)} - 1\right\}},$$

where

$$h(p) = \begin{cases} 2p & \text{for } 0 < p < 0.5 \\ 2(1-p) & \text{for } 0.5 \le p < 1. \end{cases}$$

The QF of a t-distribution simplifies when ν takes on specific values. For example,

$$Q_Y(p;2) = \frac{2p-1}{\sqrt{2p-2p^2}}$$

and

$$Q_Y(p;4) = \operatorname{sgn}\left(p-\frac{1}{2}\right)\sqrt{u-4},$$

where $u = \dfrac{4}{\sqrt{\alpha}}\cos\left\{\dfrac{1}{3}\arccos(\sqrt{\alpha})\right\}$ and $\alpha = 4p(1-p)$.

Chi-squared

A random variable Y is said to follow a *chi-squared distribution* with parameter ν, $Y \sim \chi^2_\nu$, if its PDF is given by

$$f_Y(y;\nu) = \begin{cases} \dfrac{y^{\frac{\nu}{2}-1}\exp\left(-\frac{y}{2}\right)}{2^{\frac{\nu}{2}}\Gamma\left(\frac{\nu}{2}\right)} & \text{for } y > 0 \\ 0 & \text{otherwise,} \end{cases}$$

where $\nu > 0$.

The corresponding distribution function is

$$F_Y(y;\nu) = \begin{cases} 0 & \text{for } y \le 0 \\ \gamma_r\left(\frac{\nu}{2},\frac{y}{2}\right) & \text{for } y > 0. \end{cases}$$

The quantile function $F_Y^{-1}(p;\nu)$ does not have, in general, a simple representation. However, as in the case of a t-distribution, the QF of a χ^2 distribution simplifies when ν takes on specific values. For example,

$$Q_Y(p;2) = -2\log(1-p)$$

for $p \in [0,1)$.

Gamma

A random variable Y is said to follow a *gamma distribution* with parameters a and b, $Y \sim \mathcal{G}(a,b)$, if its PDF is given by

$$f_Y(y;a,b) = \begin{cases} \dfrac{b^a}{\Gamma(a)} y^{a-1}\exp(-by) & \text{for } y \geq 0 \\ 0 & \text{otherwise,} \end{cases}$$

where $a > 0$ (shape) and $b > 0$ (rate). Equivalently, the gamma distribution can be parameterised in terms of the scale, say, $\sigma = 1/b$. The distribution $\mathcal{G}(1,b) \equiv \mathcal{E}(b)$ is called exponential distribution and has density $f_Y(y;b) = b\exp(-by)$. Note that if $Y_i \sim \mathcal{E}(b)$, $i = 1,2$, with $Y_1 \perp\!\!\!\perp Y_2$, then $Y_1 - Y_2 \overset{d}{=} \mathcal{L}(0, 1/(2b))$ (hence the name 'double exponential').

The corresponding distribution function is

$$F_Y(y;a,b) = \begin{cases} 0 & \text{for } y < 0 \\ \gamma_r(a, by) & \text{for } y \geq 0. \end{cases}$$

The quantile function $F_Y^{-1}(p;a,b)$ does not have, in general, a simple representation. However, when $a = 1$, the QF of a gamma (exponential) distribution has a simple form

$$Q_Y(p;1,b) = -\frac{1}{b}\log(1-p)$$

for $p \in [0,1)$.

Gumbel

A *Gumbel distribution* (or type I extreme value distribution) with parameters μ and σ belongs to the family of probability density functions defined as

$$f_Y(y;\mu,\sigma) = \frac{1}{\sigma}\exp\left\{-\frac{y-\mu}{\sigma} - \exp\left(-\frac{y-\mu}{\sigma}\right)\right\}$$

for $y \in \mathbb{R}$, where $\mu \in \mathbb{R}$ and $\sigma > 0$. The Gumbel distribution is often referred to as the double exponential distribution. We will reserve the latter name to the Laplace distribution.

The corresponding CDF and QF are, respectively,

$$F_Y(y;\mu,\sigma) = \exp\left\{-\exp\left(-\frac{y-\mu}{\sigma}\right)\right\}$$

and

$$Q_Y(p;\mu,\sigma) = \mu - \sigma\log\{-\log(p)\}$$

for $p \in (0,1)$. Note that $Q_Y(p;0,1)$ is known as *complementary log-log* (cloglog) function.

Weibull

A random variable Y is said to follow a (2 parameter) *Weibull distribution* (or type III extreme value distribution) with parameters a and b, $Y \sim \mathcal{W}(a,b)$, if its PDF is given by

$$f_Y(y;a,b) = \begin{cases} ab(by)^{a-1}\exp\{-(by)^a\} & \text{for } y \geq 0 \\ 0 & \text{otherwise,} \end{cases}$$

where $a > 0$ and $b > 0$. The 2-parameter Weibull distribution can be extended to include a location parameter. Note that $\mathcal{W}(1,b)$ is an exponential distribution.

The corresponding CDF and QF are, respectively,

$$F_Y(y;a,b) = \begin{cases} 0 & \text{for } y < 0 \\ 1 - \exp\{-(by)^a\} & \text{for } y \geq 0 \end{cases}$$

and

$$Q_Y(p;a,b) = \frac{1}{b}\{-\log(1-p)\}^{\frac{1}{a}}$$

for $p \in [0,1)$.

Generalised extreme value

A random variable Y is said to follow a *generalised extreme value distribution* with parameters μ, σ, and γ, $Y \sim \text{GEV}(\mu,\sigma,\gamma)$, if its PDF is given by

$$f_Y(y;\mu,\sigma,\gamma) = \begin{cases} \frac{1}{\sigma}\exp\left\{-\left(1+\gamma\frac{y-\mu}{\sigma}\right)^{-\frac{1}{\gamma}}\right\}\left(1+\gamma\frac{y-\mu}{\sigma}\right)^{-1-\frac{1}{\gamma}} \\ \qquad\qquad \text{for } 1+\gamma\frac{y-\mu}{\sigma} > 0,\ \gamma \neq 0 \\ \frac{1}{\sigma}\exp\left\{-\exp\left(-\frac{y-\mu}{\sigma}\right) - \frac{y-\mu}{\sigma}\right\} \\ \qquad\qquad \text{for } y \in \mathbb{R},\ \gamma = 0, \end{cases}$$

where $\mu \in \mathbb{R}$, $\sigma > 0$, and $\gamma \in \mathbb{R}$.

The corresponding CDF and QF are, respectively,

$$F_Y(y;\mu,\sigma,\gamma) = \begin{cases} \exp\left\{-\left(1+\gamma\frac{y-\mu}{\sigma}\right)^{-\frac{1}{\gamma}}\right\} & \text{for } 1+\gamma\frac{y-\mu}{\sigma} > 0,\ \gamma \neq 0 \\ \exp\left(-e^{-\frac{y-\mu}{\sigma}}\right) & \text{for } y \in \mathbb{R},\ \gamma = 0 \end{cases}$$

and

$$Q_Y(p;\mu,\sigma,\gamma) = \begin{cases} \mu + \frac{\sigma}{\gamma}\{(-\log p)^{-\gamma} - 1\} & \text{for } \gamma \neq 0 \\ \mu - \sigma\log(-\log p) & \text{for } \gamma = 0 \end{cases}$$

for $p \in (0,1)$.

Generalised Pareto

A random variable Y is said to follow a *generalised Pareto distribution* with parameters μ, σ, and γ, $Y \sim \text{GP}(\mu,\sigma,\gamma)$, if its PDF is given by

$$f_Y(y;\mu,\sigma,\gamma) = \begin{cases} \frac{1}{\sigma}\left(1+\gamma\frac{y-\mu}{\sigma}\right)^{-\frac{1}{\gamma}-1} & \text{for } 1+\gamma\frac{y-\mu}{\sigma} > 0,\ \gamma \neq 0 \\ \frac{1}{\sigma}\exp\left(-\frac{y-\mu}{\sigma}\right) & \text{for } y \in \mathbb{R},\ \gamma = 0 \end{cases}$$

for $\mu \in \mathbb{R}$, $\sigma > 0$, and $\gamma \in \mathbb{R}$. Additionally, $y \geq \mu$ when $\gamma \geq 0$, and $\mu \leq y \leq \mu - \sigma/\gamma$ when $\gamma < 0$.

The corresponding CDF and QF are, respectively,

$$F_Y(y; \mu, \sigma, \gamma) = \begin{cases} 1 - \left(1 + \gamma \frac{y-\mu}{\sigma}\right)^{-\frac{1}{\gamma}} & \text{for } \gamma \neq 0 \\ 1 - \exp\left(-\frac{y-\mu}{\sigma}\right) & \text{for } \gamma = 0 \end{cases}$$

and

$$Q_Y(p; \mu, \sigma, \gamma) = \begin{cases} \mu + \frac{\sigma}{\gamma}\{(1-p)^{-\gamma} - 1\} & \text{for } \gamma \neq 0 \\ \mu - \sigma \log(1-p) & \text{for } \gamma = 0 \end{cases}$$

for $p \in (0, 1)$.

B

Appendix B

B.1 Multivariate distributions

B.1.1 Joint and marginal distributions

Let $Y_1, \ldots, Y_k$ be k random variables defined on the same probability space. We call $\boldsymbol{Y} = (Y_1, \ldots, Y_k)$ a k-dimensional or multivariate random variable.

Definition 18 (Joint cumulative distribution function). *The joint cumulative distribution function of $Y_1, \ldots, Y_k$, denoted by $F_{Y_1,\ldots,Y_k}(\cdot)$, is a function whose domain and codomain are the k-fold Cartesian product $\mathbb{R}^k$ and the unit interval $[0, 1]$, respectively. The joint CDF of $Y_1, \ldots, Y_k$ is defined as $F_{Y_1,\ldots,Y_k}(y_1, \ldots, y_k) = \Pr(Y_1 \leq y_1, \ldots, Y_k \leq y_k)$ for every $(y_1, \ldots, y_k) \in \mathbb{R}^k$.*

The distinction between joint discrete and continuous random variables is as follows.

Definition 19 (Joint discrete random variables). *The k-dimensional random variable $\boldsymbol{Y} = (Y_1, \ldots, Y_k)$ is defined to be a k-dimensional discrete random variable if it can assume values only at a countable number of points $\boldsymbol{y} = (y_1, \ldots, y_k)$ in k-dimensional real space. We also say that the random variables $Y_1, \ldots, Y_k$ are joint discrete random variables.*

Analogously to the univariate case, we now introduce the probability function for multiple variables. If $Y_1, \ldots, Y_k$ are joint discrete random variables taking on $m_1, \ldots, m_k$ distinct values $y_{j_1}, \ldots, y_{j_k}$, where $\{j_1 = 1, \ldots, m_1; \ldots; j_k = 1, \ldots, m_k\}$, then the function

$$m_{\boldsymbol{Y}}(\boldsymbol{y}) = \begin{cases} \Pr(Y_1 = y_{j_1}, \ldots, Y_k = y_{j_k}) & \text{for } \boldsymbol{y} = (y_{j_1}, \ldots, y_{j_k}) \\ 0 & \text{for } \boldsymbol{y} \neq (y_{j_1}, \ldots, y_{j_k}) \end{cases}$$

is called *joint probability (mass) function* of $\boldsymbol{Y}$. The relationship between the CDF and the PMF of joint discrete variables is given by the following expression:

$$F_{\boldsymbol{Y}}(\boldsymbol{y}) = \sum_{j:\boldsymbol{y}_j \leq \boldsymbol{y}} m_{\boldsymbol{Y}}(\boldsymbol{y}_j),$$

where $\boldsymbol{y} = (y_1, \ldots, y_k)$, $j = (j_1, \ldots, j_k)$ and $\boldsymbol{y}_j \leq \boldsymbol{y}$ if and only if each component of $\boldsymbol{y}_j$ is less than or equal to the corresponding component of $\boldsymbol{y}$.

Definition 20 (Joint continuous random variables). *The k-dimensional random variable $\boldsymbol{Y} = (Y_1, \ldots, Y_k)$ is defined to be a k-dimensional continuous random variable if and only if there exists a function $f_{Y_1,\ldots,Y_k}$ such that $F_{Y_1,\ldots,Y_k}(y_1, \ldots, y_k) = \int_{-\infty}^{y_1} \cdots \int_{-\infty}^{y_k} f_{Y_1,\ldots,Y_k}(t_1, \ldots, t_k)\, \mathrm{d}t_1 \ldots \mathrm{d}t_k$ for all $(y_1, \ldots, y_k)$. We also say that the random variables $(Y_1, \ldots, Y_k)$ are joint continuous random variables.*

The function $f_{\boldsymbol{Y}}(\cdot)$ is called *joint probability density function*, which satisfies $\int_{-\infty}^{\infty} \int_{-\infty}^{\infty} \cdots \int_{-\infty}^{\infty} f_{Y_1,\ldots,Y_k}(t_1, \ldots, t_k)\, \mathrm{d}t_1 \ldots \mathrm{d}t_k = 1$.

DOI: 10.1201/9781003326410-B

Knowledge of the joint distribution of $\boldsymbol{Y} = (Y_1, Y_2, \ldots, Y_k)$ implies knowledge of the *marginal* distribution of each variable Y_j, $j = 1, 2, \ldots, k$. For example, let $\boldsymbol{Y} = (Y_1, Y_2)$ be a bivariate continuous random variable with joint PDF $f_{Y_1,Y_2}(y_1, y_2)$. Then,

$$f_{Y_1}(y_1) = \int f_{Y_1,Y_2}(y_1, y_2)\, \mathrm{d}y_2 \quad \text{and} \quad f_{Y_2}(y_2) = \int f_{Y_1,Y_2}(y_1, y_2)\, \mathrm{d}y_1.$$

In the discrete case, the above integrals are replaced by summations.

B.1.2 Conditional distributions

Definition 21 (Conditional discrete cumulative distribution function)**.** *Let* $\boldsymbol{Y} = (\boldsymbol{Y}_1, \boldsymbol{Y}_2)$ *be a partition of a discrete random vector. The conditional discrete CDF of* $\boldsymbol{Y}_1$ *given* $\boldsymbol{Y}_2 = \boldsymbol{y}_2$ *is defined as*

$$F_{\boldsymbol{Y}_1|\boldsymbol{Y}_2=\boldsymbol{y}_2}(\boldsymbol{y}_1) = \Pr(\boldsymbol{Y}_1 \le \boldsymbol{y}_1 \mid \boldsymbol{Y}_2 = \boldsymbol{y}_2),$$

for values $\boldsymbol{y}_2$ *such that* $\Pr(\boldsymbol{Y}_2 = \boldsymbol{y}_2) > 0$.

The associated conditional PMF is

$$m_{\boldsymbol{Y}_1|\boldsymbol{Y}_2=\boldsymbol{y}_2}(\boldsymbol{y}_1) = \frac{m_{\boldsymbol{Y}_1,\boldsymbol{Y}_2}(\boldsymbol{y}_1, \boldsymbol{y}_2)}{m_{\boldsymbol{Y}_2}(\boldsymbol{y}_2)},$$

provided the denominator is positive.

Definition 22 (Conditional continuous cumulative distribution function)**.** *Let* $\boldsymbol{Y} = (\boldsymbol{Y}_1, \boldsymbol{Y}_2)$ *be a partition of a continuous random vector. The conditional continuous CDF of* $\boldsymbol{Y}_1$ *given* $\boldsymbol{Y}_2 = \boldsymbol{y}_2$ *is defined as*

$$F_{\boldsymbol{Y}_1|\boldsymbol{Y}_2=\boldsymbol{y}_2}(\boldsymbol{y}_1) = \int_{-\infty}^{y_1} \cdots \int_{-\infty}^{y_r} f_{\boldsymbol{Y}_1|\boldsymbol{Y}_2=\boldsymbol{y}_2}(t_1, \ldots, t_r)\, \mathrm{d}t_1 \cdots \mathrm{d}t_r.$$

The associated conditional PDF is

$$f_{\boldsymbol{Y}_1|\boldsymbol{Y}_2=\boldsymbol{y}_2}(\boldsymbol{y}_1) = \frac{f_{\boldsymbol{Y}_1,\boldsymbol{Y}_2}(\boldsymbol{y}_1, \boldsymbol{y}_2)}{f_{\boldsymbol{Y}_2}(\boldsymbol{y}_2)},$$

whenever $f_{\boldsymbol{Y}_2}(\boldsymbol{y}_2) > 0$.

In the continuous case, the expression

$$\Pr(\boldsymbol{Y}_1 \le \boldsymbol{y}_1 \mid \boldsymbol{Y}_2 = \boldsymbol{y}_2)$$

cannot be interpreted through the elementary ratio formula for conditional probability, because $\Pr(\boldsymbol{Y}_2 = \boldsymbol{y}_2) = 0$. The conditional distribution must therefore be defined through densities, regular conditional probabilities, or equivalent constructions. In most regression settings, however, the conditioning value $\boldsymbol{y}_2$ is treated as given and the formal difficulty is rarely consequential in practice.

B.2 Further results on conditional quantiles

B.2.1 Rosenblatt representation

The following result extends the uniform transformation rule (Proposition 1) to a bivariate random variable.

Proposition 3 (Rosenblatt representation, bivariate case). *Let X and Y be continuous random variables defined on the same probability space with marginal CDFs F_X and F_Y, and joint CDF $F_{X,Y}$. Assume F_X is continuous and strictly increasing (so that F_X^{-1} is well-defined and single-valued). Then the random vector $(X,Y) \stackrel{d}{=} \left(F_X^{-1}(U), F_{Y|X}^{-1}(V \mid X = F_X^{-1}(U))\right)$, where U and V are independent standard uniform random variables, and $F_{Y|X}^{-1}(\cdot \mid x)$ denotes a (generalised) quantile function of the conditional distribution of Y given $X = x$.*

Proof. Let U and V be independent standard uniform random variables. Define

$$\tilde{X} = F_X^{-1}(U), \qquad \tilde{Y} = F_{Y|X}^{-1}(V \mid \tilde{X}),$$

where $F_{Y|X}^{-1}(\cdot \mid x)$ is a quantile function of the conditional distribution of Y given $X = x$.

We show that the joint CDF of $(\tilde{X}, \tilde{Y})$ equals $F_{X,Y}(x, y)$. Compute

$$\begin{aligned}
\Pr(\tilde{X} \leq x, \tilde{Y} \leq y) &= \Pr\big(F_X^{-1}(U) \leq x,\ F_{Y|X}^{-1}(V \mid F_X^{-1}(U)) \leq y\big) \\
&= \Pr\big(U \leq F_X(x),\ V \leq F_{Y|X}(y \mid F_X^{-1}(U))\big).
\end{aligned}$$

Since U and V are independent,

$$\begin{aligned}
\Pr(\tilde{X} \leq x, \tilde{Y} \leq y) &= \mathrm{E}\left\{I(U \leq F_X(x)) \cdot \Pr\big(V \leq F_{Y|X}(y \mid F_X^{-1}(U)) \mid U\big)\right\} \\
&= \mathrm{E}\left\{I(U \leq F_X(x)) \cdot F_{Y|X}(y \mid F_X^{-1}(U))\right\} \\
&= \int_0^{F_X(x)} F_{Y|X}(y \mid F_X^{-1}(u))\, du.
\end{aligned}$$

Under the assumption that F_X is continuous and strictly increasing, the change of variables $t = F_X^{-1}(u)$ is valid (with $du = f_X(t)\, dt$ if a density exists, or more generally via the Lebesgue–Stieltjes measure). The integral becomes

$$\int_{-\infty}^{x} F_{Y|X=t}(y)\, \mathrm{d}F_X(t) = F_{X,Y}(x, y).$$

The last equality holds by the definition of the joint CDF via conditioning on X. Thus $(\tilde{X}, \tilde{Y}) \stackrel{d}{=} (X, Y)$. □

Proposition 3 provides a sequential representation of the joint distribution of (X, Y) in terms of independent uniform variables. Essentially, this construction is the bivariate case of the inverse Rosenblatt transform introduced by Murray Rosenblatt (Rosenblatt, 1952). If X and Y are continuous and F_X is strictly increasing, the construction holds directly. If F_X is continuous but has flat intervals, the result still holds provided the quantile function is interpreted as the generalised inverse. With discrete or mixed marginals, one must use the conditional distribution function instead of the strict inverse. More generally, the representation remains valid whenever suitable conditional quantile functions exist.

B.2.2 Conditional quantile representation

We now show a result which follows from the direct application of the Q-transformation rule but that is also related to the comparison distribution of F_Y and $F_{Y|X}$.

Proposition 4 (Conditional quantile representation). *Let X and Y be two random variables defined on the same probability space, and let F_Y denote the marginal CDF of Y. Define $U = F_Y(Y)$. Then the conditional quantile function of Y given $X = x$ can be written as (Parzen, 2004)*

$$Q_{Y|X=x}(p) = Q_Y(s), \tag{B.1}$$

where $s = Q_{U|X=x}(p)$.

Proof. Let $s = Q_{U|X=x}(p)$. Then, by inversion,

$$p = F_{U|X=x}(s) = \Pr(U \leq s \mid X = x) = \Pr(F_Y(Y) \leq s \mid X = x).$$

Since Q_Y is nondecreasing,

$$\Pr(F_Y(Y) \leq s \mid X = x) = \Pr(Y \leq Q_Y(s) \mid X = x) = F_{Y|X=x}(Q_Y(s)).$$

Hence

$$F_{Y|X=x}(Q_Y(s)) = p,$$

which shows that

$$Q_{Y|X=x}(p) = Q_Y(s).$$

□

The relationship between p and s in Equation (B.1) is a special case of the comparison cumulative distribution

$$v(s; F_Y, F_{Y|X=x}) = F_{Y|X=x}(F_Y^{-1}(s)).$$

This relationship can be verified directly. By definition of the comparison distribution,

$$p = v(s; F_Y, F_{Y|X=x}) = F_{Y|X=x}(F_Y^{-1}(s)).$$

Inverting with respect to s yields

$$s = v^{-1}(p; F_Y, F_{Y|X=x}) = F_Y(Q_{Y|X=x}(p)).$$

Applying Q_Y to both sides gives

$$Q_Y(s) = Q_Y\big(v^{-1}(p; F_Y, F_{Y|X=x})\big) = Q_{Y|X=x}(p),$$

which is equivalent to (B.1).

Equation (B.1) shows that a conditional quantile can be expressed as a marginal quantile evaluated at a 'distorted' probability level $s = Q_{U|X=x}(p)$, where $U = F_Y(Y)$ is the marginal rank of Y. In other words, the conditional quantile curve is obtained by reparameterising the horizontal axis of the marginal quantile function according to the conditional distribution of Y given $X = x$.

A related result concerns conditional quantile functions and *copulas*, which is discussed in the next section.

B.2.3 Copulas

Copulas provide a bridge between joint distributions and their marginals. Let us begin with the definition of copula function (we restrict our attention to a bivariate random variable but the multivariate extension is also possible).

Definition 23 (Copula function). *A bivariate copula is a function* $C : [0,1]^2 \to [0,1]$ *such that:*

(a) for all $a \in [0,1]$,

$$C(a,1) = C(1,a) = a, \qquad C(a,0) = C(0,a) = 0;$$

(b) C *is nondecreasing in each argument and is 2-increasing on* $[0,1]^2$.

In statistical terms, a copula function is a joint CDF of variables each of which has a marginal uniform distribution. The following theorem elucidates the role of copulas in the relationship between joint and marginal CDFs.

Theorem 9 (Sklar's theorem). *Let* X *and* Y *be random variables with CDFs* F_X *and* F_Y, *and joint CDF* $F_{X,Y}$. *Then there exists a copula* C *such that*

$$F_{X,Y}(x,y) = C(F_X(x), F_Y(y)) \tag{B.2}$$

for all $(x,y) \in \mathbb{R}^2$. *If* F_X *and* F_Y *are continuous, then* C *is unique.*

The proof of Theorem 9 makes use of the uniform transformation rule.
When F_X and F_Y are continuous, the copula can be written as

$$C(u_x, u_y) = F_{X,Y}(F_X^{-1}(u_x), F_Y^{-1}(u_y)). \tag{B.3}$$

We now introduce a theorem, due to Bouyé and Salmon (2009) and also discussed in Koenker (2005, p. 266), that provides the link between copula functions and conditional quantile functions.

Proposition 5. *Let* $C(u_x, u_y) = F_{X,Y}(F_X^{-1}(u_x), F_Y^{-1}(u_y))$ *be a bivariate copula for continuous* X *and* Y. *Then*

$$\frac{\partial C(u_x, u_y)}{\partial u_x} = F_{Y|X}\left(F_Y^{-1}(u_y) \mid X = F_X^{-1}(u_x)\right), \tag{B.4}$$

whenever the derivative exists.

Consequently,

$$Q_Y(p \mid X = x) = Q_Y(c_u^{-1}(p)), \qquad x = F_X^{-1}(u_x), \tag{B.5}$$

where c_u^{-1} denotes the generalised inverse with respect to u_y of the function

$$c_u(u_y) = \frac{\partial C(u_x, u_y)}{\partial u_x}.$$

Hence, the copula representation makes explicit that conditional quantiles are marginal quantiles evaluated at probability levels transformed by the dependence structure. Note that in Equation (B.4) we find the comparison distribution

$$\upsilon(u_y; F_Y, F_{Y|X=x}) = F_{Y|X=x}\left(F_Y^{-1}(u_y)\right).$$

Bibliography

Abramowitz, M. and Stegun, I. A. (1972). *Handbook of mathematical functions with formulas, graphs, and mathematical tables*, volume Applied Mathematics Series 55. National Bureau of Standards, Washington, DC, tenth edition.

Abrevaya, J. (2001). The effects of demographics and maternal behavior on the distribution of birth outcomes. *Empirical Economics*, 26(1):247–257.

Abrevaya, J. and Dahl, C. M. (2008). The effects of birth inputs on birthweight: Evidence from quantile estimation on panel data. *Journal of Business & Economic Statistics*, 26(4):379–397.

Agresti, A. (2013). *Categorical data analysis.* Wiley, New York, NY, third edition.

Aitchison, J. and Brown, J. (1957). *The lognormal distribution.* Cambridge University Press, Cambridge, UK.

Alkan Ozdemir, S., Ozer, E. A., Kose, S., Ilhan, O., Ozturk, C., and Sutcuoglu, S. (2016). Reference values of serum IgG and IgM levels in preterm and term newborns. *The Journal of Maternal-Fetal & Neonatal Medicine*, 29(6):972–976.

Alvarez, F. J. R. (2020). Letter to the editor 'on quantile-based asymmetric family of distributions: Properties and inference'. *International Statistical Review*, 88(3):793–796.

Anderson, C. W. (1970). Extreme value theory for a class of discrete distributions with applications to some stochastic processes. *Journal of Applied Probability*, 7(1):99–113.

Anderson, C. W. (1980). Local limit theorems for the maxima of discrete random variables. *Mathematical Proceedings of the Cambridge Philosophical Society*, 88(1):161–165.

Aranda-Ordaz, F. J. (1981). On two families of transformations to additivity for binary response data. *Biometrika*, 68(2):357–363.

Azzalini, A. and Bowman, A. W. (1990). A look at some data on the Old Faithful Geyser. *Journal of the Royal Statistical Society C*, 39(3):357–365.

Bahadur, R. R. (1966). A note on quantiles in large samples. *The Annals of Mathematical Statistics*, 37(3):577–580.

Balanda, K. P. and MacGillivray, H. L. (1988). Kurtosis: A critical review. *The American Statistician*, 42(2):111–119.

Bapat, R. B. and Beg, M. I. (1989). Order statistics for nonidentically distributed variables and permanents. *Sankhyā: The Indian Journal of Statistics, Series A*, 51(1):79–93.

Bassett, G. and Koenker, R. (1978). Asymptotic theory of least absolute error regression. *Journal of the American Statistical Association*, 73(363):618–622.

Bassett, G. W. and Koenker, R. W. (1992). A note on recent proposals for computing l_1 estimates. *Computational Statistics and Data Analysis*, 14(2):207–211.

Begun, J. M., Hall, W. J., Huang, W.-M., and Wellner, J. A. (1983). Information and asymptotic efficiency in parametric–nonparametric models. *The Annals of Statistics*, 11(2):432–452.

Beirlant, J., Goegebeur, Y., Teugels, J., and Segers, J. (2004). *Statistics of extremes: Theory and applications*. Wiley, Chichester, UK.

Benjamini, Y. and Krieger, A. M. (1996). Concepts and measures for skewness with data-analytic implications. *Canadian Journal of Statistics*, 24(1):131–140.

Bera, A. K., Galvao, A. F., Montes-Rojas, G. V., and Park, S. Y. (2015). Asymmetric Laplace regression: Maximum likelihood, maximum entropy and quantile regression. *Journal of Econometric Methods*, 5(1):79–101.

Berkelaar, M. (2024). *lpSolve: Interface to 'lp_solve' v. 5.5 to solve linear/integer programs*. R package version 5.6.23.

Berman, A. (1973). *Cones, matrices and mathematical programming*. Springer-Verlag, New York, NY.

Bickel, P. J. and Freedman, D. A. (1981). Some asymptotic theory for the bootstrap. *The Annals of Statistics*, 9(6):1196–1217.

Birch, J. M., Geraci, M., Alston, R., Wallingford, S., and Kelsey, A. (2010). Low birthweight and aetiology of childhood liver tumours in North West England. *Pediatric Blood & Cancer*, 55(5):932–932.

Bloomfield, P. and Steiger, W. L. (1983). *Least absolute deviations: Theory, applications, and algorithms*. Birkhäuser, Boston, MA.

Bofinger, E. (1975). Estimation of a density function using order statistics. *Australian Journal of Statistics*, 17(1):1–7.

Boghossian, N. S., Geraci, M., Edwards, E. M., and Horbar, J. D. (2018). Morbidity and mortality in small for gestational age infants at 22 to 29 weeks' gestation. *Pediatrics*, 141(2):e20172533.

Boghossian, N. S., Geraci, M., Edwards, E. M., Morrow, K. A., and Horbar, J. D. (2016). Anthropometric charts for infants born between 22 and 29 weeks' gestation. *Pediatrics*, 138(6):e20161641.

Bolker, B. (2023). *bbmle: Tools for general maximum likelihood estimation*. R package version 1.0.25.1.

Bosch, R. J., Ye, Y., and Woodworth, G. G. (1995). A convergent algorithm for quantile regression with smoothing splines. *Computational Statistics & Data Analysis*, 19(6):613–630.

Boscovich, R. J. (1757). De litteraria expeditione per pontificiam ditionem. In *De Bononiensi Scientiarum et Artium Instituto Atque Academia Commentarii*, volume IV, pages 353–396. Bononiae.

Bottai, M., Orsini, N., and Geraci, M. (2015). A gradient search maximization algorithm for the asymmetric Laplace likelihood. *Journal of Statistical Computation and Simulation*, 85(10):1919–1925.

Bouyé, E. and Salmon, M. (2009). Dynamic copula quantile regressions and tail area dynamic dependence in Forex markets. *The European Journal of Finance*, 15(7-8):721–750.

Bowley, A. L. (1901). *Elements of statistics*. P. S. King & Son, London, first edition.

Box, G. E. P. and Cox, D. R. (1964). An analysis of transformations. *Journal of the Royal Statistical Society B*, 26(2):211–252.

Boyd, S. P. and Vandenberghe, L. (2004). *Convex optimization*. Cambridge University Press, Cambridge, UK.

Buchinsky, M. (1995). Quantile regression, Box–Cox transformation model, and the US wage structure, 1963-1987. *Journal of Econometrics*, 65(1):109–154.

Cameron, A. C. and Trivedi, P. K. (1990). Regression-based tests for overdispersion in the Poisson model. *Journal of Econometrics*, 46(3):347–364.

Canguilhem, G. (1989). *The normal and the pathological*. The MIT Press, New York, NY.

Canty, A. and Ripley, B. D. (2025). *boot: Bootstrap R (S-Plus) functions*. R package version 1.3-32.

Carroll, R. J. and Ruppert, D. (1988). *Transformation and weighting in regression*. Chapman and Hall, London, UK.

Casella, G. and Berger, R. (2002). *Statistical inference*. Duxbury, Pacific Grove, CA, second edition.

Chamberlain, G. (1994). Quantile regression, censoring, and the structure of wages. In Sims, C. A., editor, *Advances in Econometrics: Sixth World Congress*, volume 1 of *Econometric Society Monographs*, chapter 5, pages 171–210. Cambridge University Press, Cambridge, UK.

Chen, C. (2007). A finite smoothing algorithm for quantile regression. *Journal of Computational and Graphical Statistics*, 16(1):136–164.

Chen, C. and Mangasarian, O. L. (1996). A class of smoothing functions for nonlinear and mixed complementarity problems. *Computational Optimization and Applications*, 5(2):97–138.

Chernozhukov, V. and Fernández-Val, I. (2011). Inference for extremal conditional quantile models, with an application to market and birthweight risks. *Review of Economic Studies*, 78(2):559–589.

Chernozhukov, V., Fernández-Val, I., and Galichon, A. (2010). Quantile and probability curves without crossing. *Econometrica*, 78(3):1093–1125.

Chernozhukov, V., Fernández-Val, I., Melly, B., and Wüthrich, K. (2020). Generic inference on quantile and quantile effect functions for discrete outcomes. *Journal of the American Statistical Association*, 115(529):123–137.

Clark, D. and Osborne, M. (1986). Finite algorithms for Huber's M-estimator. *SIAM Journal on Scientific and Statistical Computing*, 7(1):72–85.

Clarke, F. H. (1990). *Optimization and nonsmooth analysis.* SIAM, Philadelphia, PA.

Cline, D. B. H. (1988). Admissibile kernel estimators of a multivariate density. *The Annals of Statistics*, 16(4):1421–1427.

Cole, T. J. (1988). Fitting smoothed centile curves to reference data. *Journal of the Royal Statistical Society A*, 151(3):385–418.

Cole, T. J. and Green, P. J. (1992). Smoothing reference centile curves: The LMS method and penalized likelihood. *Statistics in Medicine*, 11(10):1305–1319.

Coles, S. (2001). *An introduction to statistical modeling of extreme values.* Springer, London, UK.

Conn, A. R., Scheinberg, K., and Vicente, L. N. (2009). *Introduction to derivative-free optimization.* SIAM, Philadelphia, PA.

Conroy, R. M. (2012). What hypotheses do "nonparametric" two-group tests actually test? *The Stata Journal*, 12(2):182–190.

Cook, C. D. and Hamann, J. F. (1961). Relation of lung volumes to height in healthy persons between the ages of 5 and 38 years. *The Journal of Pediatrics*, 59(5):710–714.

Cox, D. R. (1984). Interaction. *International Statistical Review*, 52(1):1–24.

Cox, D. R. (2006). *Principles of statistical inference.* Cambridge University Press, New York, NY.

Cox, D. R. and Hinkley, D. V. (1997). *Theoretical statistics.* Chapman and Hall, London, UK.

Csörgő, M. (1983). *Quantile processes with statistical applications*, volume 42. SIAM, Philadelphia, PA.

Darmois, G. (1935). Sur les lois de probabilité à estimation exhaustive. *Comptes rendus hebdomadaires des séances de l'Académie des sciences*, 200:1265–1266.

DasGupta, A. (2008). *Asymptotic theory of statistics and probability.* Springer, New York, NY.

David, H. A. (1995). First (?) occurrence of common terms in mathematical statistics. *The American Statistician*, 49(2):121–133.

David, H. A. and Nagaraja, H. N. (2003). *Order statistics.* Wiley, Hoboken, NJ, third edition.

Davino, C., Furno, M., and Vistocco, D. (2013). *Quantile regression: Theory and applications.* Wiley, Chichester, UK.

Davison, A. C. and Hinkley, D. V. (1997). *Bootstrap methods and their applications.* Cambridge University Press, Cambridge, UK.

De Angelis, D., Hall, P., and Young, G. A. (1993). A note on coverage error of bootstrap confidence intervals for quantiles. *Mathematical Proceedings of the Cambridge Philosophical Society*, 114(3):517–531.

de Haan, L. (1970). *On regular variation and its application to the weak convergence of sample extremes*, volume 32 of *Mathematical Centre Tracts*. Mathematisch Centrum, Amsterdam, NL.

de Haan, L. and Ferreira, A. (2006). *Extreme value theory*. Springer, New York, NY.

Dehbi, H.-M., Cortina-Borja, M., and Geraci, M. (2016). Aranda-Ordaz quantile regression for student performance assessment. *Journal of Applied Statistics*, 43(1):58–71.

Delignette-Muller, M. L. and Dutang, C. (2015). fitdistrplus: An R package for fitting distributions. *Journal of Statistical Software*, 64(4):1–34.

Di Lorenzo, P. (2025). *usmap: US maps including Alaska and Hawaii*. R package version 1.0.0.

Dobson, A. J. (1983). *An introduction to statistical modelling*. Chapman and Hall, London, UK.

Doksum, K. (1974). Empirical probability plots and statistical inference for nonlinear models in the two-sample case. *The Annals of Statistics*, 2(2):267–277.

Doksum, K. A. and Sievers, G. L. (1976). Plotting with confidence: Graphical comparisons of two populations. *Biometrika*, 63(3):421–434.

Dunn, C. L. (1991). Precise simulated percentiles in a pinch. *The American Statistician*, 45(3):207–211.

Efron, B. (1979). Bootstrap methods: Another look at the jackknife. *The Annals of Statistics*, 7(1):1–26.

Efron, B. (1982). *The jackknife, the bootstrap, and other resampling plans*. SIAM, Philadelphia, PA.

Ellis, S. P. (1998). Instability of least squares, least absolute deviation and least median of squares linear regression. *Statistical Science*, 13(4):337–344.

Embrechts, P. and Hofert, M. (2013). A note on generalized inverses. *Mathematical Methods of Operations Research*, 77(3):423–432.

Embrechts, P., Klüppelberg, C., and Mikosch, T. (1997). *Modelling extremal events for insurance and finance*. Springer, Berlin.

Epanechnikov, V. A. (1969). Non-parametric estimation of a multivariate probability density. *Theory of Probability & Its Applications*, 14(1):153–158.

Eubank, R. L. (1988). *Spline smoothing and nonparametric regression*. Dekker, New York, NY, first edition.

Eubank, R. L. (1999). *Nonparametric regression and spline smoothing*. Dekker, New York, NY, second edition.

Everitt, B. S. and Skrondal, A. (2010). *The Cambridge dictionary of statistics*. Cambridge University Press, New York, NY, fourth edition.

Expert Group on Prematurity (1950). Final report, Geneva, 17–21 April 1950. Technical Report 27, World Health Organization.

Falk, M. (1986). On the estimation of the quantile density function. *Statistics & Probability Letters*, 4(2):69–73.

Falk, M. and Kaufmann, E. (1991). Coverage probabilities of bootstrap-confidence intervals for quantiles. *The Annals of Statistics*, 19(1):485–495.

Falk, M. and Marohn, F. (1993). Von Mises conditions revisited. *The Annals of Probability*, 21(3):1310–1328.

Farcomeni, A. and Geraci, M. (2024). Quantile ratio regression. *Statistics and Computing*, 34(2):15.

Fasiolo, M., Wood, S. N., Zaffran, M., Nedellec, R., and Goude, Y. (2021). Fast calibrated additive quantile regression. *Journal of the American Statistical Association*, 116(535):1402–1412.

Fechner, G. T. (1897). *Kollektivmasslehre*. Engelmann, Leipzig.

Feng, X., He, X., and Hu, J. (2011). Wild bootstrap for quantile regression. *Biometrika*, 98(4):995–999.

Firpo, S., Fortin, N. M., and Lemieux, T. (2009). Unconditional quantile regressions. *Econometrica*, 77(3):953–973.

Fitzenberger, B., Wilke, R. A., and Zhang, X. (2010). Implementing Box–Cox quantile regression. *Econometric Reviews*, 29(2):158–181.

Fox, M. and Rubin, H. (1964). Admissibility of quantile estimates of a single location parameter. *The Annals of Mathematical Statistics*, 35(3):1019–1030.

Fu, L. and Wang, Y.-G. (2012). Quantile regression for longitudinal data with a working correlation model. *Computational Statistics & Data Analysis*, 56(8):2526–2538.

Gabaix, X. (1999). Zipf's law for cities: An explanation. *The Quarterly Journal of Economics*, 114(3):739–767.

Galilei, G. (1632). *Dialogo sopra i due massimi sistemi del mondo—Tolemaico e Copernicano*. Giovanni Battista Landini, Firenze.

Galton, F. (1875). IV. Statistics by intercomparison, with remarks on the law of frequency of error. *Philosophical Magazine Series 4*, 49(322):33–46.

Galton, F. (1882). Report of the anthropometric committee. In *Report of the 51st Meeting of the British Association for the Advancement of Science, 1881*, pages 245–260, London. John Murray.

Galton, F. (1885a). Anthropometric percentiles. *Nature*, 31(793):223–225.

Galton, F. (1885b). Some results of the anthropometric laboratory. *The Journal of the Anthropological Institute of Great Britain and Ireland*, 14:275–287.

Galvao, A. F. (2011). Quantile regression for dynamic panel data with fixed effects. *Journal of Econometrics*, 164(1):142–157.

Galvao, A. F. and Montes-Rojas, G. V. (2010). Penalized quantile regression for dynamic panel data. *Journal of Statistical Planning and Inference*, 140(11):3476–3497.

Genton, M. G., Ma, Y., and Parzen, E. (2006). Discussion of "sur une limitation très générale de la dispersion de la médiane" by M. Fréchet. *Journal de la Société Française de Statistique*, 147(2):51–60.

Geraci, M. (2014). Linear quantile mixed models: The lqmm package for Laplace quantile regression. *Journal of Statistical Software*, 57(13):1–29.

Geraci, M. (2016a). Estimation of regression quantiles in complex surveys with data missing at random: An application to birthweight determinants. *Statistical Methods in Medical Research*, 25(4):1393–1421.

Geraci, M. (2016b). Qtools: A collection of models and tools for quantile inference. *R Journal*, 8(2):117–138.

Geraci, M. (2019a). Additive quantile regression for clustered data with an application to children's physical activity. *Journal of the Royal Statistical Society C*, 68(4):1071–1089.

Geraci, M. (2019b). Letter to the editor. *Statistics and Its Interface*, 12(1):71–75.

Geraci, M. (2019c). Modelling and estimation of nonlinear quantile regression with clustered data. *Computational Statistics & Data Analysis*, 136:30–46.

Geraci, M. (2020). *nlqmm: Nonlinear quantile mixed models.* R package version 1.2.

Geraci, M. (2026). *Qtools: Utilities for quantiles.* R package version 1.6.1.

Geraci, M., Boghossian, N. S., Farcomeni, A., and Horbar, J. D. (2019). Quantile contours and allometric modelling for risk classification of abnormal ratios with an application to asymmetric growth-restriction in preterm infants. *Statistical Methods in Medical Research*, 29(7):1769–1786.

Geraci, M. and Bottai, M. (2007). Quantile regression for longitudinal data using the asymmetric Laplace distribution. *Biostatistics*, 8(1):140–154.

Geraci, M. and Bottai, M. (2014). Linear quantile mixed models. *Statistics and Computing*, 24(3):461–479.

Geraci, M. and Cortina-Borja, M. (2018). The Laplace distribution. *Significance*, 15(5):10–11.

Geraci, M. and Farcomeni, A. (2022). Mid-quantile regression for discrete responses. *Statistical Methods in Medical Research*, 31(5):821–838.

Geraci, M. and Jones, M. C. (2015). Improved transformation-based quantile regression. *Canadian Journal of Statistics*, 43(1):118–132.

Ghosh, J. K. (1971). A new proof of the Bahadur representation of quantiles and an application. *The Annals of Mathematical Statistics*, 42(6):1957–1961.

Gijbels, I., Karim, R., and Verhasselt, A. (2019). On quantile-based asymmetric family of distributions: Properties and inference. *International Statistical Review*, 87(3):471–504.

Gilbert, P. and Varadhan, R. (2019). *numDeriv: Accurate numerical derivatives.* R package version 2016.8-1.1.

Gilchrist, W. (2000). *Statistical modelling with quantile functions.* Chapman & Hall/CRC, Boca Raton, FL.

Gilchrist, W. (2008). Regression revisited. *International Statistical Review*, 76(3):401–418.

Gnedenko, B. (1943). Sur la distribution limite du terme maximum d'une série aléatoire. *Annals of Mathematics*, 44(3):423–453.

Gneiting, T. (2011). Quantiles as optimal point forecasts. *International Journal of Forecasting*, 27(2):197–207.

Gory, J. J., Craigmile, P. F., and MacEachern, S. N. (2021). A class of generalized linear mixed models adjusted for marginal interpretability. *Statistics in Medicine*, 40(2):427–440.

Green, P. J. and Silverman, B. W. (1994). *Nonparametric regression and generalized linear models.* Chapman & Hall/CRC, Boca Raton, FL.

Greenwald, M. and Khanna, S. (2001). Space-efficient online computation of quantile summaries. *SIGMOD Record*, 30(2):58–66.

Griffiths, D. A. (1980). Interval estimation for the three-parameter lognormal distribution via the likelihood function. *Journal of the Royal Statistical Society C*, 29(1):58–68.

Groeneveld, R. A. (1998). A class of quantile measures for kurtosis. *The American Statistician*, 52(4):325–329.

Groeneveld, R. A. and Meeden, G. (1984). Measuring skewness and kurtosis. *Journal of the Royal Statistical Society D*, 33(4):391–399.

Gutenbrunner, C. and Jurečková, J. (1992). Regression rank scores and regression quantiles. *The Annals of Statistics*, 20(1):305–330.

Hájek, J. and Šidák, Z. (1967). *Theory of rank tests.* Academia, Prague.

Hald, A. (1998). *A history of mathematical statistics from 1750 to 1930.* Wiley, New York, NY.

Hald, A. (2003). *A history of probability and statistics and their applications before 1750.* Wiley, New York, NY.

Hall, P. (1988). Theoretical comparison of bootstrap confidence intervals. *The Annals of Statistics*, 16(3):927–953.

Hall, P. and Sheather, S. J. (1988). On the distribution of a Studentized quantile. *Journal of the Royal Statistical Society B*, 50(3):381–391.

Hallin, M., Paindaveine, D., and Šiman, M. (2010). Multivariate quantiles and multiple-output regression quantiles: From l_1 optimization to halfspace depth. *The Annals of Statistics*, 38(2):635–639.

Hampel, F. R. (1974). The influence curve and its role in robust estimation. *Journal of the American Statistical Association*, 69(346):383–393.

Handcock, M. S. (2023). *reldist: Relative distribution methods.* R package version 1.7-2.

Handcock, M. S. and Morris, M. (1999). *Relative distribution methods in the social sciences.* Springer, New York, NY.

Hayfield, T. and Racine, J. S. (2008). Nonparametric econometrics: The np package. *Journal of Statistical Software*, 27(5):1–32.

He, X. (1997). Quantile curves without crossing. *The American Statistician*, 51(2):186–192.

He, X., Pan, X., Tan, K. M., and Zhou, W.-X. (2023a). *conquer: Convolution-type smoothed quantile regression.* R package version 1.3.3.

He, X., Pan, X., Tan, K. M., and Zhou, W.-X. (2023b). Smoothed quantile regression with large-scale inference. *Journal of Econometrics*, 232(2):367–388.

He, X. M. and Zhu, L. X. (2003). A lack-of-fit test for quantile regression. *Journal of the American Statistical Association*, 98(464):1013–1022.

Hepper, N. G. G., Fowler, W. S., and Helmholz, F. (1960). Relationship of height to lung volume in healthy men. *Diseases of the Chest*, 37(3):314–320.

Heyde, C. C. and Seneta, E. (2001). *Statisticians of the centuries.* Springer, New York, NY.

Hill, B. M. (1963). The three-parameter lognormal distribution and Bayesian analysis of a point-source epidemic. *Journal of the American Statistical Association*, 58(301):72–84.

Hinkley, D. V. (1975). On power transformations to symmetry. *Biometrika*, 62(1):101–111.

Horn, P. S. (1983). A measure for peakedness. *The American Statistician*, 37(1):55–56.

Horowitz, J. L. (1992). A smoothed maximum score estimator for the binary response model. *Econometrica*, 60(3):505–531.

Horowitz, J. L. and Lee, S. (2005). Nonparametric estimation of an additive quantile regression model. *Journal of the American Statistical Association*, 100(472):1238–1249.

Huber, P. J. (1964). Robust estimation of a location parameter. *The Annals of Mathematical Statistics*, 35(1):73–101.

Huber, P. J. (1981). *Robust statistics.* Wiley, New York, NY.

Hunter, D. R. and Lange, K. (2000). Quantile regression via an MM algorithm. *Journal of Computational and Graphical Statistics*, 9(1):60–77.

Hunter, D. R. and Lange, K. (2004). A tutorial on MM algorithms. *American Statistician*, 58(1):30–37.

Hyndman, R. J. and Fan, Y. (1996). Sample quantiles in statistical packages. *The American Statistician*, 50(4):361–365.

International Agency for Research on Cancer (2014). *World cancer report 2014.* International Agency for Research on Cancer, Lyon. Edited by B. W. Stewart and C. P. Wild.

Isaacs, D., Altman, D. G., Tidmarsh, C. E., Valman, H. B., and Webster, A. D. (1983). Serum immunoglobulin concentrations in preschool children measured by laser nephelometry: reference ranges for IgG, IgA, IgM. *Journal of Clinical Pathology*, 36(10):1193.

Jentsch, C. and Leucht, A. (2016). Bootstrapping sample quantiles of discrete data. *Annals of the Institute of Statistical Mathematics*, 68(3):491–539.

Joag-Dev, K. (1989). MAD property of a median: A simple proof. *The American Statistician*, 43(1):26–27.

Johnson, N. L. and Kotz, S. (1970). *Continuous univariate distributions*, volume I. Houghton Mifflin, Boston, MA.

Jones, M. C. (1988). Discussion of the paper by Cole 'Fitting smoothed centile curves to reference data'. *Journal of the Royal Statistical Society A*, 151(3):412–413.

Jones, M. C. (1992). Estimating densities, quantiles, quantile densities and density quantiles. *Annals of the Institute of Statistical Mathematics*, 44(4):721–727.

Jones, M. C. (1993). Simple boundary correction for kernel density estimation. *Statistics and Computing*, 3(3):135–146.

Jones, M. C. (1994). Expectiles and M-quantiles are quantiles. *Statistics & Probability Letters*, 20(2):149–153.

Jones, M. C., Rosco, J. F., and Pewsey, A. (2011). Skewness-invariant measures of kurtosis. *The American Statistician*, 65(2):89–95.

Karlsson, A. (2008). Nonlinear quantile regression estimation of longitudinal data. *Communications in Statistics – Simulation and Computation*, 37(1):114–131.

Kendall, M. G. (1940). Note on the distribution of quantiles for large samples. *Supplement to the Journal of the Royal Statistical Society*, 7(1):83–85.

Keynes, J. M. (1911). The principal averages and the laws of error which lead to them. *Journal of the Royal Statistical Society*, 74(3):322–331.

Kiefer, J. (1967). On Bahadur's representation of sample quantiles. *The Annals of Mathematical Statistics*, 38(5):1323–1342.

Kleiber, C. and Zeileis, A. (2008). *Applied econometrics with R*. Springer, New York, NY.

Kleiber, C. and Zeileis, A. (2026). *Applied econometrics with R*. R package version 1.2-16.

Knuth, D. E. (1998). *The art of computer programming. Volume III: Sorting and searching*. Addison-Wesley, Reading, MA, second edition.

Koenker, R. (2004). Quantile regression for longitudinal data. *Journal of Multivariate Analysis*, 91(1):74–89.

Koenker, R. (2005). *Quantile regression*. Cambridge University Press, New York, NY.

Koenker, R. (2011). Additive models for quantile regression: Model selection and confidence bandaids. *Brazilian Journal of Probability and Statistics*, 25(3):239–262.

Koenker, R. (2013). Discussion: Living beyond our means. *Statistical Modelling*, 13(4):323–333.

Koenker, R. (2025). *quantreg: Quantile regression*. R package version 6.1.

Koenker, R. and Bassett, G. (1978). Regression quantiles. *Econometrica*, 46(1):33–50.

Koenker, R. and Bassett, G. (1985). On Boscovich's estimator. *The Annals of Statistics*, 13(4):1625–1628.

Koenker, R., Chernozhukov, V., He, X., and Peng, L. (2018). *Handbook of quantile regression*. Handbook of Modern Statistical Methods. Chapman & Hall/CRC, Boca Raton, FL.

Koenker, R. and D'Orey, V. (1987). Algorithm AS 229: Computing regression quantiles. *Journal of the Royal Statistical Society C*, 36(3):383–393.

Koenker, R. and Geling, O. (2001). Reappraising medfly longevity. *Journal of the American Statistical Association*, 96(454):458–468.

Koenker, R. and Hallock, K. F. (2001). Quantile regression. *Journal of Economic Perspectives*, 15(4):143–156.

Koenker, R. and Machado, J. A. F. (1999). Goodness of fit and related inference processes for quantile regression. *Journal of the American Statistical Association*, 94(448):1296–1310.

Koenker, R., Ng, P., and Portnoy, S. (1994). Quantile smoothing splines. *Biometrika*, 81(4):673–680.

Koenker, R. and Park, B. J. (1996). An interior point algorithm for nonlinear quantile regression. *Journal of Econometrics*, 71(1-2):265–283.

Koenker, R. and Xiao, Z. J. (2002). Inference on the quantile regression process. *Econometrica*, 70(4):1583–1612.

Kong, L. and Mizera, I. (2012). Quantile tomography: Using quantiles with multivariate data. *Statistica Sinica*, 22(4):1589–1610.

Koopman, B. O. (1936). On distributions admitting a sufficient statistic. *Transactions of the American Mathematical Society*, 39(3):399–409.

Kordas, G. (2006). Smoothed binary regression quantiles. *Journal of Applied Econometrics*, 21(3):387–407.

Kotz, S., Balakrishnan, N., and Johnson, N. L. (2000). *Continuous multivariate distributions: Models and applications*, volume 1. Wiley, New York, NY, second edition.

Kotz, S., Kozubowski, T., and Podgorski, K. (2001). *The Laplace distribution and generalizations*. Birkhäuser, Boston, MA.

Kotz, S. and Nadarajah, S. (2000). *Extreme value distributions*, volume 31. Imperial College Press, London, UK.

Kuczmarski, R. J., Ogden, C. L., Guo, S. S., Grummer-Strawn, L. M., Flegal, K. M., Mei, Z., Wei, R., Curtin, L. R., Roche, A. F., and Johnson, C. L. (2002). 2000 CDC growth charts for the United States: Methods and development. Technical Report 246, National Center for Health Statistics.

Kwan, K. C., Breault, G. O., Umbenhauer, E. R., McMahon, F. G., and Duggan, D. E. (1976). Kinetics of indomethacin absorption, elimination, and enterohepatic circulation in man. *Journal of Pharmacokinetics and Biopharmaceutics*, 4(3):255–80.

Lamarche, C. (2010). Robust penalized quantile regression estimation for panel data. *Journal of Econometrics*, 157(2):396–408.

Lancaster, H. O. (1961). Significance tests in discrete distributions. *Journal of the American Statistical Association*, 56(294):223–234.

Laplace, P.-S. (1774). Mémoire sur la probabilité des causes par les évènements. *Mémoires de l'Académie Royale des Sciences de Paris (Savants étrangers)*, 6:621–656.

Lau, L.-C. and Boos, D. D. (1994). A fast and low-storage algorithm for finding quantiles. Technical report, Department of Statistics, North Carolina State University.

Lawless, J. F. (1980). Inference in the generalized gamma and log gamma distributions. *Technometrics*, 22(3):409–419.

Lee, M.-j. (1992). Median regression for ordered discrete response. *Journal of Econometrics*, 51(1):59–77.

Lehmann, E. L. (1975). *Nonparametrics: Statistical methods based on ranks.* Holden-Day, San Francisco, CA.

Li, Q. and Racine, J. S. (2008). Nonparametric estimation of conditional CDF and quantile functions with mixed categorical and continuous data. *Journal of Business & Economic Statistics*, 26(4):423–434.

Lindsay, B. G. and Basak, P. (2000). Moments determine the tail of a distribution (but not much else). *The American Statistician*, 54(4):248–251.

Lipsitz, S. R., Fitzmaurice, G. M., Molenberghs, G., and Zhao, L. P. (1997). Quantile regression methods for longitudinal data with drop-outs: Application to CD4 cell counts of patients infected with the human immunodeficiency virus. *Journal of the Royal Statistical Society C*, 46(4):463–476.

Liu, R. Y. and Singh, K. (1997). Notions of limiting p values based on data depth and bootstrap. *Journal of the American Statistical Association*, 92(437):266–277.

Ma, Y., Genton, M. G., and Parzen, E. (2011). Asymptotic properties of sample quantiles of discrete distributions. *Annals of the Institute of Statistical Mathematics*, 63(2):227–243.

Machado, J. A. F. (1993). Robust model selection and M-estimation. *Econometric Theory*, 9(3):478–493.

Machado, J. A. F. and Mata, J. (2000). Box–Cox quantile regression and the distribution of firm sizes. *Journal of Applied Econometrics*, 15(3):253–274.

Machado, J. A. F. and Santos Silva, J. M. C. (2005). Quantiles for counts. *Journal of the American Statistical Association*, 100(472):1226–1237.

Madsen, K. and Nielsen, H. B. (1993). A finite smoothing algorithm for linear l_1 estimation. *SIAM Journal on Optimization*, 3(2):223–235.

Mann, H. B. and Whitney, D. R. (1947). On a test of whether one of two random variables is stochastically larger than the other. *The Annals of Mathematical Statistics*, 18(1):50–60.

Manski, C. F. (1975). Maximum score estimation of the stochastic utility model of choice. *Journal of Econometrics*, 3(3):205–228.

Manski, C. F. (1985). Semiparametric analysis of discrete response: Asymptotic properties of the maximum score estimator. *Journal of Econometrics*, 27(3):313–333.

Mazucheli, J., Alves, B., Menezes, A. F. B., and Leiva, V. (2022). An overview on parametric quantile regression models and their computational implementation with applications to biomedical problems including COVID-19 data. *Computer Methods and Programs in Biomedicine*, 221:106816.

McCullagh, P. (1994). Does the moment-generating function characterize a distribution? *The American Statistician*, 48(3):208–208.

Melly, B., Chernozhukov, V., Fernandez-Val, I., and Wüthrich, K. (2025). *discreteQ: Inference on quantile and quantile effect functions for discrete outcomes.* R package version 0.4.4.

Menezes, A. F. B. and Mazucheli, J. (2023). *unitquantreg: Parametric quantile regression models for bounded data.* R package version 0.0.6.

Mersmann, O. (2024). *microbenchmark: Accurate timing functions.* R package version 1.5.0.

Millard, S. P. (2013). *EnvStats: An R package for environmental statistics.* Springer, New York, NY.

Mizera, I. (2018). Quantile regression: Penalized. In Koenker, R., Chernozhukov, V., He, X., and Peng, L., editors, *Handbook of quantile regression*, Handbook of Modern Statistical Methods, pages 21–39. Chapman & Hall/CRC, Boca Raton, FL.

Mood, A. M. F., Graybill, F. A., and Boes, D. C. (1974). *Introduction to the theory of statistics.* McGraw-Hill, New York, NY, third edition.

Morgenthaler, S. and Tukey, J. W. (2000). Fitting quantiles: Doubling, HR, HQ, and HHH distributions. *Journal of Computational and Graphical Statistics*, 9(1):180–195.

Mosteller, F. and Tukey, J. W. (1977). *Data analysis and regression: A second course in statistics.* Addison-Wesley, Reading, MA.

Mu, Y. M. and He, X. M. (2007). Power transformation toward a linear regression quantile. *Journal of the American Statistical Association*, 102(477):269–279.

Muir, P. R., Wallace, C. C., Done, T., and Aguirre, J. D. (2015). Limited scope for latitudinal extension of reef corals. *Science*, 348(6239):1135–1138.

Nassiri, V. and Loris, I. (2013). A generalized quantile regression model. *Journal of Applied Statistics*, 40(5):1090–1105.

Noufaily, A. and Jones, M. C. (2013). Parametric quantile regression based on the generalized gamma distribution. *Journal of the Royal Statistical Society C*, 62(5):723–740.

Nyblom, J. (1992). Note on interpolated order statistics. *Statistics & Probability Letters*, 14(2):129–131.

Nychka, D., Gray, G., Haaland, P., Martin, D., and O'Connell, M. (1995). A nonparametric regression approach to syringe grading for quality improvement. *Journal of the American Statistical Association*, 90(432):1171–1178.

Oberhofer, W. (1982). The consistency of nonlinear regression minimizing the L_1-norm. *The Annals of Statistics*, 10(1):316–319.

Oberhofer, W. and Haupt, H. (2005). The asymptotic distribution of the unconditional quantile estimator under dependence. *Statistics & Probability Letters*, 73(3):243–250.

Oh, H.-S., Lee, T. C. M., and Nychka, D. W. (2011). Fast nonparametric quantile regression with arbitrary smoothing methods. *Journal of Computational and Graphical Statistics*, 20(2):510–526.

Oja, H. (1981). On location, scale, skewness and kurtosis of univariate distributions. *Scandinavian Journal of Statistics*, 8(3):154–168.

Parente, P. and Santos Silva, J. (2016). Quantile regression with clustered data. *Journal of Econometric Methods*, 5(1):1–15.

Park, S. (1996). Fisher information in order statistics. *Journal of the American Statistical Association*, 91(433):385–390.

Parzen, E. (1979). Nonparametric statistical data modeling. *Journal of the American Statistical Association*, 74(365):105–121.

Parzen, E. (1993). Change PP plot and continous sample quantile function. *Communications in Statistics – Theory and Methods*, 22(12):3287–3304.

Parzen, E. (2004). Quantile probability and statistical data modeling. *Statistical Science*, 19(4):652–662.

Parzen, M. I., Wei, L., and Ying, Z. (1994). A resampling method based on pivotal estimating functions. *Biometrika*, 81(2):341–350.

Pearson, K. (1905). "Das fehlergesetz und seine verallgemeinerungen durch fechner und pearson." A rejoinder. *Biometrika*, 4(1/2):169–212.

Pfanzagl, J. (1976). Investigating the quantile of an unknown distribution. In Ziegler, W. J., editor, *Contributions to Applied Statistics*. Springer, Basel, CH.

Pinheiro, J. C. and Bates, D. M. (2000). *Mixed-effects models in S and S-PLUS*. Springer-Verlag, New York, NY.

Pitman, E. J. G. (1936). Sufficient statistics and intrinsic accuracy. *Mathematical Proceedings of the Cambridge Philosophical Society*, 32(4):567–579.

Pollard, D. (1984). *Convergence of stochastic processes*. Springer-Verlag, New York, NY.

Portnoy, S. (1991). Asymptotic behavior of the number of regression quantile breakpoints. *SIAM Journal on Scientific and Statistical Computing*, 12(4):867–883.

Portnoy, S. and Koenker, R. (1997). The Gaussian hare and the Laplacian tortoise: Computability of squared-error versus absolute-error estimators. *Statistical Science*, 12(4):279–300.

Portnoy, S. and Mizera, I. (1998). Comment on "Instability of least squares, least absolute deviation and least median of squares linear regression". *Statistical Science*, 13(4):344–347.

Powell, J. L. (1991). Estimation of monotonic regression models under quantile restrictions. In Barnett, W., Powell, J., and Tauchen, G., editors, *Nonparametric and semiparametric methods in econometrics and statistics: Proceedings of the fifth international Symposium on Economic Theory and Econometrics*, pages 357–384. Cambridge University Press, New York, NY.

Powell, J. L. (1994). Estimation of semiparametric models. In Engle, R. F. and McFadden, D. L., editors, *Handbook of Eeonometrics*, volume 4, chapter 41, pages 2443–2521. Elsevier, Amsterdam.

Prendergast, L. A. and Staudte, R. G. (2016). Exploiting the quantile optimality ratio in finding confidence intervals for quantiles. *Stat*, 5(1):70–81.

Pruim, R. (2015). *NHANES: Data from the US National Health and Nutrition Examination Study*. R package version 2.1.0.

Quetelet, A. (1835). *Sur l'homme et le développement de ses facultés, ou Essai de physique sociale*, volume 2. Bachelier, Paris.

R Core Team (2026). *R: A language and environment for statistical computing*. R Foundation for Statistical Computing, Vienna, Austria.

Racine, J. S. and Hayfield, T. (2026). *Nonparametric kernel smoothing methods for mixed data types*. R package version 0.70-1.

Redden, D. T., Fernandez, J. R., and Allison, D. B. (2004). A simple significance test for quantile regression. *Statistics in Medicine*, 23(16):2587–2597.

Reed, W. J. (2002). On the rank-size distribution for human settlements. *Journal of Regional Science*, 42(1):1–17.

Reiss, R. D. (1989). *Approximate distributions of order statistics: With applications to nonparametric statistics*. Springer, New York, NY.

Rigby, R. A. and Stasinopoulos, D. M. (2005). Generalized additive models for location, scale and shape. *Journal of the Royal Statistical Society C*, 54:507–544.

Ripley, B. and Venables, B. (2025). *MASS: Support functions and datasets for Venables and Ripley's MASS*. R package version 7.3-65.

Ritz, J. and Spiegelman, D. (2004). Equivalence of conditional and marginal regression models for clustered and longitudinal data. *Statistical Methods in Medical Research*, 13(4):309–323.

Rockafellar, R. T. (1990). Nonsmooth analysis and parametric optimization. In Ekeland, I., Marcellini, P., Marino, A., Tosques, M., Olech, C., Pianigiani, G., Rockafeller, T., Valadier, M., and Cellina, A., editors, *Methods of Nonconvex Analysis*, pages 137–151, Heidelberg. Springer.

Rooth, G. (1980). Low birthweight revised. *The Lancet*, 315(8169):639–641.

Rosenblatt, M. (1952). Remarks on a multivariate transformation. *The Annals of Mathematical Statistics*, 23(3):470–472.

Rousseeuw, P. J. and Bassett, G. W. (1990). The remedian: A robust averaging method for large data sets. *Journal of the American Statistical Association*, 85(409):97–104.

Rubin, D. B. (1981). The Bayesian bootstrap. *The Annals of Statistics*, 9(1):130–134.

Ruggles, S., Flood, S., Sobek, M., Backman, D., Cooper, G., Rivera Drew, J. A., Richards, S., Rodgers, R., Schroeder, J., and Williams, K. C. W. (2025). *IPUMS USA: Version 16.0 [dataset]*. Minneapolis, MN: IPUMS, 2025.

Ruppert, D. (1987). What is kurtosis?: An influence function approach. *The American Statistician*, 41(1):1–5.

Ruppert, D., Sheather, S. J., and Wand, M. P. (1995). An effective bandwidth selector for local least squares regression. *Journal of the American Statistical Association*, 90(432):1257–1270.

Ruppert, D., Wand, M., and Carroll, R. (2003). *Semiparametric regression.* Cambridge University Press, New York.

Sakia, R. M. (1992). The Box–Cox transformation technique: A review. *Journal of the Royal Statistical Society D*, 41(2):169–178.

Scharf, H. R., Lu, X., Williams, P. J., and Hooten, M. B. (2022). Constructing flexible, identifiable and interpretable statistical models for binary data. *International Statistical Review*, 90(2):328–345.

Schmidt, M., Fung, G., and Rosales, R. (2007). Fast optimization methods for L1 regularization: A comparative study and two new approaches. In Kok, J. N., Koronacki, J., de Mantaras, R. L., Matwin, S., Mladenič, D., and Skowron, A., editors, *Machine learning: ECML 2007*, pages 286–297, Heidelberg. Springer.

Sen, P. K. (1968). Asymptotic normality of sample quantiles for m-dependent processes. *The Annals of Mathematical Statistics*, 39(5):1724–1730.

Serfling, R. J. (2002). *Approximation theorems of mathematical statistics.* Wiley, New York, NY.

Shaw, W. T. (2006). Sampling Student's T distribution—use of the inverse cumulative distribution function. *Journal of Computational Finance*, 9(4):37–73.

Sheather, S. J. and Jones, M. C. (1991). A reliable data-based bandwidth selection method for kernel density estimation. *Journal of the Royal Statistical Society B*, 53(3):683–690.

Shi, X. (1991). Some asymptotic results for jackknifing the sample quantile. *The Annals of Statistics*, 19(1):496–503.

Shou, Y. and Smithson, M. (2019). cdfquantreg: An R package for CDF-quantile regression. *Journal of Statistical Software*, 88(1):1–30.

Silverman, B. W. (1986). *Density estimation for statistics and data analysis.* Chapman and Hall, London, UK.

Smithson, M. and Shou, Y. (2017). CDF-quantile distributions for modelling random variables on the unit interval. *British Journal of Mathematical and Statistical Psychology*, 70(3):412–438.

Soliman, S. A., Christensen, G. S., and Rouhi, A. (1988). A new technique for curve fitting based on minimum absolute deviations. *Computational Statistics & Data Analysis*, 6(4):341–351.

Sorrentino, D., Avellini, C., Geraci, M., Dassopoulos, T., Zafiri, D., Vadalà, S., and Benevento, G. (2014). Tissue studies in screened first degree relatives reveal a distinct Crohn's disease phenotype. *Inflammatory Bowel Diseases*, 20(6):1049–1056.

Stacy, E. W. (1962). A generalization of the gamma distribution. *The Annals of Mathematical Statistics*, 33(3):1187–1192.

Staudte, R. G. (2017). Inference for quantile measures of kurtosis, peakedness, and tail weight. *Communications in Statistics – Theory and Methods*, 46(7):3148–3163.

Steinbrecher, G. and Shaw, W. T. (2008). Quantile mechanics. *European Journal of Applied Mathematics*, 19(2):87–112.

Stephenson, A. G. (2002). evd: Extreme value distributions. *R News*, 2(2):31–32.

Stevens, W. L. (1950). Fiducial limits of the parameter of a discontinuous distribution. *Biometrika*, 37(1/2):117–129.

Stigler, S. M. (1977). Fractional order statistics, with applications. *Journal of the American Statistical Association*, 72(359):544–550.

Stigler, S. M. (1986). *The history of statistics: The measurement of uncertainty before 1900.* Harvard University Press, Cambridge, MA.

Stuart, A., Ord, K., and Arnold, S. (1999). *Kendall's advanced theory of statistics, classical inference and the linear model*, volume 2A. Arnold, London, UK, sixth edition.

Stülp, P., Bazán, J. L., and Valdivieso Serrano, L. H. (2025). A new quantile regression model for bounded responses with applications. *Computational Statistics*, 40(8):4081–4113.

Tanner, M. A. (1996). *Tools for statistical inference.* Springer, New York, NY, third edition.

Thompson, W. R. (1936). On confidence ranges for the median and other expectation distributions for populations of unknown distribution form. *The Annals of Mathematical Statistics*, 7(3):122–128.

Tsybakov, A. B. (2009). *Introduction to nonparametric estimation.* Springer, New York, NY.

Tukey, J. W. (1965). Which part of the sample contains the information? *Proceedings of the National Academy of Sciences of the United States of America*, 53(1):127–134.

U.S. Energy Information Administration (2016). International energy outlook. Technical report, U.S. Department of Energy, Washington, DC.

van Zwet, W. R. (1964). Convex transformations: A new approach to skewness and kurtosis. *Statistica Neerlandica*, 18(4):433–441.

Venables, W. N. and Ripley, B. D. (2002). *Modern applied statistics with S.* Springer, New York, NY, fourth edition.

von Mises, R. (1936). La distribution de la plus grande de n valeurs. *Revue mathématique de l'Union Interbalcanique*, 1:141–160.

Walker, A. M. (1968). A note on the asymptotic distribution of sample quantiles. *Journal of the Royal Statistical Society B*, 30(3):570–575.

Wallis, K. F. (2014). The two-piece normal, binormal, or double Gaussian distribution: Its origin and rediscoveries. *Statistical Science*, 29(1):106–112.

Wand, M. P. (2025). *KernSmooth: Functions for kernel smoothing supporting Wand & Jones (1995).* R package version 2.23-26.

Wand, M. P. and Jones, M. C. (1995). *Kernel smoothing.* Springer, New York, NY.

Wang, D. and Hutson, A. D. (2011). A fractional order statistic towards defining a smooth quantile function for discrete data. *Journal of Statistical Planning and Inference*, 141(9):3142–3150.

Wang, J. D. (1995). Asymptotic normality of L_1-estimators in nonlinear regression. *Journal of Multivariate Analysis*, 54(2):227–238.

Wang, Y. (2011). *Smoothing splines: Methods and applications*. Chapman & Hall/CRC, New York, NY.

Warnes, G. R., Bolker, B., and Lumley, T. (2023). *gtools: Various R programming tools*. R package version 3.9.5.

Westfall, P. H. (2014). Kurtosis as peakedness, 1905–2014. R.I.P. *The American Statistician*, 68(3):191–195.

White, H. (1982). Maximum likelihood estimation of misspecified models. *Econometrica*, 50(1):1–25.

Wickham, H. (2016). *ggplot2: Elegant graphics for data analysis*. Springer, New York, NY.

Wilcox, R. R., Erceg-Hurn, D. M., Clark, F., and Carlson, M. (2014). Comparing two independent groups via the lower and upper quantiles. *Journal of Statistical Computation and Simulation*, 84(7):1543–1551.

Wilcoxon, F. (1945). Individual comparisons by ranking methods. *Biometrics Bulletin*, 1(6):80–83.

Wilks, S. S. (1942). Statistical prediction with special reference to the problem of tolerance limits. *The Annals of Mathematical Statistics*, 13(4):400–409.

Wilson, E. B. (1923). First and second laws of error. *Journal of the American Statistical Association*, 18(143):841–851.

Wright, S. E. (2024). A note on computing maximum likelihood estimates for the three-parameter asymmetric Laplace distribution. *Applied Mathematics and Computation*, 464:128381.

Ylppö, A. (1919). Pathologisch-anatomische studien bei frühgeborenen. *Zeitschrift für Kinderheilkunde*, 20(1):212–431.

Yu, K. and Jones, M. C. (1998). Local linear quantile regression. *Journal of the American Statistical Association*, 93(441):228–237.

Yu, K. and Lu, Z. (2004). Local linear additive quantile regression. *Scandinavian Journal of Statistics*, 31(3):333–346.

Yu, K. M. and Zhang, J. (2005). A three-parameter asymmetric Laplace distribution and its extension. *Communications in Statistics – Theory and Methods*, 34(9-10):1867–1879.

Zhao, Q. S. (2000). Restricted regression quantiles. *Journal of Multivariate Analysis*, 72(1):78–99.

Zheng, S. (2011). Gradient descent algorithms for quantile regression with smooth approximation. *International Journal of Machine Learning and Cybernetics*, 2(3):191.

Zieliński, R. (1998). Uniform strong consistency of sample quantiles. *Statistics & Probability Letters*, 37(2):115–119.

Subject Index

Author Index

For Product Safety Concerns and Information please contact our EU
representative GPSR@taylorandfrancis.com
Taylor & Francis Verlag GmbH, Kaufingerstraße 24, 80331 München, Germany

www.ingramcontent.com/pod-product-compliance
Ingram Content Group UK Ltd.
Pitfield, Milton Keynes, MK11 3LW, UK
UKHW020131161026
476707UK00005B/152
* 9 7 8 1 0 3 2 3 5 3 3 0 2 *